Lecture Notes in Artificial Intelligence   4929

Edited by J. G. Carbonell and J. Siekmann

Subseries of Lecture Notes in Computer Science

Malte Helmert

Understanding Planning Tasks

Domain Complexity
and Heuristic Decomposition

 Springer

Series Editors

Jaime G. Carbonell, Carnegie Mellon University, Pittsburgh, PA, USA
Jörg Siekmann, University of Saarland, Saarbrücken, Germany

Author

Malte Helmert
Albert-Ludwigs-Universität Freiburg, Institut für Informatik
Georges-Köhler-Allee, Geb. 052, 79110 Freiburg, Germany
E-mail: helmert@informatik.uni-freiburg.de

Library of Congress Control Number: 2007943037

CR Subject Classification (1998): I.2.8, F.1.3, F.2, E.1, G.1.6

LNCS Sublibrary: SL 7 – Artificial Intelligence

ISSN   0302-9743
ISBN-10   3-540-77722-9 Springer Berlin Heidelberg New York
ISBN-13   978-3-540-77722-9 Springer Berlin Heidelberg New York

Springer is a part of Springer Science+Business Media

springer.com

© Springer-Verlag Berlin Heidelberg 2008
Printed in Germany

Typesetting: Camera-ready by author, data conversion by Markus Richter, Heidelberg
Printed on acid-free paper   SPIN: 12215133   06/3180   5 4 3 2 1 0

Foreword

Action planning is an area that has played a central role in Artificial Intelligence since its beginning. Given a description of the current situation, a description of possible actions and a description of the goals, the task is to identify a sequence of actions, a plan, that transforms the current situation into one that satisfies the goal description.

Even if we restrict this problem to a setting where the environment is completely observable, all actions are deterministic, and there are no exogenous events, the planning problem is computationally very difficult. For planning formalisms that are considered these days, such as PDDL, the problem is EXPSPACE-complete, and even for a purely propositional setting the planning problem is still PSPACE-complete. For this reason, it is very unlikely that we will ever come up with planning algorithms that can solve arbitrary planning tasks in reasonable time. Instead, the planning community has concentrated on developing methods that work well with "typical" tasks.

In order to push the advance of the state of the art, in 1998 planning researchers initiated the biennial International Planning Competitions (IPC). These competitions had two very important effects on research in planning. First of all, they led to the process of defining a quasi standard planning formalism, called the *planning domain definition language* (PDDL). Secondly, for each competition, new sets of planning problems are introduced, giving the research community a rich set of benchmark problems.

And in this context, Malte Helmert makes two important contributions. First of all, the book contains an exhaustive analysis of the computational complexity of the benchmark problems that have been used in the first four competitions. Secondly, drawing on structural similarities between a number of the benchmark problems, a new planning technique is derived. The effectiveness of this technique was demonstrated at the fourth International Planning Competition in 2004, where the planning system *Fast Downward* won the first prize in the non-optimal propositional planning track.

When using benchmark problems to evaluate algorithms, one should have a good idea of the properties of these problems. In particular, in the planning case one should have an idea of what the inherent computational complexity of the problem is and what the source of the difficulty is. So far, only one planning domain, the so-called *blocks world* had been analyzed from a computational complexity point of view. However, nobody had a look at all the other planning domains. In fact, Malte was the first one to analyze all the benchmark problems from an analytical point of view. Furthermore, he also provided an in-depth analysis of what he called routing and transportation problems, which is a recurring topic in a large number of planning problems in the literature and of the benchmark problems designed for the IPC. Additionally, he analyzed the approximability of all the benchmarks, giving an indication of how easy it may be to come up with near-optimal solutions instead of optimal solutions. All in all, this book gives the most comprehensive analysis of planning benchmarks so far and it provides a good idea of how hard the different planning domains are and where the sources for (NP- or PSPACE-)hardness lie.

One outcome of the analysis was the observation that a large number of domains have a "hierarchical structure". However, in order to exploit this structure, it is necessary to move from binary to multi-valued state variables. So, one part of this contribution is a development of a method for automatic reformulation of planning domains. On the basis of such a reformulation of the domain description, it is then possible to derive heuristics that are based on the hierarchical structure of the domain. Interestingly, this resembles planning based on abstraction with the important difference that one does not need to rely on strong refinement properties. Thus, abstractions appear to be much more useful in the context of generating heuristics.

In summary, both contributions of this book advance the state of the art in automatic planning significantly, and in particular the new method for deriving heuristics appears to be quite powerful and seems to have potential beyond what has been explored in this book.

Freiburg, November 2007   Bernhard Nebel

Preface

This volume is concerned with the *classical planning problem*, which can be informally defined as follows:

> Given a description of the current situation (an *initial state*) of an agent, the means by which the agent can alter this situation (a set of *actions*) and a description of desirable situations (a *goal*), find a sequence of actions (a *plan*) that leads from the current situation to one which is desirable.

Instances of the classical planning problem, called *planning tasks*, can model all kinds of abstract reasoning problems in areas as diverse as elevator control, the transportation of petroleum products through pipeline networks, or the solution of solitaire card games.

These applications have little in common apart from the fact that, at a suitable level of abstraction, they can be precisely modelled using the notions of initial states, actions, and goals. In the elevator example, the initial state is given by the current location of the elevator and the locations of passengers waiting at different floors to board the elevator. The set of actions comprises movements of the elevator between different floors along with the ensuing activities of passengers boarding and leaving the elevator. The goal specifies a destination floor for each passenger. In the pipeline example, the initial state describes the initial contents of the pipelines and of the areas they connect. Actions model the changes in the contents of a pipeline as products are pumped through it, and a typical goal requires a certain amount of petroleum product to be available in a certain area. In the card game example, the initial state is given by a randomly dealt card tableau. The set of actions models the different ways of moving cards between piles that are allowed by the rules of the game. The goal consists of achieving a certain arrangement of the cards.

Because planning is not limited to a particular application area, or indeed any finite set of application areas, it is an example of *general problem solving*, and in fact the planning problem was first introduced to the Artificial Intelligence community under that name. Classical planning has been an active research area for about half a century, with Newell and Simon's work on the General Problem Solver [94] usually seen as a starting point. A historical perspective on planning research is provided by a collection of classical papers edited by Allen et al. [1] and a more recent survey by Weld [112].

One of the well-established facts about the planning problem is that it is *hard*. In the very general variants commonly studied in the early days of planning research, such as the original STRIPS formalism [40,84], it is known to be undecidable [39]. In the more restricted formalisms typically considered today, the problem is still at least PSPACE-hard [19].

How do we solve such a hard problem? In his 1945 classic *How to Solve It*, mathematician George Pólya describes a four-step strategy to problem solving. Here is the first and most important step:

"First, you have to *understand* the problem." [99, p. 5]

This is solid general advice. To solve the planning problem, that is to design efficient planning algorithms, it is important to understand it. Is planning difficult? Can we *prove* that it is difficult? Are there relevant special cases which are easier to solve than others?

This volume contributes to the understanding of the planning problem by formally analyzing those special cases which have attracted most attention in the past decade, namely, the standard *benchmark domains* of the International Planning Competitions [8,66,86,87,91]. This is the topic of Part I, *Planning Benchmarks*.

Because planning is general *problem* solving, planning tasks are commonly called planning *problems* in the literature. This implies that Pólya's recommendation is equally applicable to *their* solution: To solve a planning task, one has to understand it. Without any kind of intuition of which actions are useful for achieving the goals in a certain situation, the problem solver is more or less limited to blindly exploring the space of possible solutions, which is usually a fruitless endeavour. Given that we are interested in *algorithmic* approaches to planning, this "understanding" must be arrived at algorithmically. One well-established approach to *informed* planning algorithms is the use of heuristic search techniques [16,68,90]. (Not entirely coincidentally, Pólya's work is also responsible for introducing the word *heuristic* into modern scientific discourse, although not quite with the meaning in which it is generally used in Artificial Intelligence these days.)

This volume contributes to the practice of solving planning tasks by presenting a new approach to heuristic planning based on two central ideas: reformulating planning tasks into a form in which its logical structure is more apparent than in the original specification, and exploiting the information encoded in the *causal graphs* and *domain transition graphs* of these reformulated tasks. This is the topic of Part II, *Fast Downward*.

This volume is a revised version of my doctoral thesis, *Solving Planning Tasks in Theory and Practice* [61], submitted to Albert-Ludwigs-Universität Freiburg in June 2006. It has been a long time in the making, with the first ideas conceived in 1998. Over the years, many people have contributed to the work in some way or other, and I would like to use this opportunity to thank them.

Throughout my PhD studies, which began in 2001, Bernhard Nebel served as my advisor. In addition to providing a perfect work environment, his advice has always been very helpful. While giving me lots of freedom to pursue the scientific topics I was interested in, he pushed me at the right times and with the right amount of force to actually get the work done, which is a crucial contribution.

Sylvie Thiébaux served as the second reviewer of the thesis and gave me some exceptional feedback. I very strongly doubt that anyone else will ever read the thesis as closely as her, myself included.

This work has greatly benefited from feedback and discussions with many people in the AI planning community and some people outside. In addition to Bernhard and Sylvie, I particularly want to thank Carmel Domshlak, Stefan Edelkamp, Maria Fox, Inge Li Gørtz, Jörg Hoffmann, Derek Long, Silvia Richter, Jussi Rintanen and Menkes van den Briel for their scientific input.

Many of the results in Chaps. 4 and 5 build on the work of former students whose work I supervised. I want to thank Michael Drescher, Robert Mattmüller and Gabi Röger for their contributions.

I already mentioned the perfect work environment, and I want to use the opportunity to thank all my former and current colleagues at the Foundations of Artificial Intelligence group at Albert-Ludwigs-Universität Freiburg for useful discussions and for having a lot of fun together. Special thanks go to my officemates throughout the years, in chronological order: Rudi Triebel, Yacine Zemali, Christian Köhler, Yuliya Lierler, Marco Ragni, Yannis Dimopoulos and Sebastian Kupferschmid.

For their help – and patience – in the preparation of this volume, I want to thank Ursula Barth and Frank Holzwarth from Springer. Extra special thanks go to Gabi Röger, who undertook the very laborious task of preparing the keyword index.

Last but certainly not least, for moral support my heartfelt thanks goes to Stefan Franck, Sebastian Kupferschmid, Silvia Richter, Anna Seesjärvi and Libor Valevsky and to my family Gundula, Michael and Volker Helmert.

Freiburg, November 2007   Malte Helmert

Contents

Part I Planning Benchmarks

1 The Role of Benchmarks 3
   1.1 Evaluating Planner Performance 3
        1.1.1 Worst-Case Evaluation 4
        1.1.2 Average-Case Evaluation 5
   1.2 Planning Benchmarks Are Important 7
   1.3 Theoretical Analyses of Planning Benchmarks 8
        1.3.1 Why Theoretical Analyses Are Useful 8
        1.3.2 Published Results on Benchmark Complexity 9
   1.4 Standard Benchmarks 9
   1.5 Summary and Overview 11

2 Defining Planning Domains 13
   2.1 Optimization Problems 13
        2.1.1 Minimization Problems 14
        2.1.2 Approximation Algorithms 15
        2.1.3 Approximation Classes 16
        2.1.4 Reductions 18
   2.2 Formalizing Planning Domains 21
   2.3 General Results and Reductions 24
        2.3.1 Upper Bounds 24
        2.3.2 Shortest Plan Length 25
        2.3.3 Approximation Classes of Limited Interest 26
        2.3.4 Relating Planning and (Bounded) Plan Existence ... 28
        2.3.5 Generalization and Specialization 29

3 The Benchmark Suite 31
   3.1 Defining the Competition Domains 31
   3.2 The Benchmark Suite 32
        3.2.1 IPC1 Domains 32

    3.2.2 IPC2 Domains 34
    3.2.3 IPC3 Domains 34
    3.2.4 IPC4 Domains 35
  3.3 Domains and Domain Families 36

4 **Transportation and Route Planning** 39
  4.1 TRANSPORT and ROUTE 39
    4.1.1 The TRANSPORT Domain 41
    4.1.2 The ROUTE Domain 43
    4.1.3 Special Cases and Hierarchy 44
  4.2 General Results 46
  4.3 Plan Existence 52
  4.4 Hardness of Optimization............................... 54
  4.5 Constant Factor Approximation 59
  4.6 Hardness of Constant Factor Approximation 62
  4.7 Summary 68
  4.8 Beyond TRANSPORT and ROUTE 71

5 **IPC Domains: Transportation and Route Planning** 75
  5.1 GRIPPER 75
  5.2 MYSTERY and MYSTERYPRIME 76
  5.3 LOGISTICS 78
  5.4 ZENOTRAVEL 83
  5.5 DEPOTS 85
  5.6 MICONIC-10 88
  5.7 ROVERS 93
  5.8 GRID 98
  5.9 DRIVERLOG 103
  5.10 AIRPORT 108
  5.11 Summary 111

6 **IPC Domains: Others** 113
  6.1 ASSEMBLY 113
  6.2 BLOCKSWORLD 117
  6.3 FREECELL 117
  6.4 MOVIE 126
  6.5 PIPESWORLD 127
  6.6 PROMELA 132
  6.7 PSR 138
  6.8 SATELLITE 142
  6.9 SCHEDULE 145
  6.10 Summary 149

7   Conclusions...151
    7.1   Ten Conclusions.......................................151
    7.2   Going Further...154

Part II Fast Downward

8   Solving Planning Tasks Hierarchically 157
    8.1   Introduction 157
    8.2   Related Work 163
          8.2.1   Causal Graphs and Abstraction 164
          8.2.2   Causal Graphs and Unary STRIPS Operators.........165
          8.2.3   Multi-Valued Planning Tasks 167
    8.3   Architecture and Overview 168

9   Translation 171
    9.1   PDDL and Multi-valued Planning Tasks..................171
    9.2   Translation Overview 175
    9.3   Normalization...176
          9.3.1   Compiling Away Types 177
          9.3.2   Simplifying Conditions 177
          9.3.3   Simplifying Effects 179
          9.3.4   Normalization Result 179
    9.4   Invariant Synthesis 180
          9.4.1   Initial Candidates 182
          9.4.2   Proving Invariance 183
          9.4.3   Refining Failed Candidates 186
          9.4.4   Examples 188
          9.4.5   Related Work..................................188
    9.5   Grounding...190
          9.5.1   Overview of Horn Exploration 191
          9.5.2   Generating the Logic Program 191
          9.5.3   Translating the Logic Program to Normal Form 193
          9.5.4   Computing the Canonical Model 195
          9.5.5   Axiom and Operator Instantiation..............197
    9.6   Multi-valued Planning Task Generation 197
          9.6.1   Variable Selection 198
          9.6.2   Converting the Initial State..................199
          9.6.3   Converting Operator Effects 200
          9.6.4   Converting Conditions 201
          9.6.5   Computing Axiom Layers 202
          9.6.6   Generating the Output.........................202
    9.7   Performance Notes 203
          9.7.1   Relative Performance Compared to MIPS Translator... 203
          9.7.2   Absolute Performance..........................205

10 Knowledge Compilation 207
   10.1 Overview 207
   10.2 Domain Transition Graphs 208
   10.3 Causal Graphs 213
      10.3.1 Acyclic Causal Graphs 214
      10.3.2 Generating and Pruning Causal Graphs 215
      10.3.3 Causal Graph Examples 217
   10.4 Successor Generators and Axiom Evaluators 220
      10.4.1 Successor Generators 220
      10.4.2 Axiom Evaluators 221

11 Search 223
   11.1 Overview 223
   11.2 The Causal Graph Heuristic 224
      11.2.1 Conceptual View of the Causal Graph Heuristic 225
      11.2.2 Computation of the Causal Graph Heuristic 226
      11.2.3 States with Infinite Heuristic Value 228
      11.2.4 Helpful Transitions 229
   11.3 The FF Heuristic 230
   11.4 Greedy Best-First Search in Fast Downward 231
      11.4.1 Preferred Operators 231
      11.4.2 Deferred Heuristic Evaluation 232
   11.5 Multi-heuristic Best-First Search 233
   11.6 Focused Iterative-Broadening Search 234

12 Experiments 239
   12.1 Experiment Design 239
      12.1.1 Benchmark Set....................................240
      12.1.2 Experiment Setup 242
      12.1.3 Translation and Knowledge Compilation vs. Search 243
   12.2 STRIPS Domains from IPC1–3.........................243
   12.3 ADL Domains from IPC1–3 246
   12.4 Domains from IPC4 248
   12.5 Conclusions from the Experiment........................251

13 Discussion 253
   13.1 Summary..253
   13.2 Major Contributions 254
      13.2.1 Multi-valued Representations 254
      13.2.2 Task Decomposition Heuristics 256
   13.3 Minor Contributions 257
   13.4 Going Further...258

References..259

Index 267

Part I

Planning Benchmarks

1

The Role of Benchmarks

Given the theoretical hardness of classical planning, the chance of developing practically usable solution algorithms appears slim. However, modern planning systems developed since the 1990s such as Graphplan [14], SAT-PLAN [75–78], HSP [16] and FF [68] have demonstrated their ability to solve planning tasks of considerable size. Their efficiency, or – put a bit more carefully – their *perceived* efficiency is somewhat at odds with the theoretical hardness of planning.

To understand why we can observe such good planner performance, we must consider the methods with which the performance of planning systems is evaluated. This is the topic of the following Sect. 1.1, which will lead us to a discussion of *planning benchmarks* in Sect. 1.2 and their theoretical properties in Sect. 1.3. In the penultimate Sect. 1.4, we briefly introduce a set of *standard* benchmarks. The chapter concludes with an overview of Part I in Sect. 1.5.

1.1 Evaluating Planner Performance

Planning systems are usually evaluated empirically by measuring their runtime on a number of planning tasks. In most evaluations, these tasks are not generated completely randomly, but taken from a number of so-called *benchmark domains*, families of related planning tasks often modelled after real-world planning activities such as vehicle routing, machine-shop scheduling, or assembly problems.

Benchmark domains are limited: They do not span the complete space of planning tasks definable in the PDDL language. Before we turn to a detailed discussion of these domains, it is thus worth asking whether planner evaluations based on such a restricted set of instances is a good practice.

For this purpose, let us contrast the planning problem with the classical algorithmic problem of sorting a sequence of numbers. For the sorting problem, it would be quite unusual to evaluate an algorithm based on its performance on a set of arbitrary predefined "benchmark sequences". Instead, one would typically use one or both of the following modes of evaluation:

- *Worst-case evaluation*: The algorithm is evaluated by determining the *maximal* runtime for a given instance size. Worst-case evaluations are usually theoretical analyses using the asymptotic O- and Θ-notations [25].
- *Average-case evaluation*: The algorithm is evaluated by determining or estimating the *average* runtime for a given instance size. In some cases, average-case evaluations can be conducted theoretically, but for more complicated algorithms, one often resorts to experimental studies involving statistical sampling of the problem space.

Worst-case evaluations and average-case evaluations are not limited to the sorting problem, of course. They are general metrics for algorithm performance. Would they not make better alternatives for planner evaluations than considering benchmark tasks?

Indeed, it can be argued that evaluating planners by using tasks from benchmark domains is a flawed approach [102]. A benchmark task taken from one of the classical domains like BLOCKSWORLD or LOGISTICS is not at all representative of an average planning task drawn randomly from the full problem space. In particular, for many of these benchmark domains the problem of deciding whether a given task admits a solution is easy (i. e., polynomially solvable), in stark contrast to the fact that (non-temporal, non-numeric) PDDL planning is EXPSPACE-complete and planning in a fixed domain (i. e., with a fixed set of predicates) is PSPACE-complete in general [39]. How then can an evaluation which is limited to "simple" special cases accurately assess the performance of a planning algorithm? To address this question, let us discuss the two alternatives of worst-case and average-case evaluation.

1.1.1 Worst-Case Evaluation

To evaluate a sorting algorithm, one would normally start by analyzing its worst-case behaviour. If it is worse than the optimal (for *comparison sorts* [25]) $O(n \log n)$ bound, the algorithm would typically be rejected immediately. On the other hand, a propositional planning algorithm with exponential space requirements in the worst case is considered acceptable, even though polynomial space is sufficient. Indeed, no commonly used complete planning algorithm is guaranteed to occupy sub-exponential space.

Similar to space bounds, worst-case time bounds of planning algorithms are considered equally unimportant in practice. For example, it is not difficult to construct planning tasks for which a typical "planning as satisfiability" [75–78] algorithm requires time which is doubly exponential in the number of state variables, while a naive breadth-first search algorithm with duplicate checking runs in (singly) exponential time. Yet the former approach is considered to be state of the art for optimal propositional planning [66], while no-one would seriously consider the latter.

Indeed, people are even willing to (grudgingly) accept *incomplete* planning techniques which will (necessarily or subject to some randomness in the algorithm) fail in some cases [47, 69, 80, 93], while such behaviour would be completely unacceptable for a sorting algorithm.

How can this discrepancy between sorting and planning be explained? The key to understanding why worst-case analyses of planning algorithms are not very useful is that *in the worst case, all planning algorithms are bad.* Propositional planning is PSPACE-complete, general (non-numerical) PDDL planning even EXPSPACE-complete. It is thus extremely unlikely that polynomial algorithms exist for the former, and for the latter, this possibility can even be ruled out theoretically. Exponentially scaling algorithms will always be too slow for some instances. Even though they might be asymptotically optimal, they will always be *practically incomplete* in the sense that there will be relevant planning tasks that they cannot solve within reasonable time bounds. Therefore, for the planning problem, even asymptotically optimal algorithms are of limited usefulness unless they can improve on the worst-case behaviour in practice.

1.1.2 Average-Case Evaluation

If we cannot make our algorithms work well for *all* planning tasks, the next best thing is making them work well *most of the time.* Unfortunately, "most of the time" is an ill-defined concept. Given that there are infinitely many planning tasks, any planning algorithm will be efficient for infinitely many instances and inefficient for infinitely many instances. Indeed, the only exceptions to this rule are those algorithms which are inefficient for almost all instances and thus not really worthy of consideration. *Average case analyses* offer a way of comparing "typical" algorithm performance despite these difficulties.

For an average case analysis, a measure for the *size* of an instance is required. For the sorting problem, the size of an instance is usually considered to be the length of the input sequence. For the planning problem, one could use the number of state variables or operators (or some combination thereof) or the encoding length in a planning formalism such as PDDL as a measure of size.

Having decided on a size measure, the algorithm is evaluated by averaging some performance criterion for all instances of a certain size. In practice, there are typically too many instances of any given size to consider, so the real performance criterion is approximated by statistical sampling, considering only a subset of inputs. For the sorting problem, the usual performance criteria are the number of comparisons performed by the algorithm or its runtime; for planning, runtime is typically considered.

One difficulty in average-case evaluations of planning algorithms is that the distribution of runtimes is often heavy-tailed. If some tasks take exponentially longer to solve than others, they will dominate the average runtime to the extent that other tasks contribute very little to the overall result. In such situations, it is questionable whether the evaluation result accurately captures "typical" algorithm performance. Even worse, some tasks might not be solvable at all within practical resource limits, which makes averaging over runtime impossible. Therefore, it is also common to consider the median runtime or count the percentage of tasks solved within a given timeout instead of taking averages.

To evaluate the overall performance of an algorithm, one would then observe how performance scales with increasing input size, usually by estimating performance for a limited number of sizes and extrapolating a curve from that. In addition to such statistical techniques, it is sometimes possible to determine the average case behaviour of an algorithm with analytical methods, but this is rare for non-trivial planning algorithms.

How useful are average case analyses? For the sorting problem, very useful. Indeed, analyzing the average case is one of the most important techniques for the evaluation of modern sorting algorithms. By averaging running time (or number of comparisons) for uniformly chosen permutations of a certain fixed input sequence, one usually obtains a measurement of algorithm performance which predicts the suitability to practical applications fairly well. Of course, even for the sorting problem, a careful algorithm analysis would also need to consider inputs which are *not* completely random, such as sequences which are already sorted, or presorted to some degree. Still, no-one would doubt the usefulness of pure average-case experiments.

Planning algorithms can be analysed in a similar fashion, and such analyses have been conducted by Bylander [20] and Rintanen [102]. However, their results are somewhat disappointing. In particular, the vast majority of planning tasks in Bylander's model can be solved by the following very simple algorithm:

1. If there is a goal condition that is not satisfied in the initial state and does not appear as the effect of any operator, report that the task has no solution.
2. Otherwise, starting from the initial state, repeatedly apply some operator which increases the number of satisfied goals, until no such operator can be found.
3. If a goal state was reached, return the sequence of operators that were applied in the previous step; otherwise fail.

It is evident that this algorithm is practically useless for all interesting applications of classical planning despite being "good on average".

To obtain more practically significant results, Rintanen's study refines Bylander's model somewhat and focuses on "hard" instances in the *phase transition* region of the problem space [23]. Concentrating on this hard region, the study compares the performance of three prototypical planning systems (FF [68], LPG [44] and a SATPLAN-like planner [104]) on these tasks, finding that the satisfiability planner vastly outperforms FF and LPG as the problem instances grow in size. Rintanen notices that this observed behaviour does not reflect their relative performance on typical benchmark tasks from planning domains like BLOCKSWORLD or LOGISTICS.

Rintanen argues that this shows that typical planning benchmarks do not reflect the structure of "hard" planning problems. He reasons that the prevalence of the planning benchmarks has led to an inordinate focus on heuristic techniques, which are of limited usefulness for hard planning tasks, as the comparatively bad performance on FF on the randomly generated instances shows. However, this argument can also be turned around. Even though they are often derided as "toy problems", the planning benchmarks are the closest available approximations to real-world applications of classical planning. On these benchmarks, FF usually outperforms SATPLAN by a huge margin, not least due to the fact that for many of these domains, FF is provably polynomial [64, 65] while SATPLAN solves an NP-hard problem, as we will show in the following chapters. Thus, the outcome of the average-case analysis runs counter to conventional wisdom in planner evaluation, which we may interpret as a sign that random tasks do not match typical application tasks very well, and are thus of limited applicability in practice.

The truth is somewhere in between these two positions. One particular caveat when evaluating modern planning systems such as FF against the classical planning benchmarks is that these systems have been developed with the benchmarks in mind. For example, the Goal Agenda Management technique used in IPP and FF [68,81] is motivated by the BLOCKSWORLD domain, and the new heuristics we describe in Part II is inspired by "transportation domains" such as LOGISTICS and MYSTERY. Evaluating a planning system on the same set of benchmarks that were instrumental to its development is clearly a dangerous practice.

On the other hand, heuristic planners like FF have also demonstrated good performance on benchmark domains that were introduced *after* the planner's conception, which indicates that the lessons learned from the earlier benchmarks that have factored into the planning system's design may be similarly applicable to other practically interesting planning domains.

1.2 Planning Benchmarks Are Important

From the preceding discussion, it should be apparent that whether or not the community's focus on benchmark domains constitutes a healthy trend is a matter of some debate. In addition to Rintanen's criticism that they represent an uncharacteristically easy fragment of the general planning problem, it can be argued that current practice in classical planning research focuses too much on raw benchmark performance, and too little on original ideas. This has led to a certain uniformity of planning approaches, with three of the five best-performing planners in the non-optimizing propositional track of the 4th International Planning Competition [66] being some variation of Hoffmann's FF.

But irrespective of which side one takes in this debate, it is a fact that the benchmark domains *do* play an important role in evaluating planning algorithms. It has become a very rare occurrence that a paper discussing new techniques for classical planning is published without presenting performance results for some of the benchmarks.

Moreover, a number of benchmark domains are relevant for other reasons besides their use as a testing vehicle. Domains such as MICONIC-10 and PSR are modelled after application problems which are intrinsically useful to solve, although the classical planning variant of PSR is somewhat restricted in scope compared to the original non-deterministic formulation [82, 110]. As another example, the operations research community has been studying vehicle routing problems like the capacitated dial-a-ride problem for a long time due to its significance for logistics applications [11]. These problems are closely related to transportation planning domains acting as common benchmarks such as LOGISTICS, MYSTERY or GRID.

1.3 Theoretical Analyses of Planning Benchmarks

In summary, empirical experiments with planning benchmarks have become the standard for performance evaluation in classical planning. Running time on tasks from domains like LOGISTICS or BLOCKSWORLD is often used for comparing the relative merits of planning techniques, or, put a bit more provocatively, to draw the line between good and bad ones. However, without further knowledge about these benchmark domains, it is very difficult to gauge the efficiency of a planning system in *absolute* terms. If no planning system performs well in a given domain, does that mean that they are all poor, or is that domain intrinsically hard? If they all perform well, is this because of their strength or because of the simplicity of the task?

1.3.1 Why Theoretical Analyses Are Useful

To address such questions, planning domains can be analysed from the viewpoint of computational complexity.

Theoretical knowledge of the complexity of planning in a particular domain is not only useful for judging the runtime behaviour of planning systems. It also helps in assessing the trade-off between planners that plan quickly and planners that generate short plans. For the general planning problem, finding an optimal plan is just as easy – or just as difficult – as finding an arbitrary plan. For a particular benchmark domain, these problems might well be of different complexity. In domains where generating optimal plans is a problem that can be solved in polynomial time, there are fewer reasons to be content with non-optimal plans and generating overly long plans can be viewed as a deficiency of a planning algorithm. On the other hand, in domains where there is a difference in computational complexity between non-optimal and optimal planning, striving for optimality clearly demands a price in runtime performance, which the user of a planning system might not always be willing to pay. We will prove that most common benchmark domains belong to the latter category, which goes a long way towards explaining why heuristic planners like FF dramatically outperform optimizing planners like Graphplan in many domains.

Another point in favour of theoretical analyses is their potential to expose *sources* of hardness in planning domains. For instance, if we discovered that in a hypothetical PAC-MAN domain, plans can be generated in polynomial time if there is a single ghost, while the corresponding problem with multiple ghosts is NP-hard, then this would allow us to draw the conclusion that one source of hardness in this domain is the presence of multiple ghosts.

1.3.2 Published Results on Benchmark Complexity

Because of these issues it is rather surprising that so far, there is relatively little research aimed at understanding the standard benchmark domains, with the only notable exception being the BLOCKSWORLD domain.

Gupta and Nau analyse the complexity of BLOCKSWORLD planning both in its standard form and in several variants, such as ones with blocks of different size. They prove that while non-optimal BLOCKSWORLD planning is a simple polynomial problem, finding optimal plans is NP-hard [52]. Selman refines these results by showing that it is possible, in polynomial time, to generate plans that are as most a constant factor as long as the shortest one, but that this factor cannot be brought arbitrarily close to 1 unless P = NP [107]. Slaney and Thiébaux provide a deep and thorough analysis of the algorithmics and empiricism of BLOCKSWORLD planning, including the identification of efficient non-optimal and optimal planning algorithms and an investigation of phase transitions [108].

However, for other planning benchmark domains, previous work on computational complexity appears to be non-existent.

1.4 Standard Benchmarks

Seeing that domain-specific complexity results for planning problems appear to be useful, the question arises *which* domains are of particular interest. It is evident that it is neither feasible nor desirable to investigate every single planning domain that has been considered in the literature. So which domains should we focus on? Ten years ago, this question would have been difficult to answer. Many publications would introduce their own planning domains which would never be considered again in another paper, and apart from a few classical planning domains like BLOCKSWORLD or TOWERSOFHANOI,

IPC1	ASSEMBLY	IPC2	BLOCKSWORLD
	GRID		FREECELL
	GRIPPER		LOGISTICS
	LOGISTICS		MICONIC-10
	MOVIE		SCHEDULE
	MYSTERY		
	MYSTERYPRIME		
IPC3	DEPOTS	IPC4	AIRPORT
	DRIVERLOG		PIPESWORLD
	FREECELL		PROMELA
	ROVERS		PSR
	SATELLITE		SATELLITE
	ZENOTRAVEL		

Fig. 1.1. Domains from the International Planning Competitions

there would be little or no consensus about which planning domains could be considered "standard".

However, with the advent of the International Planning Competitions (abbreviated as IPC in the following), this situation has changed [8,66,86,87,91]. Instigated in 1998 and held as a biennial event since then, the competitions have led to a consolidation of classical planning benchmarks to the extent that most current research on classical PDDL planning exclusively relies on the competition domains (Fig. 1.1) for benchmarking. The proceedings of the major planning conferences since the first planning competition clearly show this:

– The ECP 1999 proceedings [12] contain eight papers on classical planning techniques. All of these use at least the BLOCKSWORLD or LOGISTICS domain in their evaluations, but only two use IPC domains exclusively. The other six papers use a number of other planning domains mostly from the UCPOP [97] suite.
– The AIPS 2000 proceedings [24] contain nine papers on classical planning techniques. All of these use at least two IPC domains, with LOGISTICS, BLOCKSWORLD, GRIPPER and GRID considered three or more times. However, six of the papers also include one non-IPC domain (most commonly ROCKET), and one paper uses two non-IPC domains (FERRY and TOWERSOFHANOI).
– The ECP 2001 proceedings [21] contain six papers on classical planning techniques. IPC domains are predominantly used in all papers, with only one paper using more than one other domain (however, that paper also uses all the IPC domains available at the time).
– The AIPS 2002 proceedings [48] contain seven papers on classical planning techniques, three of which use IPC domains exclusively for evaluation

purposes. The other four papers use IPC domains predominantly, but also consider a variety of other domains.

- The ICAPS 2003 proceedings [50] contain two papers on classical planning techniques, both of which use IPC domains exclusively for evaluation purposes.
- The ICAPS 2004 proceedings [115] contain six papers on classical planning techniques, five of which use IPC domains exclusively for evaluation purposes. The remaining paper uses five IPC domains and BRIEFCASE.
- The ICAPS 2005 proceedings [13] contain eight papers on classical planning techniques, seven of which use IPC domains exclusively for evaluation purposes. The remaining paper uses two IPC domains and TOWER-SOFHANOI.

1.5 Summary and Overview

Summarizing this chapter, we have made the following observations:

- *Planning algorithms are usually evaluated empirically, using benchmark domains.* Other modes of evaluation, like general worst-case or average-case analyses are altogether less appropriate, and thus rare.
- *Benchmark domains are important.* Besides their role for performance measurements, benchmark domains are also used for identifying short-comings of planning systems or as inspiration for new planning techniques. Moreover, some of them model relevant application problems and are thus intrinsically interesting.
- *Little is known about most benchmark domains.* With the exception of BLOCKSWORLD, the benchmark domains are not formally analysed in the planning literature.
- *The IPC domains form a set of standard benchmarks.* Most papers published in recent years use domains from the International Planning Competitions exclusively for performance evaluations.

Following these observations, we dedicate the rest of this part to a formal analysis of the IPC benchmark domains. The following chapters are organized as follows:

- In Chap. 2, we develop the formal background necessary for our analysis. In particular, we define a formal notion of *planning domain* and introduce the *plan existence* and *bounded plan existence* decision problems as well as the *planning* optimization problem. We also provide some reductions between these problems.
- In Chap. 3, we informally introduce the IPC planning domains.
- In Chap. 4, we introduce two families of planning domains related to transportation planning and route planning and present detailed complexity results for these.

- In Chap. 5, we present detailed complexity results for a number of IPC planning domains related to transportation or route planning.
- In Chap. 6, we present detailed complexity results for those IPC planning domains which are not related to transportation or route planning.
- In Chap. 7, we summarize the findings of this part and draw conclusions.

Most of the material presented in this part is based on work that we have published previously. In particular, the decision complexity results for the planning domains from the first two International Planning Competitions and the TRANSPORT domain family were first presented in the author's Master's thesis [54] and published in an article in the Artificial Intelligence Journal [57]. Part of these results are also included in an article presented at ECP 2001 [55]. The complexity results for the domains from IPC4 are included in an article presented at ICAPS 2006 [60]. Those approximation results in the planning competition domains (but not in the general TRANSPORT and ROUTE domain families) which are *not* implied by the decision complexity results are included in an article presented at ECAI 2006 [62]. The complexity results for the domains from IPC3 and for the ROUTE domain family as well as all approximation results for the TRANSPORT and ROUTE domain families are previously unpublished.

The approximation results in the GRID domain and in some of the domains in the TRANSPORT family are the result of joint work with Robert Mattmüller [89] and Michael Drescher [31].

2

Defining Planning Domains

To be able to state precise results for the planning benchmarks, we first need to formalize them. In this chapter, we lay the foundations for such formalizations. The following Sect. 2.1 provides a brief introduction into the theory of *minimization problems* and their formal properties, in particular their classification into *approximation classes*. In Sect. 2.2, planning domains are introduced as examples of minimization problems. Finally, Sect. 2.3 contains some general classification results and reductions applying to all planning domains.

2.1 Optimization Problems

We assume that the reader is familiar with the basic notions of complexity theory, such as decision problems and the complexity classes P, NP and PSPACE, and refer to the literature for definitions [43].

Classical complexity theory is concerned with yes/no-questions. Does a given planning task have a solution? Does it have a solution with certain properties (e. g., using no more than a certain number of operators)? In many cases, analyzing the theoretical difficulty of answering such questions already provides a clear picture of the hardness of planning in a certain domain. However, in some cases it does not. For example there are domains where it is easy to find *some* solution in polynomial time, but difficult to find an *optimal* solution. In such a situation, it is natural to ask just how close to optimality we can get without sacrificing polynomial runtime.

For addressing such questions, the tools of classical complexity theory are somewhat limited. A more adequate framework is provided by the theory of *approximation algorithms*, which we will now introduce. We will keep our discussion brief, referring to the textbook by Ausiello et al. [7] for a more thorough treatment. For the most part, our presentation follows theirs.

2.1.1 Minimization Problems

The central concept in the theory of approximation algorithms is that of
optimization problems, which play the same role that decision problems do in
classical complexity theory. Optimization problems are either *minimization
problems* or *maximization problems*.

Definition 2.1.1. *Minimization Problem*
*A **minimization problem** is a 3-tuple $\langle \mathcal{I}, \mathcal{S}, m \rangle$ with the following components (Σ and Γ are finite alphabets):*

- *$\mathcal{I} \subseteq \Sigma^*$ is a polynomial-time recognizable set of **instances**.*
- *$\mathcal{S} : \mathcal{I} \to 2^{\Gamma^*}$ is a **solution function**, which maps instances $I \in \mathcal{I}$ to a
  (possibly empty) set of **solutions** of I.*
- *$m : \bigcup_{I \in \mathcal{I}}(\{I\} \times \mathcal{S}(I)) \to \mathbb{N}_0$ is a polynomial-time computable **measure
  function**, which maps an instance $I \in \mathcal{I}$ and a solution S for I to a
  natural number called its **measure**.*

*An instance $I \in \mathcal{I}$ is called **solvable** if $\mathcal{S}(I) \neq \emptyset$, and **unsolvable** otherwise.*

Maximization problems are similarly defined. However, we do not need them
for this volume, so we do not introduce them formally.

Similar to common notation in complexity theory, in practice the instances
and solutions of a minimization problem are described by graphs, sets or functions rather than words over an alphabet, and it is assumed that a "reasonable
encoding" is used for representing such mathematical structures in contexts
that require "proper" words, for example as inputs to Turing Machines. Such
leniency is justified by the fact that problems that differ only in encoding
share the same approximability (and complexity) properties if the encoding
lengths are polynomially related. For all the minimization problems we will
introduce, recognizing instances and computing measures is an easy task, so
we do not describe these when defining a problem. Although measure functions are formally defined to take both the instance and the solution as an
argument, we usually omit the instance and write $m(S)$ instead of $m(I, S)$ to
simplify notation, when the instance is clear from context.

Our definition of minimization problems generalizes the usual definition [7]
(called the "standard definition" in the following) in two ways. First, we do not
require that the lengths of solutions are polynomially bounded in the length
of the instance. Solutions to planning tasks typically do not satisfy this polynomial length bound property because it is often possible to "waste time"
by applying inconsequential actions indefinitely. More importantly, there exist planning tasks for which no short solutions exist. To model the planning
problem faithfully, we thus need the ability to specify exponentially long solutions. The second extension is that we allow for solutions of measure 0 –
it is more common to demand that measure functions map to the positive

integers. We will see shortly how these extensions influence the definition of the common approximation classes.

Two examples of minimization problems are given by the MINIMUM VERTEX COVER and MINIMUM SET COVER problems, which we will use for a number of reductions in the following chapters.

Problem 2.1.2. MINIMUM VERTEX COVER
The MINIMUM VERTEX COVER problem is defined as follows:

INSTANCE: A graph $\langle V, E \rangle$.
SOLUTION: A subset of vertices $U \subseteq V$ such that for all edges $\{u, v\} \in E$,
    $u \in U$ or $v \in U$.
MEASURE: $|U|$.

Problem 2.1.3. MINIMUM SET COVER
The MINIMUM SET COVER problem is defined as follows:

INSTANCE: A finite set S and collection C of subsets of S.
SOLUTION: A set cover of S, i.e., a subset $D \subseteq C$ such that each element of
    S is contained in some set in D.
MEASURE: $|D|$.

2.1.2 Approximation Algorithms

Informally, the algorithmic task associated with a minimization problem is that of computing, for a given instance, a solution with measure as small as possible. If an instance has no solution, the algorithm must detect this. We will now formalize these notions, starting with algorithms that do not provide any guarantee regarding the quality of the generated solutions.

Definition 2.1.4. *Approximation Algorithm*
Let $\mathcal{P} = \langle \mathcal{I}, \mathcal{S}, m \rangle$ be a minimization problem.

*An **approximation algorithm** $\mathcal{A}$ for $\mathcal{P}$ is a Turing Machine which, given an instance $I \in \mathcal{I}$, recognizes whether or not it is solvable, and if this is the case, computes some solution $S \in \mathcal{S}(I)$.*

*Depending on the properties of $\mathcal{A}$, we distinguish **deterministic** and **non-deterministic** approximation algorithms, and **polynomial-time** and **exponential-time** approximation algorithms.*

We assume a strong notion of non-deterministic computation where it is allowed for a Turing Machine to compute different solutions if different non-deterministic choices are made. (There are weaker notions of non-deterministic computation where the non-deterministic choices may only influence whether or not the machine produces a solution at all, but not which solution is produced [95].) When we speak of *the* solution computed by a non-deterministic

Turing Machine, we mean *one* (arbitrarily chosen) solution of minimal measure among those that it may compute. As usual in complexity theory, *exponential time* is understood as "$O(2^{n^k})$ for some $k \in \mathbb{N}_0$".

Before we can introduce formal notions for approximation algorithms of different quality, we need to define the concept of *performance ratio*.

Definition 2.1.5. *Performance Ratio*
Let $\langle \mathcal{I}, \mathcal{S}, m \rangle$ be a minimization problem, and let $I \in \mathcal{I}$ be a solvable instance.
  *The **optimal measure** for I is defined as $m^*(I) = \min_{S \in \mathcal{S}(I)} m(S)$.*
  *An **optimal solution** of I is a solution $S^* \in \mathcal{S}(I)$ with $m(S^*) = m^*(I)$.*
  *For any solution $S \in \mathcal{S}(I)$, the **performance ratio** of S is defined as the fraction $\frac{m(S)}{m^*(I)}$, where $\frac{0}{0} = 1$ and $\frac{k}{0} = \infty$ for $k \neq 0$.*

For minimization problems, our objective is to generate solutions with small measure. We are thus chiefly interested in algorithms which can guarantee a certain upper bound on the performance ratios of the solutions they compute.

Definition 2.1.6. *f-Approximation, c-Approximation, Optimal Algorithm*
*Let $f : \mathbb{N}_0 \to \mathbb{R}^+$ be a function. An approximation algorithm $\mathcal{A}$ for some minimization problem $\mathcal{P} = \langle \mathcal{I}, \mathcal{S}, m \rangle$ is called an f-**approximation** if, for all solvable instances $I \in \mathcal{I}$, the performance ratio of the solution generated by $\mathcal{A}$ is at most $f(|I|)$.*

*If f is a constant function $f : n \mapsto c$ ($c \in \mathbb{R}$), then $\mathcal{A}$ is also called a c-**approximation**. A 1-approximation is also called an **optimal algorithm**.*

*A problem $\mathcal{P}$ is called f-**approximable** (c-**approximable**, **optimizable**, **solvable**) by a deterministic (non-deterministic) polynomial (exponential) algorithm if there exists a deterministic (non-deterministic) polynomial-time (exponential-time) f-approximation (c-approximation, optimal algorithm, approximation algorithm) for $\mathcal{P}$. The qualifier "deterministic" may be omitted.*

2.1.3 Approximation Classes

We can now define a number of *approximation classes*, which are the optimization problem counterparts of complexity classes in classical complexity theory.

Definition 2.1.7. *Approximation Classes*
*The **approximation classes** PO, FPTAS, PTAS, APX, poly-APX, exp-APX, PS, NPO, NPS, EXPO and EXPS contain a minimization problem $\mathcal{P}$ under the following conditions:*

- *for PO (P-optimizable):*
  $\mathcal{P}$ is optimizable by a polynomial algorithm.

- *for* **FPTAS** *(fully polynomial-time approximation scheme):*
  $\mathcal{P}$ *is c-approximable for all real numbers $c > 1$ by an algorithm with running time polynomial both in the instance size and in $\frac{1}{c-1}$.*
- *for* **PTAS** *(polynomial-time approximation scheme):*
  $\mathcal{P}$ *is c-approximable by a polynomial algorithm for all real numbers $c > 1$.*
- *for* **APX** *(approximable):*
  $\mathcal{P}$ *is c-approximable by a polynomial algorithm for some real number $c > 1$.*
- *for* **poly-APX** *(poly-approximable):*
  $\mathcal{P}$ *is p-approximable by a polynomial algorithm for some polynomial $p \in \mathbb{R}[x]$.*
- *for* **exp-APX** *(exp-approximable):*
  $\mathcal{P}$ *is f-approximable by a polynomial algorithm for some exponential function $f \in O(2^{n^k})$ $(k \in \mathbb{N}_0)$.*
- *for* **PS** *(P-solvable):*
  $\mathcal{P}$ *is solvable by a polynomial algorithm.*
- *for* **NPO** *(NP-optimizable):*
  $\mathcal{P}$ *is optimizable by a non-deterministic polynomial algorithm.*
- *for* **NPS** *(NP-solvable):*
  $\mathcal{P}$ *is solvable by a non-deterministic polynomial algorithm.*
- *for* **EXPO** *(EXP-optimization):*
  $\mathcal{P}$ *is optimizable by an exponential algorithm.*
- *for* **EXPS** *(EXP-solvable):*
  $\mathcal{P}$ *is solvable by an exponential algorithm.*

Note that for class **PTAS**, unlike **FPTAS**, the c-approximating algorithm may have arbitrary dependence on c (i. e., it may grow very fast as c approaches 1).

Our definition includes four unusual approximation classes not found in the book by Ausiello et al. [7], namely **PS**, **NPS**, **EXPS** and **EXPO**. We now explain why they are necessary for our analysis.

If we followed the standard definition of minimization problems, classes **exp-APX** and **PS** would be identical, so that there would be no need to define the latter. To see this, consider a problem $\mathcal{P} \in$ **PS**. Membership in **PS** implies that there exists a deterministic polynomial-time approximation algorithm which solves $\mathcal{P}$. Running in polynomial-time, such an algorithm can only generate polynomially large solutions. Moreover, the measure function must be computable in polynomial time in the size of that solution, and thus also in polynomial time in the instance size. In polynomial time, it is only possible to compute exponentially large numbers (rather than, for example, doubly exponentially large numbers), and thus the measure of any solution computed by the algorithm is at most exponential in the instance size. According to the usual definition of minimization problems, this would imply that the performance ratio of the solution is at most exponential in the task size because the optimal solution must have a measure of at least 1. However, we allow for solutions of measure 0, so that this argument does not work. We thus distinguish between **exp-APX** and **PS**.

The approximation class NPS is not required under the standard definition of minimization problems because it would be identical to NPO: With the requirement that the size of any solution is polynomially bounded in the size of the instance, *all* minimization problems belong to NPO, as it is always possible to guess an optimal solution in this setting. By contrast, our more general definition allows for the possibility that optimal solutions are exponentially long, while non-optimal solutions are short. Problems with this property belong to NPS, but not to NPO. (Note that the set of solutions of a given instance is always polynomial-time recognizable *in the size of the solution*. This follows from the polynomial-time computability requirement for measure functions.)

Finally, classes EXPO and EXPS are obvious extensions once exponentially long solutions are introduced, and should not require further comment.

It is easy to see that $PO \subseteq PTAS \subseteq APX \subseteq poly\text{-}APX \subseteq exp\text{-}APX \subseteq PS \subseteq NPS$ and $PO \subseteq NPO \subseteq NPS$ and that all these classes are identical if $P = NP$. More interestingly, all these inclusions are strict if $P \neq NP$. For example, if $P \neq NP$, then MINIMUM VERTEX COVER belongs to $APX \setminus PTAS$ and MINIMUM SET COVER belongs to $poly\text{-}APX \setminus APX$ [7].

Beyond polynomial time, we have $NPO \subseteq EXPO \subseteq EXPS$ and $NPS \subseteq EXPS$. All these inclusions are strict: To see $NPO \not\supseteq EXPO$ and $NPS \not\supseteq EXPS$, consider problems which only have exponentially long solutions. To see $EXPO \not\supseteq EXPS$, consider problems which have exponentially long non-optimal solutions and doubly exponentially long optimal solutions.

An inclusion which holds in the standard theory of approximation algorithms, but *not* in our setting is $exp\text{-}APX \subseteq NPO$. To see this, note that our definition allows for problems where non-optimal solutions can be generated in polynomial time by deterministic algorithms, but optimal solutions cannot be generated in polynomial time by non-deterministic algorithms because they are exponentially long.

2.1.4 Reductions

Similar to classical complexity theory, the notion of *reducibility* between problems is central to the theory of approximation algorithms. Before we introduce reductions between minimization problems, let us briefly review the definition of *Karp reductions* (also called *polynomial many-one reductions*), which are the basis of most results in classical complexity theory.

Definition 2.1.8. *Karp Reduction*
*Let P and P' be decision problems. A **Karp reduction** between P and P' is a polynomial-time computable function mapping positive instances of P to positive instances of P' and negative instances of P to negative instances of P'. If there exists a Karp reduction between P and P', we say that P is **Karp-reducible** to P', in symbols $P \leq_p P'$.*

In classical complexity theory, Karp reductions are sufficient for most purposes. Unfortunately, there is no such unifying notion of reducibility for approximation algorithms. As a case in point, Crescenzi identifies no fewer than nine types of reductions between optimization problems in his overview article [26], and these exhibit widely differing properties. Although all of these have been used in the literature, one of them – *approximation-preserving reductions*, or *AP-reductions* for short – have appeared to gain the widest support in recent years, and might emerge as a kind of "standard" reducibility for approximation problems. We thus exclusively use AP-reductions throughout this work, despite the fact that some results might be somewhat easier to prove using other notions of reducibility, such as *L-reductions* [26].

Definition 2.1.9. AP-Reduction
Let $\mathcal{P} = \langle \mathcal{I}, \mathcal{S}, m \rangle$ and $\mathcal{P}' = \langle \mathcal{I}', \mathcal{S}', m' \rangle$ be minimization problems. An **AP-reduction** or **approximation-preserving reduction** between $\mathcal{P}$ and $\mathcal{P}'$ is a 3-tuple $\langle \alpha, f, g \rangle$ with the following components:

- $\alpha \geq 1$ is a real-valued parameter.
- f is a function which, given an instance $I \in \mathcal{I}$ and a real parameter $r > 1$, generates an instance $f(I, r) \in \mathcal{I}'$ such that $f(I, r)$ is solvable iff I is solvable.
- g is a function which, given a solvable instance $I \in \mathcal{I}$, a real parameter $r > 1$, and some solution $S' \in \mathcal{S}'(f(I, r))$ with performance ratio at most r, generates a solution $g(I, r, S') \in \mathcal{S}(I)$ to the original instance with performance ratio at most $1 + \alpha(r - 1)$.

For each fixed value of r, the functions f and g are polynomial-time computable.

If there exists an AP-reduction between $\mathcal{P}$ and $\mathcal{P}'$, we say that $\mathcal{P}$ is **AP-reducible** to $\mathcal{P}'$, in symbols $\mathcal{P} \leq_{\text{AP}} \mathcal{P}'$.

AP-reducibility is a generalization of Karp-reducibility in a certain sense. In particular, function f maps instances of the "easier" problem to an instance of the "harder" problem in the same way and subject to the same constraints as a Karp reduction does. However, this is not enough to provide a reduction for an optimization problem: We are not just interested in the solvability of an instance, but also in the quality of the solutions generated by an approximation algorithm. Therefore, the reduction needs another component, which maps solutions to the instance generated by the reduction back to solutions of the original instance in such a way that "good quality" solutions are mapped to "good quality" solutions. This mapping is provided by function g, and the quality guarantee is provided by the condition that solutions with performance ratio r are mapped back to solutions with performance ratio at most $1 + \alpha(r - 1)$.

As an example, we can use the following approximation-preserving reduction to show MINIMUM VERTEX COVER $\leq_{\text{AP}}$ MINIMUM SET COVER:

- The parameter α is set to 1.
- Function f maps MINIMUM VERTEX COVER instances (graphs) to MINIMUM SET COVER instances as follows: Given graph $\langle V, E \rangle$, the edge set E is used as the finite set that must be covered, and the collection of subsets to use for covering includes one set for each vertex $v \in V$, consisting of the set of edges incident to v. In this case, the function f is independent of the real parameter $r > 1$, which is a common situation.

  This function clearly satisfies the requirement that $f(I, r)$ is solvable iff I is solvable. To be more precise, both I and $f(I, r)$ are always solvable, as every graph has a trivial vertex cover consisting of all vertices.
- Function g maps MINIMUM VERTEX COVER instances (graphs) and MINIMUM SET COVER solutions (set covers) to MINIMUM VERTEX COVER solutions (vertex covers) as follows: Given graph $\langle V, E \rangle$ and set cover D', for each set $E' \in D'$, add a vertex $v \in V$ whose set of adjacent edges equals E' to the vertex cover. Such a vertex exists by construction of the MINIMUM SET COVER instance. The choice of $v \in V$ is usually unique except for the case where E' is a singleton set consisting of an edge which connects two vertices of degree 1. In this case, either vertex can be selected. Again, the mapping is independent of the parameter $r > 1$.

  It is apparent that g maps set covers D' to vertex covers U. Moreover, it is easy to see that $|U| = |D'|$, and hence both solutions have the same measure. Finally, the *optimal* measures for both instances are identical, so that the performance ratio of U equals the performance ratio of D'. In other words, if D' has performance ratio r, then U has performance ratio at most (actually, equal to) $r = 1 + 1 \cdot (r - 1) = 1 + \alpha(r - 1)$, satisfying the last requirement for an AP-reduction.

Observe that the mapping from MINIMUM VERTEX COVER instances to MINIMUM SET COVER instances in the example reduction is identical to the one that is commonly used to establish Karp-reducibility between the related decision problems. While there are many cases in which Karp reductions can easily be extended to AP-reductions, there are also many counter-examples.

AP-reductions satisfy the properties that would normally be expected from reductions. In particular, AP-reducibility is transitive and preserves membership in most of the optimization classes defined earlier, most importantly PTAS and APX [7]. However, it does *not* necessarily preserve membership in PO and FPTAS: Unless P = NP, there exist problems $\mathcal{P}$ and $\mathcal{P}'$ such that $\mathcal{P} \leq_{\mathrm{AP}} \mathcal{P}'$ and $\mathcal{P}' \in$ PO (or $\mathcal{P}' \in$ FPTAS), but $\mathcal{P} \notin$ PO ($\mathcal{P} \notin$ FPTAS).

There are stricter notions of reducibility which do satisfy these properties, but we do not need them for our purposes. However, we introduce one slight extension to AP-reductions (not taken from the literature) that we will make use of in many of our reductions.

Definition 2.1.10. OP-Reduction
Let $R = \langle \alpha, f, g \rangle$ be an AP-reduction between minimization problems $\mathcal{P}$ and $\mathcal{P}'$, let $\mathcal{I}$ be the set of instances of $\mathcal{P}$, and let $\mathcal{I}'$ be the set of instances of $\mathcal{P}'$.

*R is called **optimization-preserving** iff there exists a constant $r > 1$ and a polynomial-time computable function $b_r : \mathcal{I} \times \mathbb{N}_0 \to \mathbb{N}_0$ such that for all instances $I \in \mathcal{I}$ and all natural numbers $K \in \mathbb{N}_0$:*

$$m^*(I) \leq K \text{ iff } m^*(f(I,r)) \leq b_r(I,K).$$

*In this case, we say that R is an OP-reduction **for the approximation factor** r **via the bound function** b_r. (In cases where f and b_r do not depend on r, we do not mention the approximation factor.)*

*If there exists an OP-reduction between $\mathcal{P}$ and $\mathcal{P}'$, we say that $\mathcal{P}$ is **OP-reducible** to $\mathcal{P}'$, in symbols $\mathcal{P} \leq_{\mathrm{OP}} \mathcal{P}'$.*

The purpose of OP-reductions is to allow an AP-reduction between $\mathcal{P}$ and $\mathcal{P}'$ to double as a Karp reduction between their *associated decision problems* which ask whether or not an instance of $\mathcal{P}$ (or $\mathcal{P}'$) has a solution of measure at most $K \in \mathbb{N}_0$. In particular, if $\langle \alpha, f, g \rangle$ is an OP-reduction between $\mathcal{P}$ and $\mathcal{P}'$ for the approximation factor r via the bound function b_r, then $\langle I, K \rangle \mapsto \langle f(I,r), b_r(I,K) \rangle$ is a Karp reduction between their associated decision problems.

2.2 Formalizing Planning Domains

For our purposes, planning is simply a special kind of minimization problem. The *syntactical* aspects of specifying planning tasks are of little importance to our analysis; we only care about being able to define planning domains in a way that is reasonably concise and accessible. To provide a clear idea of the *semantical* aspects of planning, we first define the notion of *state spaces*.

Definition 2.2.1. State Space
*A **state space** is a 5-tuple $\mathcal{S} = \langle S, s_0, S_*, O, w \rangle$ with the following components:*

- *S is a finite set of **states**.*
- *$s_0 \in S$ is the **initial state**.*
- *$S_* \subseteq S$ is the set of **goal states**.*
- *O is a finite set of **operators**, functions mapping states from a subset of S to S. An operator is called **applicable** in a given state iff it is defined on that state. Operators are also called **actions**.*
- *$w : O \to \mathbb{N}_0$ is the **operator cost** function.*

*S defines the **state transition graph** $G(\mathcal{S})$, a labelled, weighted digraph with vertex set S and an arc labelled o with weight $w(o)$ from s to $o(s)$ for all operators o and all states s for which o is defined.*

In practice, we will define states in a structured way, for example as tuples of functions, and describe operators only in terms of the *changes* to the state in which they are applied. For example, we might define states as pairs of functions $s = \langle f, g \rangle$ and describe an operator as "o changes $g(x)$ to y". By this we mean that $o(s) = \langle f, g' \rangle$ where $g'(x) = y$ and $g'(x') = g(x)$ for all $x' \neq x$.

In most cases – in particular when we do not mention operator costs – we assume a unit cost model, where the operator cost function is a constant function mapping to 1. However, there are some settings in which we find it useful to generalize this model, and thus we allow for arbitrary operator cost functions, even including operators of cost 0.

We can now define what we mean by a planning domain.

Definition 2.2.2. *Planning Domain*
*A **planning domain** $\mathcal{D}$ is a function that maps words T (called **planning tasks**) over some finite alphabet Σ to state spaces $\mathcal{D}(T) = \langle S, s_0, S_\star, O, w \rangle$ with the following properties:*

- *States $s \in S$ can be represented in space polynomially bounded by $|T|$.*
- *$|O|$ is polynomially bounded by $|T|$.*
- *Operator costs are exponentially bounded by $|T|$.*
- *There exists a polynomial algorithm which takes T, $o \in O$ and $s \in S$ as inputs, decides if s is applicable in o and computes $o(s)$ if this is the case.*
- *There exists a polynomial algorithm which takes T and $s \in S$ as inputs and decides if $s \in S_\star$.*

In practice, the encoding language for planning tasks is usually PDDL [38, 42, 91]. However, since we are not interested in representational issues here, we will describe planning tasks in terms of structures that are natural to the domain at hand (such as roadmap graphs or fuel functions) rather than encode them in propositional or first-order logic. For our results to be applicable to planning in PDDL, we must make sure that encoding lengths of planning tasks in these two different representations are polynomially equivalent. This is the case for all the domains we will investigate, and we will not explicitly mention encoding lengths in our definitions of planning domains.

The formal restrictions for planning domains capture the properties of classical (i. e., non-numerical, non-temporal) PDDL-style planning with schematic operators in a fixed domain. They ensure that the difficulty of planning in a given domain does not grow without bounds: In particular, they ensure that we can always plan by exhaustively searching the state space of a task. If, for example, we would drop the first restriction, there could be infinitely many states, as in *numerical planning*, which is generally undecidable [56].

We can now finally define what a plan is.

Definition 2.2.3. *Plan*
*Let $\mathcal{D}$ be a planning domain, and let T be one of its tasks. A **plan** π for T is a path in the state space $\mathcal{D}(T)$ leading from the initial state to some goal state.*

*The **cost** or **measure** of $\pi = \pi_1 \ldots \pi_n$, in symbols $m(\pi)$, is defined as $m(\pi) = \sum_{i=1}^{n} w(\pi_i)$, where w is the operator cost function of $\mathcal{D}(T)$.*

In the following, we will identify a task with its state space. For example, we will speak of the states of a *task*, rather than of its state space. Moreover, we will use the notation $T \in \mathcal{D}$ to denote that T is a task of domain $\mathcal{D}$, although formally $\mathcal{D}$ is actually a *mapping* defined on a set of tasks, rather than a set of tasks.

For a given planning domain, we are chiefly interested in the *planning problem* for that domain: Given a task of the domain, compute a (preferably "cheap") plan, or prove that no plan exists. The theory of minimization problems provides an adequate formalism for this.

Problem 2.2.4. Planning
Let $\mathcal{D}$ be a planning domain. The **planning problem** PLAN-$\mathcal{D}$ for the domain $\mathcal{D}$ is the following minimization problem:

INSTANCE: A planning task $T \in \mathcal{D}$.
SOLUTION: A plan π for T.
MEASURE: The cost of the plan, $m(\pi)$.

In addition to the planning problem proper, we will also consider two related decision problems. Decision problems have a longer history than optimization problem, so that it is possible to utilize a larger body of work for reductions. Moreover, as we will see in the following section, analyzing the complexity of these decision problems is often all that is needed to classify the approximation complexity of the planning problem in a given domain.

Problem 2.2.5. Plan Existence
Let $\mathcal{D}$ be a planning domain. The **plan existence problem** PLANEX-$\mathcal{D}$ for the domain $\mathcal{D}$ is the following decision problem:

GIVEN: A planning task $T \in \mathcal{D}$.
QUESTION: Is there a plan for T?

Plan existence is closely related to the problem of *generating a plan* for the task. If the plan existence problem is easy to solve, it is typically easy to generate a plan; the converse is always true.

Problem 2.2.6. Bounded Plan Existence
Let $\mathcal{D}$ be a planning domain. The **bounded plan existence problem** PLANLEN-$\mathcal{D}$ for the domain $\mathcal{D}$ is the following decision problem:

GIVEN: A planning task $T \in \mathcal{D}$ and a natural number $K \in \mathbb{N}_0$.
QUESTION: Is there a plan π for T with $m(\pi) \leq K$?

Bounded plan existence is related to *generating an optimal plan* for a task. If bounded plan existence is easy to solve, it is typically easy to generate an optimal plan; the converse is always true.

2.3 General Results and Reductions

Before moving on to the planning domains we are interested in, we now present some general results which hold for *all* planning domains. This reduces the busywork required for classifying the complexity of PLAN-$\mathcal{D}$, PLANEX-$\mathcal{D}$ and PLANLEN-$\mathcal{D}$.

2.3.1 Upper Bounds

We first observe some general upper bounds.

Theorem 2.3.1. PLAN-$\mathcal{D}$ ∈ EXPO

PLAN-$\mathcal{D}$ ∈ EXPO, *for any planning domain* $\mathcal{D}$.

Proof. Let $\mathcal{D}$ be a planning domain. From Definition 2.2.2, there exist polynomials p_S and p_O such that for any planning task $T \in \mathcal{D}$, the number of states of T is bounded by $2^{p_S(|T|)}$ and the number of operators is bounded by $p_O(|T|)$. Moreover, there is an algorithm with polynomial run-time bound $p_a(|T|)$ which determines for each state s and operator o whether o is applicable in s and what the resulting state $o(s)$ is. Finally, there is an algorithm with polynomial run-time bound $p_\star(|T|)$ which determines whether or not a given state is a goal state.

We can thus explicitly create the state transition graph for T in time $O(2^{p_S(|T|)} \cdot p_O(|T|) \cdot p_a(|T|))$ (for each state, determine the set of applicable operators and introduce arcs for them) and mark all goal states in time $O(2^{p_S(|T|)} \cdot p_\star(|T|))$. Standard graph search techniques [25] find a shortest path in that graph leading from the initial state to some goal state in time which is polynomial in the size of the graph. Given such a shortest path, a plan is easy to generate. Combining the individual run-time bounds, we see that the total running time of this explicit search algorithm is bounded by $O(2^{q(|T|)})$ for some polynomial q, proving that PLAN-$\mathcal{D}$ ∈ EXPO. □

In general, we cannot do better than exponential time for PLAN-$\mathcal{D}$, because there are planning tasks for which all solutions are exponentially long in the task size. Therefore, the bound from the previous proof is tight. For the decision problems, however, we can prove a slightly better result.

Theorem 2.3.2. PLANLEN-$\mathcal{D}$ ∈ PSPACE

PLANLEN-$\mathcal{D}$ ∈ PSPACE, *for any planning domain* $\mathcal{D}$.

Proof. Let $\mathcal{D}$ be a planning domain. The following non-deterministic algorithm decides PLANLEN-$\mathcal{D}$ given a task $T \in \mathcal{D}$ and cost bound $K \in \mathbb{N}_0$:

1. *Set s to the initial state of the task and k to K.*
2. *While s is not a goal state:*
   - *Non-deterministically choose an operator o which is applicable in s. Fail if no such operator exists.*
   - *Set s to $o(s)$ and reduce k by the cost of o. Fail if $k < 0$.*
3. *Succeed.*

All steps of the plan (maintaining state s and counter k, testing whether a state is a goal state, testing for applicability of an operator and applying it) can clearly be computed in polynomial space. Moreover, it is obvious that the algorithm can succeed iff the input task has a plan of measure at most K. Thus, we have provided a non-deterministic polynomial-space algorithm for PLANLEN-$\mathcal{D}$. *Savitch's Theorem [95] implies that* PSPACE = NPSPACE, *so that indeed* PLANLEN-$\mathcal{D}$ ∈ PSPACE. □

The following result bounds the complexity of plan existence by the complexity of bounded plan existence. This supports the intuition that the one problem relates to finding arbitrary plans while the other relates to finding optimal plans, a harder problem.

Theorem 2.3.3. PLANEX-$\mathcal{D}$ $\leq_p$ PLANLEN-$\mathcal{D}$
PLANEX-$\mathcal{D}$ $\leq_p$ PLANLEN-$\mathcal{D}$, *for any planning domain* $\mathcal{D}$.

Proof. For a given planning domain $\mathcal{D}$, let p_S and p_w be polynomials such that for all tasks $T \in \mathcal{D}$ the number of states is bounded by $2^{p_S(|T|)}$ and operator costs are bounded by $2^{p_w(|T|)}$. We can then provide the following reduction: Given a planning task $T \in \mathcal{D}$ (a PLANEX-$\mathcal{D}$ *instance), we compute the* PLANLEN-$\mathcal{D}$ *instance consisting of the same planning task T and the length bound $K = (2^{p_S(|T|)} - 1) \cdot 2^{p_w(|T|)}$. This is a polynomial mapping, since a binary representation of K has only $p_S(|T|) + p_w(|T|)$ digits. Clearly, if T has a solution of cost at most K, then T has a solution, so the resulting instance is solvable only if the original instance is solvable. On the other hand, if the original instance is solvable by some plan, then it must also be solvable by a plan which does not traverse the same state several times (parts of a plan between repeated traversals of the same state can be omitted). Plans which do not traverse the same state twice cannot have more than $2^{p_S(|T|)} - 1$ actions, each of which cannot cost more than $2^{p_w(|T|)}$. Therefore, the mapping is indeed a polynomial reduction.* □

This gives us our final upper bound:

Theorem 2.3.4. PLANEX-$\mathcal{D}$ ∈ PSPACE
PLANEX-$\mathcal{D}$ ∈ PSPACE, *for any planning domain* $\mathcal{D}$.

Proof. This follows immediately from the previous two results. □

2.3.2 Shortest Plan Length

One evident observation is that we cannot efficiently generate plans if the plans themselves are too long to be written down in reasonable time. In the framework of classical complexity, this can lead to somewhat unintuitive results in the sense that planning is "provably easy" (plan existence, and maybe even bounded plan existence are polynomial problems), but actually generating a plan requires exponential algorithms. One class of planning tasks exhibiting such a property has been studied by Jonsson and Bäckström [71,73].

In the context of minimization problems, however, we can capture such anomalies more cleanly. To do so, we start with the following definition.

Definition 2.3.5. *Admitting Short (Optimal) Plans*
*A planning domain $\mathcal{D}$ **admits short plans** if there exists a polynomial p such that for all solvable tasks $T \in \mathcal{D}$, the length of some plan for T is at most $p(|T|)$.*

*A planning domain $\mathcal{D}$ **admits short optimal plans** if there exists a polynomial p such that for all solvable tasks $T \in \mathcal{D}$, the length of some optimal plan for T is at most $p(|T|)$.*

*In both cases, p is called a **length bounding polynomial** for $\mathcal{D}$.*

Of course, a planning domain that admits short optimal plans also admits short plans. The converse is not necessarily true for domains with super-polynomial operator costs. However, it *is* true for all the domains we will consider.

Whether or not a planning domain admits short (optimal) plans leads to a clear distinction in approximation complexity.

Theorem 2.3.6. *Short Plans and Approximation Classes*
Let $\mathcal{D}$ be a planning domain.
    If $\mathcal{D}$ admits short plans, then PLAN-$\mathcal{D} \in$ NPS.
    Otherwise, PLAN-$\mathcal{D} \in$ EXPO $\setminus$ NPS.
    If $\mathcal{D}$ admits short optimal plans, then PLAN-$\mathcal{D} \in$ NPO.
    Otherwise, PLAN-$\mathcal{D} \in$ EXPO $\setminus$ NPO.

Proof. If a planning domain admits short plans, we can, given a solvable instance, guess a plan and validate that it is indeed a solution within polynomial time, so that PLAN-$\mathcal{D} \in$ NPS. Similarly, if a domain admits short optimal plans, we can guess and verify an optimal solution in polynomial time, so that PLAN-$\mathcal{D} \in$ NPO.

On the other hand, if a domain does not admit short plans, then there cannot be an algorithm that always generates a solution to a solvable instance in polynomial time, because super-polynomial time is required for writing down the solution. Hence, PLAN-$\mathcal{D} \notin$ NPS. Similarly, if a domain does not admit short optimal plans, we must have PLAN-$\mathcal{D} \notin$ NPO.

With Theorem 2.3.1 we obtain PLAN-$\mathcal{D} \in$ EXPO $\setminus$ NPS and PLAN-$\mathcal{D} \in$ EXPO $\setminus$ NPO for the two cases. $\square$

2.3.3 Approximation Classes of Limited Interest

All of the approximation classes introduced in Definition 2.1.7 are relevant to the theory of approximation algorithms. However, using the notion of short plans, we can prove that some of the introduced approximation classes are of limited interest for planning problems. Before we can to do this, we must formalize another important property of planning domains.

Definition 2.3.7. *Having Cheap Operators*
*A planning domain $\mathcal{D}$ **has cheap operators** if there exists a polynomial p such that for all tasks $T \in \mathcal{D}$, the operator cost function is bounded by $p(|T|)$. Such a polynomial p is called a **cost bounding polynomial** for $\mathcal{D}$.*

We remarked before that we usually adopt a unit operator cost model, so that must planning domains we discuss do have cheap operators. The following proof is similar to Theorem 3.15 in the textbook by Ausiello et al. [7].

Theorem 2.3.8. FPTAS *Is of Limited Interest*
Let $\mathcal{D}$ be a planning domain having cheap operators.
If PLAN-$\mathcal{D} \in$ FPTAS, *then* PLAN-$\mathcal{D} \in$ PO.

Proof. With PLAN-$\mathcal{D} \in$ FPTAS, *Theorem 2.3.6 and* FPTAS $\subseteq$ NPS, *it follows that $\mathcal{D}$ admits short plans. Let p_L be a length bounding polynomial for $\mathcal{D}$, and let p_w be a cost bounding polynomial for $\mathcal{D}$.*

Due to membership in FPTAS, *there must be a planning algorithm for $\mathcal{D}$ which generates c-approximating solutions for tasks $T \in \mathcal{D}$ in time $q(|T|, \frac{1}{c-1})$, where q grows polynomially in both parameters.*

In particular, choose $c = 1 + \frac{1}{2p_L(|T|)p_w(|T|)}$. Then $\frac{1}{c-1} = 2p_L(|T|)p_w(|T|)$ and hence the running time of the algorithm is polynomial in $|T|$, $p_L(|T|)$ and $p_w(|T|)$, which is still polynomial in $|T|$.

Because the algorithm is c-approximating, the measure of the plan π computed by the algorithm is bounded as $m(\pi) \leq c \cdot m^*(T)$. Substituting the value of c, we get $m(\pi) \leq (1 + \frac{1}{2p_L(|T|)p_w(|T|)})m^*(\pi) = m^*(T) + \frac{1}{2} \cdot \frac{m^*(T)}{p_L(|T|)p_w(|T|)}$. Because p_L is a length bounding polynomial, there exists a plan π' with at most $p_L(|T|)$ operators, each of which costs at most $p_w(|T|)$. Thus, we must have $m^*(T) \leq p_L(|T|)p_w(|T|)$, as an optimal plan cannot have a larger measure than π'. We must thus have $m(\pi) \leq m^*(T) + \frac{1}{2}$, and hence $m(\pi) = m^*(T)$, because $m(\pi)$ and $m^*(T)$ are natural numbers.

We can thus use the fully polynomial-time approximation scheme to produce optimal solutions in polynomial time, and hence PLAN-$\mathcal{D} \in$ PO. □

This result is an instance of a general theorem about optimization problems, namely that any optimization problem in FPTAS \ PO must have a measure function which grows super-polynomially with the instance size. This is most commonly the case for problems with a strong numerical aspect; one well-known example is the MAXIMUM KNAPSACK problem, which belongs to FPTAS, but not to PO unless P = NP [7].

Note that the restriction to domains with cheap operators is critical. For example, for a given propositional formula φ of size $O(n)$, it is not difficult to construct a planning task T_φ with the following properties:

- The size of T_φ is polynomial in the size of φ, denoted by n.
- T_φ always has a trivial solution of measure $2^n + 1$.
- If φ is satisfiable, T_φ has a second solution of measure 2^n.

There exist trivial fully-polynomial time approximation schemes for such planning tasks, but existence of a polynomial-time optimal planning algorithm would imply P = NP.

At the other end of the approximation complexity spectrum, we can prove a similar result.

Theorem 2.3.9. PS *and* exp-APX *Are of Limited Interest*
Let $\mathcal{D}$ be a planning domain having cheap operators for which there exists a polynomial-time algorithm which determines whether or not $m^*(T) = 0$ for a given task $T \in \mathcal{T}$ and computes a plan of measure 0 if this is the case.
If PLAN-$\mathcal{D} \in$ PS, then PLAN-$\mathcal{D} \in$ poly-APX.

Proof. Let $\mathcal{D}$ be a planning domain such that PLAN-$\mathcal{D} \in$ PS. This means that we must have a polynomial-time planning algorithm for $\mathcal{D}$. A polynomial-time planning algorithm can only generate polynomial-size plans, and polynomial-size plans in a domain having cheap operators have polynomial measure. Thus, the algorithm guarantees a polynomial performance ratio for all input tasks with $m^*(T) \neq 0$. Moreover, input tasks with $m^*(T) = 0$ can be optimally solved in polynomial time. Together, this implies PLAN-$\mathcal{D} \in$ poly-APX. □

Note that the requirements of the proof are satisfied in particular for planning domains with unit operator costs: These clearly have cheap operators, and tasks can have optimal measure 0 only if they are solved by the empty plan, which is trivial to determine.

Again, the restriction to cheap operators is necessary, and indeed in the wider world of optimization, there are problems which belong to exp-APX, but not to poly-APX unless P = NP, even when measures of 0 are forbidden. One example of such a problem which is easily expressible as a planning domain is (a modified version of) MINIMUM WEIGHTED SATISFIABILITY [7].

It is also easy to see that the requirement to be able to deal with zero-measure solutions is important. It is easy to define a planning domain which can be solved by a plan with measure 0 iff a given propositional formula is satisfiable, and which can always be solved by a plan with measure 1. The planning problem for such a domain clearly belongs to PS, but not to exp-APX or poly-APX unless P = NP.

2.3.4 Relating Planning and (Bounded) Plan Existence

We have claimed before that there is a relationship between the planning problem and the plan existence and bounded plan existence decision problems which makes the latter two relevant to our study. Here, we state this relationship formally.

Theorem 2.3.10. PLAN-$\mathcal{D}$ *vs.* PLANEX-$\mathcal{D}$ *and* PLANLEN-$\mathcal{D}$
For any planning domain $\mathcal{D}$, the following relationships hold:

- *If* PLAN-$\mathcal{D}$ $\in$ PO, *then* PLANEX-$\mathcal{D}$ $\in$ P *and* PLANLEN-$\mathcal{D}$ $\in$ P.
- *If* PLAN-$\mathcal{D}$ $\in$ PS, *then* PLANEX-$\mathcal{D}$ $\in$ P.
- *If* PLAN-$\mathcal{D}$ $\in$ NPO, *then* PLANEX-$\mathcal{D}$ $\in$ NP *and then* PLANLEN-$\mathcal{D}$ $\in$ NP.
- *If* PLAN-$\mathcal{D}$ $\in$ NPS, *then* PLANEX-$\mathcal{D}$ $\in$ NP.
- *If* PLANLEN-$\mathcal{D}$ *is* NP-*hard, then* PLAN-$\mathcal{D}$ $\notin$ PO *unless* P = NP.
- *If* PLANEX-$\mathcal{D}$ *is* NP-*hard, then* PLAN-$\mathcal{D}$ $\notin$ PS *unless* P = NP.

Proof. For the first statement, an optimal polynomial time planning algorithm can be used for polynomially deciding plan existence and bounded plan existence: Run the algorithm and test if it generates a solution (for plan existence) or a solution satisfying the given measure bound (for bounded plan existence).

Similarly, any polynomial time planning algorithm can be used for polynomially deciding plan existence: Run the algorithm, check if it generates a plan, and accept the input iff it does.

The same arguments works for the third and fourth statement, using non-deterministic algorithms in place of deterministic ones.

We show the fifth statement by contradiction. Assume that PLANLEN-$\mathcal{D}$ *were* NP-*hard and* PLAN-$\mathcal{D}$ *were in* PO. *By the first statement, we have* PLANLEN-$\mathcal{D}$ $\in$ P. *Together with the fact that this problem is* NP-*hard, this implies* P = NP.

The sixth statement follows similarly by contradiction from the second. $\square$

Generally, the theorem states that hardness of the decision problems carries over to hardness of the minimization problem, and easiness of minimization carries over to easiness of the decision problems. The converse is not necessarily true. For example, plan existence might be easy, but the argument showing that a plan exists might be non-constructive and not easily translatable to an approximation algorithm. Moreover, the minimization problem may be hard simply because the domain does not admit short plans; however, this does not preclude the possibility of easy decision problems. (Consider the TOWERSOFHANOI domain as an example.)

To provide the maximum amount of information, we will thus only consider the decision problems for proving *hardness* results, and always work directly with the planning problems for proving *easiness* results.

2.3.5 Generalization and Specialization

The final result we want to present in this chapter is quite simple. Nevertheless, it is fundamental to the following analysis and thus worth stating explicitly.

Theorem 2.3.11. *Generalization and Specialization*
Let $\mathcal{D}$ and $\mathcal{D}'$ be planning domains such that all tasks of $\mathcal{D}$ are tasks of $\mathcal{D}'$, and for all tasks $T \in \mathcal{D}$, we have $\mathcal{D}(T) = \mathcal{D}'(T)$. Then:

- PLAN-$\mathcal{D} \leq_{\text{OP}}$ PLAN-$\mathcal{D}'$.
- PLANEX-$\mathcal{D} \leq_{\text{p}}$ PLANEX-$\mathcal{D}'$.
- PLANLEN-$\mathcal{D} \leq_{\text{p}}$ PLANLEN-$\mathcal{D}'$.

Proof. All proofs are simply by embedding: For the AP-reduction, $\alpha = 1$ and f maps tasks to themselves and g maps solutions to themselves. To see that this is also an OP-reduction, consider the bound function which maps a given bound to itself, independent of the instance.

For plan existence and bounded plan existence, the polynomial reductions are by the identity function. □

This concludes the chapter. We have presented sufficient formal background to conduct complexity analyses for planning domains, and shall now give a brief overview of the domains we are interested in.

3

The Benchmark Suite

Having formalized planning domains, the question arises how the planning domains from the International Planning Competitions fit into this framework. We discuss this issue in Sect. 3.1, then informally introduce the competition domains in Sect. 3.2. The concluding Sect. 3.3 gives an outlook on the following chapters, in which we present the main technical results of this part.

3.1 Defining the Competition Domains

How are the planning competition domains defined? This question is not as straight-forward to answer as it might at first appear. The semantics of a benchmark domain – how a task is mapped to a state space – is obvious from their PDDL definitions [8, 66, 86, 91] and causes no further difficulty, maybe apart from the fact that some of the benchmark domains contain a few quirks or errors in their formalization that we might want to address.

The real question is: Which tasks should be considered part of a given domain? Again, the available PDDL definitions help somewhat: Tasks which cannot be expressed with the official PDDL domain specifications should not be considered part of the domain. However, being expressible in PDDL with the given domain specification is a criterion which we consider necessary, but not sufficient for deciding whether or not a given task belongs to a certain domain. In many domains, important pieces of information are not (and cannot be made) explicit in the PDDL domain specification. For example, the fact that a block cannot sit on top of itself is an important property of BLOCKSWORLD tasks, yet the domain file does not require or imply this property. Similarly, in domains with a transportation theme like LOGISTICS, it is usually assumed but not made explicit that locations, carriers and portable objects are disjoint classes, and hence carriers cannot pick up other carriers or move locations around.

So there must be a second, restricting criterion to narrow the choice of tasks. Unfortunately, for most domains this is not formalized in the literature.

Therefore, we must find some way of identifying the "intended" set of tasks in a given domain. Identifying intent appropriately is of special relevance if our results are to be employed for judging the performance of planning systems on the existing benchmark suites. As an example, consider the GRIPPER domain, where a robot moves objects between different rooms. Although the PDDL definition allows for more general specifications, in all GRIPPER tasks which were used for benchmarking, there are only two rooms, all objects are initially located in the same room as the robot, and all objects need to be moved to the other room. What is more, the term GRIPPER task is generally used only in this restricted sense. For this reason, it makes more sense for our formal definition of the domain to mirror these implicit constraints and *not* exploit the full potential of the PDDL specification of the domain.

So how do we identify an appropriate set of tasks? For the benchmarks of IPC1 and IPC2, we rely on two sources. The first of these is the set of benchmark tasks that were used during the competitions. In many cases, as in the GRIPPER example above, these already give some very strong indications of intent. The second source is provided by the informal domain descriptions made available by the competition organizers [8, 91]. For example, the requirement that the roadmap graph of MYSTERY tasks is planar is taken from McDermott's description and has been verified for the available MYSTERY benchmark tasks. For IPC3, the random task generators that were used by the competition organizers for generating the benchmark suite have been the major guiding criterion for our definitions. Finally, for IPC4, a very thorough article on the competition domains is available [67].

3.2 The Benchmark Suite

Altogether, the benchmark suite comprises 20 domains, some of which (namely, MICONIC-10, PIPESWORLD, PROMELA) are further divided into subdomains (e. g., MICONIC-10-SIMPLEADL, PIPESWORLD-NOTANKAGE and PROMELA-PHILOSOPHERS). We will formally introduce each of these domains as we discuss it, rather than providing all the definitions in one place. However, in order to get an impression of the scope of the benchmarks and to motivate the structure of our investigation, we now provide short, informal definitions of all the domains.

3.2.1 IPC1 Domains

The following descriptions are cited from an article by the competition organizer, Drew McDermott [91]. The only exception is the description of the MYSTERY domain, which is taken from the IPC1 website.

ASSEMBLY: The goal is to assemble a complex object made of subassemblies. There are four actions: (1) commit resource assembly, (2) release resource assembly, (3) assemble part assembly, and (4) remove part assembly. The sequence of steps must obey a given partial order. In addition, through poor engineering design, many subassemblies must be installed temporarily in one assembly, then removed and given a permanent home in another.

GRID: There is a square grid of locations. A robot can move one grid square at a time horizontally and vertically. If a square is locked, the robot can move to it only by unlocking it, which requires having a key of the same shape as the lock. The keys must be fetched and can themselves be in locked locations. Only one object can be carried at a time. The goal is to get objects from various locations to various new locations.

GRIPPER: Here, a robot must move a set of balls from one room to another, being able to grip two balls at a time, one in each gripper. There are three actions: (1) move, (2) pick, and (3) drop.

LOGISTICS: There are several cities, each containing several locations, some of which are airports. There are also trucks, which can drive within a single city, and airplanes, which can fly between airports. The goal is to get some packages from various locations to various new locations.

MOVIE: In this domain, the goal is always the same (to have lots of snacks to watch a movie). There are seven actions, including rewind-movie and get-chips, but the number of constants increases with the problem number.

MYSTERY: There is a planar graph of nodes. At each node are vehicles, cargo items, and some amount of fuel. Objects can be loaded onto vehicles (up to their capacity), and the vehicles can move between nodes; but a vehicle can leave a node only if there is a nonzero amount of fuel there, and the amount decreases by one unit. The goal is to get cargo items from various nodes to various new nodes.

MYSTERYPRIME: This is the MYSTERY domain with one extra action, the ability to squirt a unit of fuel from any node to any other node, provided the originating node has at least two units.

3.2.2 IPC2 Domains

The domains from the 2nd International Planning Competition can be described as follows (these descriptions are original):

BLOCKSWORLD: This is the classic AI planning domain where blocks stacked into towers must be rearranged by a robotic arm which can pick up the top block of a tower and place it on another tower or on the table. Table space is not limited.

FREECELL: A solitaire card game where cards are initially dealt into unordered piles and must be arranged into sorted piles. There is limited space available for moving cards, and there are constraints regarding which cards may be placed on top of which other cards in a pile. Cards may also be moved to a limited number of free cells, but these can only hold one card at a time, rather than a pile of cards.

LOGISTICS: This is the same domain as in the 1st International Planning Competition.

MICONIC-10: There is an elevator moving between the floors of a building. There are passengers waiting at various floors. The goal is to move each passenger to their destination floor. There are three variants of this domain, including one where special constraints such as VIP service restrict elevator movement.

SCHEDULE: A set of physical objects must be processed by various machines to change their physical properties, such as colour, shape and surface condition. The available equipment consists of a polisher, a roller, a lathe, a grinder, a punch, a drill press, a spray painter and an immersion painter.

3.2.3 IPC3 Domains

These descriptions (except for the FREECELL domain) are cited from an article by the competition organizers, Derek Long and Maria Fox [86]:

DEPOTS: This domain combines the transportation style problem of Logistics with the well-known Blocks domain.

DRIVERLOG: This problem involves transportation, but with the twist that vehicles must be supplied with a driver before they can move.

FREECELL: This is the same domain as in the 2nd International Planning Competition.

ROVERS: This domains was motivated by the 2003 Mars Exploration Rover (MER) missions and the planned 2009 Mars Science Laboratory (MSL) mission. The objective is to use a collection of mobile rovers to traverse between waypoints on the planet, carrying out a variety of data-collection missions and transmitting data back to a lander. The problem includes constraints on the visibility of the lander from various locations and on the ability of individual rovers to traverse between particular pairs of waypoints.

SATELLITE: This domain was inspired by the problem of scheduling satellite observations. The problems involve satellites collecting and storing data using different instruments to observe a selection of targets.

ZENOTRAVEL: Another transportation problem, inspired by a domain used in testing the ZENO planner developed by Penberthy and Weld, in which people must embark onto planes, fly between locations and then debark, with planes consuming fuel at different rates according to speed of travel.

3.2.4 IPC4 Domains

These descriptions (except for the SATELLITE domain) are again cited from an article by the competition organizers, Jörg Hoffmann and Stefan Edelkamp [66]:

AIRPORT: In the Airport domain, the planner has to control the ground traffic on an airport. The task is to find a plan that solves a specific traffic situation, specifying inbound and outbound planes along with their current and goal positions on the airport. The planes must not endanger each other, i.e., they must not both occupy the same airport "segment" (a smallest road unit), and if plane x drives behind plane y then between x and y there must be a safety distance (depending on the size of y).

PIPESWORLD: The Pipesworld domain is a PDDL adaptation of an application domain dealing with complex problems that arise when transporting oil derivative products through a pipeline system. [...] The pipelines must be filled with liquid at all times [...].

PROMELA: Promela is the input language of the model checker SPIN, used for specifying communication protocols. Communication protocols are distributed software systems, and many implementation bugs can arise, like deadlocks, failed assertions, and global invariance violations. The model checking problem is to find those errors by returning a counter-example, or to verify correctness by a complete exploration of the underlying state space. [...]

For IPC-4, two relatively simple communication protocols were selected as benchmarks – toy problem from the Model Checking area. One is the well-known Dining Philosophers protocol, the other is a larger protocol called Optical Telegraph.

PSR: PSR is short for *Power Supply Restoration*. The domain is a PDDL adaptation of an application domain investigated by Sylvie Thiébaux and other researchers, which deals with reconfiguring a faulty power distribution system to resupply customers affected by the faults.

SATELLITE: This is the same domain as in the 3rd International Planning Competition.

3.3 Domains and Domain Families

Despite their informality, the preceding descriptions should have made two things very clear:

- On the one hand, the benchmark domains are very diverse in nature, including application problems from vastly different application areas.
- On the other hand, there is one reoccurring theme for many of the benchmark domains, namely *transportation planning*.

At least nine of the planning benchmarks (GRID, GRIPPER, LOGISTICS, MYSTERY, MYSTERYPRIME, MICONIC-10, DEPOTS, DRIVERLOG and ZENO-TRAVEL) have vehicles (or robots, in the case of GRID and GRIPPER) transporting objects or people as a central theme. Two of the remaining domains, namely ROVERS and AIRPORT, include *route planning* as a central component, although there is no transportation as such. Some of these domains are very closely related to each other – this is evident for domains like MYSTERY and MYSTERYPRIME, but on closer inspection, we find similar likenesses between, for example, LOGISTICS and one of the MICONIC-10 variants.

For this reason, it is appropriate to consider the broader picture of transportation planning instead of focusing on the individual domains in isolation. Therefore, in the next chapter we investigate a *family* of related *transportation planning* domains, which includes several IPC domains but also some natural generalizations and specializations.

In addition to transportation problems, we also consider a related family of pure *route planning* domains, which can be interpreted as "degenerate" transportation planning problems, where initial and target locations of objects are so close to each other that we can focus on visiting the initial locations exclusively. While pure route planning problems are uncommon in the planning benchmark suite, they allow for some elegant reductions and show connections to related optimization problems from other areas of Computer Science such as the TRAVELLING SALESPERSON PROBLEM.

After a thorough discussion of transportation and route planning, we will return to the competition benchmark suite in Chaps. 5 and 6, where the former discusses those benchmarks related to transportation or route planning, and the latter discusses the remaining ones.

4

Transportation and Route Planning

Having established *transportation planning* and *route planning* as common concepts in the IPC benchmark suite, we now take an in-depth look at the complexity of solving such planning problems.

In the following Sect. 4.1, we introduce the general TRANSPORT and ROUTE domains. In Sect. 4.2, we present some first upper bounds and general reductions. Section 4.3 discusses the plan existence problem and shows *restricted fuel* to be a major source of hardness for transportation and route planning. Section 4.4 shows that optimal planning or polynomial-time approximation schemes are beyond reach even for quite restricted domain variants. On the positive side, we see in Sect. 4.5 that constant-factor approximations are feasible in many cases. However, in many others they are not, which is shown in Sect. 4.6. Finally, Sect. 4.7 summarizes the findings of this chapter, and Sect. 4.8 puts them into a wider context.

4.1 TRANSPORT and ROUTE

The planning domains that we subsumed under *transportation* have a number of commonalities. In particular, these are what we consider the defining properties of a prototypical transportation benchmark (terminology borrowed from Long and Fox [85]):

- There is a set of *locations* (grid squares, rooms, airports, ...), which are connected by *roads* (adjacent grid squares, doors, airways, ...), forming a *roadmap* graph.
- There is a set of *mobiles* (robots, trucks, elevators, ...), which traverse the roadmap.
- There is a set of *portables* (keys, balls, passengers, ...), which can either be at a location or carried by a mobile.
- These classes of entities are disjoint, and there are no other entities.

– The goal is to move (a subset of) the portables to their respective *final destinations*.

Most domains which we consider transportation domains satisfy all these properties. In particular, this is the case for GRID, GRIPPER, LOGISTICS, MICONIC-10, MYSTERY, MYSTERYPRIME, and ZENOTRAVEL. The DRIVER-LOG domain does not quite satisfy the last two properties, because goals can also be associated with mobiles, not just with portables, and there are *drivers* which combine some of the properties of portables and mobiles. The DEPOTS domain does not quite satisfy the fourth property, because this domain also features *hoists* for stacking and unstacking portables, which are neither mobiles nor locations or portables. Despite such variations, these five properties describe the core of a transportation domain fairly accurately.

The AIRPORT and ROVERS domains also feature mobiles traversing a roadmap of locations as a central concept. However, in these domains, there is no transportation of portables, and the AIRPORT domain in particular is very different in structure from the transportation domains due to the fact that vehicles can block each other's way. However, apart from the fact that there is no explicit transportation, ROVERS is quite similar to a typical transportation domain, and will be subsumed under the heading of *route planning* domains in our analysis.

A common analysis of transportation domains must also consider their differences. In particular, the various transportation and route planning domains from the IPC benchmark suite differ in the following ways:

– *Capacity constraints*: In GRID, mobiles can only carry one portable at a time, in GRIPPER two portables. In MYSTERY and MYSTERYPRIME, mobiles have varying capacities. In DEPOTS, DRIVERLOG, LOGISTICS, MICONIC-10 and ZENOTRAVEL, capacity is unbounded.
– *Fuel constraints*: In MYSTERY, MYSTERYPRIME and ZENOTRAVEL, fuel is consumed by and required for movement, unlike the other domains.
– *Number of mobiles*: In GRIPPER, GRID and MICONIC-10, there is only a single mobile. In the other domains, there can be several.
– *Mobile types*: In LOGISTICS, a distinction is made between trucks, which traverse roads within one city, and airplanes, which move between airports in different cities. In DRIVERLOG, trucks and drivers (which have some, but not all, properties of mobiles) use different roadmaps. In ROVERS, every rover has a different roadmap.
– *Roadmap graph types*: In the DEPOTS, GRIPPER, LOGISTICS, MICONIC-10 and ZENOTRAVEL domains, the roadmap is a complete graph. In GRID, it must be a grid, in MYSTERY and MYSTERYPRIME a planar graph. In the ROVERS domain, the roadmap of each rover must be a tree, and the roadmaps of drivers and trucks in DRIVERLOG are arbitrary connected graphs.
– *Special features*: Some of the competition domains include aspects which go beyond the common transportation theme. In DEPOTS, portables can

be stacked into towers like in the BLOCKSWORLD domain in addition to be being transported. In DRIVERLOG, trucks can only move when boarded by drivers. In GRID, new connections between locations can be made by unlocking doors. In one of the MICONIC-10 variants, the passengers inside the elevator impose different constraints on the elevator movement. In MYSTERYPRIME, fuel can be moved between locations, and in ZENO-TRAVEL, vehicles can refuel. Finally, the ROVERS domain has a host of special features not described here.

We remarked previously that we adopt a unit action cost model for most planning domains. The domains considered in this chapter are the only major exceptions to this rule, for three reasons:

- First, weighted variants of transportation or route planning domains are very natural, arguably more natural than weighted versions of most other planning domains, because action costs often have precise analogues in their application domains. For example, road distance, travel time or travel cost between two locations are natural measures for the cost of movement actions.
- Second, allowing for action weights gives us greater flexibility in specifying reductions between planning problems, which is very useful for some of the competition benchmarks. For example, the MICONIC-10-SIMPLEADL domain can be naturally linked to a variant of TRANSPORT where pickup and drop operators are free of cost.
- Third, we will be able to prove a very nice – and perhaps somewhat surprising – compilation result which allows us to reduce general action costs to unit action costs in an approximation-preserving way. With this result, we can cover the case of general actions with little additional effort.

4.1.1 The TRANSPORT Domain

We will now define a transportation domain which subsumes and generalizes most of these variations apart from the *special features*, which are too heterogeneous too be usefully integrated within a single domain. (We will see in the following chapter that for many of the domains exhibiting special features, these do not actually amount to a difference in complexity, so complexity results obtained in their absence are applicable as well.)

Definition 4.1.1. TRANSPORT **Task**
A TRANSPORT **task** *is defined by a 9-tuple* $\langle G, M, P, cap, w_m, w_p, l_0, fuel_0, l_* \rangle$ *with the following components:*

- $G = \langle V, E \rangle$ *is the **roadmap graph** or **roadmap**. Its vertices V are called **locations**, its edges **roads**.*
- *M is a finite set of **mobiles**.*
- *P is a finite set of **portables**.*

- $cap : M \to \{1, \ldots, |P|\}$ is the **capacity** function. We say that mobile m has **unbounded capacity** iff $cap(m) = |P|$.
- $w_m : M \times E \to \mathbb{N}_1 \cup \{\infty\}$ is the **movement cost** function. If $w_m(m, e) \neq \infty$, we say that mobile m **may access** road e. The **roadmap graph** or **roadmap** G_m of m is the weighted graph obtained by restricting G to those roads which m may access, weighted by the corresponding movement cost. We require that each road may be accessed by some mobile.
- $w_p : M \times P \to \mathbb{N}_0$ is the **pickup cost** function.
- $l_0 : (M \cup P) \to V$ is the **initial location** function.
- $fuel_0 : V \to \mathbb{N}_0 \cup \{\infty\}$ is the **initial fuel** function.
- $l_\star$ is a partial function from portables to locations called the **goal location** function. The set of portables on which it is defined are called **goal portables**.

The sets V, M and P must be disjoint.

The concepts of locations, roads, mobiles and portables should be clear from the preceding discussion. The capacity function bounds the number of portables a given mobile can carry at the same time. The movement and pickup cost functions can be used to define a variety of action cost models for transportation tasks.

Our definition allows movement cost functions which vary with the mobile and road, but not, for example, with the direction in which the road is traversed, or with the number of portables carried by a mobile. Movement costs must be strictly positive, while picking up and dropping portables, which both incur costs defined by the pickup cost function, may be free. Pickup cost may depend on the mobile and portable (although we will rarely need this level of generality), but not, for example, on the location where the pickup takes place.

All mobiles and portables have a specified initial location. We require mobiles to be initially empty, so we do not need a way of specifying the "initial contents" of a mobile. The fuel function bounds the number of times that a given location can be left by a mobile. Fuel is associated with locations rather than mobiles because this is the way it is handled in MYSTERY and MYSTERYPRIME. (ZENOTRAVEL associates fuel with mobiles, not locations, but in this domain fuel constraints are less important because mobiles may be refuelled.) Some portables, namely the set of goal portables, have a specified final location, given by the goal location function. The location of mobiles or non-goal portables at the end of plan execution does not matter.

We can now define the TRANSPORT planning domain.

Definition 4.1.2. TRANSPORT *Domain*
The TRANSPORT *domain maps* TRANSPORT *tasks with locations V, mobiles M and portables P to state spaces as follows:*

STATES: *Pairs $\langle l, fuel \rangle$, where $l : M \cup P \to V \cup M$ is the **location** function and fuel $: V \to \mathbb{N}_0 \cup \{\infty\}$ is the **fuel reserve** function. Only portables may have mobiles as their location.*

INITIAL STATE: *$\langle l_0, fuel_0 \rangle$, where l_0 is the initial location function and $fuel_0$ is the initial fuel function of the task.*

GOAL STATES: *Any state $\langle l, fuel \rangle$ with $l(p) = l_*(p)$ for all goal portables p, where l_* is the goal location function of the task.*

OPERATORS: *A mobile m can **move** from location v to location v' in state $\langle l, fuel \rangle$ iff $\{v, v'\}$ is a road which m may access and $fuel(v) \neq 0$. This action changes $l(m)$ to v' and reduces $fuel(v)$ by 1 (with $\infty - 1 := \infty$). The cost of this action is given by the movement cost function applied to mobile m and edge $\{v, v'\}$. A mobile m can **pick up** a portable p in state $\langle l, fuel \rangle$ iff $l(p) = l(m)$ and $l(p') = m$ for strictly less than $cap(m)$ many portables p'. This action changes $l(p)$ to m. Its cost is given by the pickup cost function applied to m and p.*
*A mobile m can **drop** a portable p in state $\langle l, fuel \rangle$ iff $l(p) = m$. This action changes $l(p)$ to $l(m)$. Its cost is given by the pickup cost function applied to m and p.*

Strictly speaking, this definition would imply an infinite number of states due to the infinite range of the fuel function. However, the number of relevant (reachable) states is finite in practice because the fuel reserve is bounded by initial fuel and locations with infinite fuel never change their fuel reserve.

4.1.2 The ROUTE Domain

Route planning domains are similar to, but simpler than transportation domains. They do not feature portables and thus there are no pick-up or drop actions. Instead, some locations are considered *target locations* and must be visited by some mobile during plan execution. It does not matter which mobile visits a given target location, or in which order they are visited. General ROUTE tasks are defined as follows.

Definition 4.1.3. ROUTE *Task*
A ROUTE **task** is defined by a 6-tuple $\langle G, M, w_m, l_0, fuel_0, V_* \rangle$ with the following components:

- $G = \langle V, E \rangle$ is the **roadmap graph** or **roadmap**. Its vertices V are called **locations**, its edges **roads**.
- M is a finite set of **mobiles**.
- $w_m : M \times E \to \mathbb{N}_1 \cup \{\infty\}$ is the **movement cost** function. If $w_m(m, e) \neq \infty$, we say that mobile m **may access** road e. The **roadmap graph** or **roadmap** G_m of m is the weighted graph obtained by restricting G to those roads which m may access, weighted by the corresponding movement cost. We require that each road may be accessed by some mobile.

– $l_0 : M \to V$ is the **initial location** function. Mobiles must be mapped to locations which are part of their respective roadmap graph.
– $fuel_0 : V \to \mathbb{N}_0 \cup \{\infty\}$ is the **initial fuel** function.
– $V_\star \subseteq V$ is the set of **target locations**.

The intuition for ROUTE tasks is similar to that of TRANSPORT tasks. The only difference is that all components pertaining to portables are removed and replaced with the set of *target locations*, which represent locations that must be visited. The definition of the ROUTE planning domain should contain no surprises.

Definition 4.1.4. ROUTE *Domain*
The ROUTE ***domain*** *maps* ROUTE *tasks with locations V and mobiles M to state spaces as follows.*

STATES: Pairs $\langle l, fuel, V^t \rangle$, where $l : M \to V$ is the **location** function, $fuel : V \to \mathbb{N}_0 \cup \{\infty\}$ is the **fuel reserve** function, and $V^t \subseteq V_\star$ is the set of **remaining targets**.

INITIAL STATE: $\langle l_0, fuel_0, V_0^t \rangle$, where l_0 is the initial location function and $fuel_0$ is the initial fuel function of the task, and V_0^t contains exactly those target locations which are not initial locations of any mobile.

GOAL STATES: Any state $\langle l, fuel, V^t \rangle$ where $V^t = \emptyset$.

OPERATORS: A mobile m can **move** from location v to location v' in state $\langle l, fuel \rangle$ iff $\{v, v'\}$ is a road which m may access and $fuel(v) \neq 0$. This action changes $l(m)$ to v' and reduces $fuel(v)$ by 1 and removes v' from V^t if present. The cost of this action is given by the movement cost function applied to mobile m and edge $\{v, v'\}$.

4.1.3 Special Cases and Hierarchy

As we saw in the description of the various planning competition domains, we usually do not need all features of TRANSPORT and ROUTE. Some benchmark domains only feature a single agent (mobile). In others, there are no fuel or capacity restrictions. All benchmark domains have strict restrictions on the movement and pickup cost functions. For this reason, we define some special cases which capture the most frequently occurring restrictions of TRANSPORT and ROUTE.

Definition 4.1.5. *Special Cases of* TRANSPORT *and* ROUTE
Let $i \in \{1, \infty, *\}$, $j \in \{1, +, *\}$, $k \in \{1, \infty, *\}$, $l \in \{1, *\}$ and $m \in \{0, 1, *\}$, and let $\mathcal{D}$ be the TRANSPORT *or* ROUTE *domain.*
    The PLAN-TRANSPORT$_{ij}^k$-$[l, m]$ *and* PLAN-ROUTE$_j^k$-$[l]$ *problems are the restrictions of* PLAN-TRANSPORT *and* PLAN-ROUTE *to planning tasks T with the following properties:*

- If $i = 1$, *the capacity of all mobiles must equal* 1.
- If $i = \infty$, *the capacity of all mobiles must be unbounded.*
- If $j = 1$, *there must be exactly one mobile in* T.
- If $j = +$, *the cost and pickup functions must be defined identically for all mobiles. (In particular, this implies that all mobiles may access the same roads.)*
- If $k = 1$, *the initial fuel of all locations must equal* 1.
- If $k = \infty$, *the initial fuel of all locations must be unbounded.*
- If $l = 1$, *the movement cost function may only map to* 1 *or* ∞. *We say that* T *has* **unit movement cost**.
- If $m = 0$, *the pickup cost function may only map to* 0. *We say that* T *has* **free pickup**.
- If $m = 1$, *the pickup cost function may only map to* 1. *We say that* T *has* **unit pickup cost**.

Parameters i and m only apply to TRANSPORT *and are omitted for* ROUTE *tasks. Parameter k may be omitted, in which case unbounded fuel is assumed. Moreover, parameters l and m may be omitted, along with the surrounding parentheses, if movement costs (and, for* TRANSPORT *tasks, pickup costs) are uniform.*

When defining restricted TRANSPORT or ROUTE tasks, we usually omit components of the task description which are implied by the restriction. For example, we may omit the definition of the capacity function for $i = 1$ or $i = \infty$ and the definition of the movement cost function for $j \neq *$ and $l = 1$.

Considering all possible combinations, this definition defines 162 variants of TRANSPORT and 18 variants of ROUTE. Fortunately, these variants are related by a generalization/specialization ordering which can be exploited for complexity classifications. In particular, for all parameters, the option $*$ is more general than the others, so that there are only two maximally general variants, TRANSPORT$^*_{**}$-$[*, *]$ and ROUTE$^*_*$-$[*]$. The number of maximally specific variants is somewhat larger, but still limited: for i and k, neither 1 nor ∞ is more specific than the other, for j and l, option 1 is most specific, and for m, neither 0 nor 1 is more specific than the other. This amounts to eight maximally specific variants for TRANSPORT and two for ROUTE. The specificity ordering is illustrated in Fig. 4.2.

i: capacity	1: one portable	∞: unbounded	$*$: varies
j: mobiles	1: one mobile	$+$: one type	$*$: many types
k: fuel units	1: one per location	∞: unbounded	$*$: varies
l: move cost	1: unit cost	$*$: varies	
m: pickup cost	0: free	1: unit cost	$*$: varies

Fig. 4.1. Summary of restrictions in TRANSPORT$^k_{ij}$-$[l, m]$ and ROUTE^k_j-$[l]$

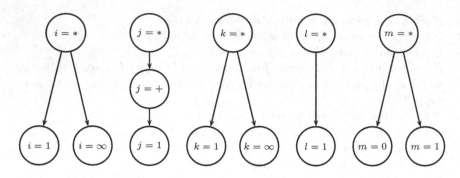

Fig. 4.2. Generalization relationships for the TRANSPORT and ROUTE domains

4.2 General Results

We begin our investigation of TRANSPORT and ROUTE by proving some general results. Along the way, we introduce some additional concepts which are important to our reductions. Our first observation is that we can limit our attention to a certain class of plans.

Definition 4.2.1. Reasonable Plans
A plan for a TRANSPORT task is called **reasonable** if it has the following properties:

– No portable is dropped at the same location twice.
– All mobiles are empty at the end of plan execution.
– The plan does not end with a movement action, and for every movement action in the plan, the next non-movement action affects the same mobile.

We first show that we only need to consider planning algorithms which generate reasonable plans: Any plan can be transformed into a reasonable one in polynomial time without increasing its measure. In particular, this implies that a task is solvable if and only if it is solvable by a reasonable plan.

Theorem 4.2.2. Reasonable Plans Are Sufficient
Given a plan π for a TRANSPORT task, a reasonable plan π' for the same TRANSPORT task with $m(\pi') \leq m(\pi)$ can be computed in time polynomial in $|\pi|$.

Proof. We need to change the plan $\pi = \pi_1 \ldots \pi_n$ so that it satisfies the conditions for reasonable plans. This is quite easy to achieve.

For any two plan steps π_i and π_j ($i < j$) where the same portable is dropped at the same location, all pick-up and drop actions for that portable in the sequence $\pi_{i+1} \ldots \pi_j$ can be omitted. By applying this transformation repeatedly, we satisfy the first condition.

If a portable is picked up but not dropped at a later step of the plan, the pick-up step can be removed. Doing this satisfies the second condition.

Finally, if a mobile is moved which is not the next mobile to pick up or drop a portable, then all its movement operators can be shifted to a later position in the plan, right before the next step where it picks up or drops a portable. If the mobile never picks up or drops a portable in the remaining plan, the movements can be removed altogether. This satisfies the third condition.

Clearly, these transformations can be performed in polynomial time. Because all transformations only remove or reorder operators, the resulting plan cannot have a higher cost than the original one.   □

The main reason why reasonable plans are important is that they cannot be overly long.

Theorem 4.2.3. *Reasonable Plans Are Short*
There exists a polynomial p such that $|\pi| \leq p(|T|)$ for all $T \in$ TRANSPORT and all reasonable plans π for T.

Proof. Because of the first restriction for reasonable plans, such a plan π cannot contain more than $|V| \cdot |P|$ drop actions, where V is the set of locations and P is the set of portables of the task. Because of the second restriction, it then cannot contain more than $|V| \cdot |P|$ pickup actions either. Because of the third restriction, there are at most $|V| - 1$ movements in between any two non-movements, or before the first non-movement, and no movement after the last non-movement. This bounds the number of movements by $(|V| - 1) \cdot 2 \cdot |V| \cdot |P|$. The sum of these three terms is $2|V|^2|P|$, which is clearly polynomial in $|T|$.   □

Combining the previous two properties, we obtain an upper bound for the complexity of TRANSPORT planning.

Theorem 4.2.4. PLAN-TRANSPORT $\in$ NPO
PLAN-TRANSPORT$^*_{**}$-$[*, *] \in$ NPO.

Proof. The two previous theorems imply that TRANSPORT admits short optimal plans. Membership in NPO follows from Theorem 2.3.6.   □

Due to the reductions presented in Chap. 2, the general result for the planning problem can be applied to the plan existence and bounded plan existence problems as well, and of course it also holds for all the special cases of the TRANSPORT domain.

We could prove an equivalent result for the ROUTE domain family by a similar line of reasoning. However, we can do better. The next result allows us to embed the ROUTE domain family in the TRANSPORT domain family. One of its implications is that the ROUTE domain also belongs to NPO, but it has much more far-reaching consequences. In particular, it allows us to prove hardness results for TRANSPORT by proving hardness results for ROUTE.

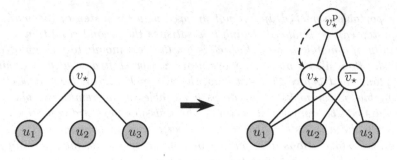

Fig. 4.3. Introducing clones and portable locations. The gray nodes indicate neighbours of target location $v_\star$; the dotted arc indicates a portable to deliver

Theorem 4.2.5. PLAN-ROUTE$_j^k$-$[l]$ $\leq_{\mathbf{OP}}$ PLAN-TRANSPORT$_{ij}^k$-$[l, m]$
For all $i, k \in \{1, \infty, *\}$, $j \in \{1, +, *\}$, $l \in \{1, *\}$ and $m \in \{0, 1, *\}$, we have:
    PLAN-ROUTE$_j^k$-$[l]$ $\leq_{\mathbf{OP}}$ PLAN-TRANSPORT$_{ij}^k$-$[l, m]$.

Proof. Let T be the given ROUTE$_j^k$-$[l]$ *task with roadmap graph $\langle V, E \rangle$ and target locations $V_\star \subseteq V$. Without loss of generality, assume that no location in $V_\star$ is the initial location of a mobile. (Target locations which are also initial locations define goals which are already satisfied in the initial state and can thus be ignored.) The reduction is independent of the performance ratio $r > 1$ and maps to a* TRANSPORT$_{ij}^k$-$[l, m]$ *task T' with the following components:*

- *The roadmap graph is $\langle V \cup V', E \cup E' \rangle$, i. e., the original roadmap graph is extended with some new locations and roads. The movement costs and fuel levels for the original locations and roads are left unchanged.*
  *The new location set V' consists of two new locations for each target location $v_\star \in V_\star$. The first, denoted by $\overline{v_\star}$ is called the **clone** of $v_\star$; the second, denoted by $v_\star^p$ is called its **portable location**. The fuel level for these locations is 1 in the restricted fuel cases ($k = 1$ or $k = *$) and unbounded otherwise.*
  The new road set E' contains one road between the clone of each target location $v_\star$ and each neighbour u of $v_\star$, with the same movement costs as the road between u and $v_\star$. Moreover, it contains roads $\{v_\star^p, v_\star\}$ and $\{v_\star^p, \overline{v_\star}\}$ connecting each portable location to the corresponding target location and its clone. The movement cost for these roads is 1 for all mobiles.
- *The set of mobiles and their initial locations are the same as in T. The mobile capacities are all equal to 1 in the bounded capacity cases ($i = 1$ or $i = *$) and unbounded otherwise.*
- *For each target location $v_\star \in V_\star$, there is a portable initially located at the corresponding portable location $v_\star^p$ with goal location $v_\star$.*
- *All pickup and drop costs are 0 if $m = 0$ or $m = *$, and 1 otherwise.*

The most important part of the reduction, the mapping of a target location of the original task to a set of three locations in the generated task, is illustrated in Fig. 4.3.

Clearly, the reduction can be performed in polynomial time. We first show that it maps solvable tasks to solvable tasks. Any plan π for T can be transformed into a plan π' for T' as follows. For each target location $v_\star \in V_\star$, identify a step in the plan where some mobile m is moved from some location $u \in V$ to $v_\star$. There must be such an action because each target location must be reached by some mobile to satisfy the original goal, and this can only be done by entering the locations because no target locations are initial locations. Replace this movement action by the following sequence of actions: move m from u to the clone of $v_\star$, then on to $v_\star^p$, pick up the portable there, move to $v_\star$, drop the portable.

It is easy to verify that this transformation generates a valid plan: All goals are satisfied, capacity constraints are met because no mobile ever carries more than one portable, access restrictions are met because all movements either mirror legal movements in the initial task or move to or from the portable locations, and fuel constraints are met because this is true for the original solution, the modifications do not increase the number of times a location $v \in V$ is left, and each clone or portable location is left only once in π'.

Considering the costs of the modified plan, we see that $m(\pi') = m(\pi) + c|V_\star|$, where $c = 2$ if pickup is free (the $m = 0$ or $m = *$ cases) and $c = 4$ otherwise: For each target location, the modifications introduce an additional cost of 2 for moving to each portable location and back and an additional cost of 2 for picking up and dropping portables unless pickup is free. In particular, this implies the relationship $m^*(T') \leq m^*(T) + c|V_\star|$ for the optimal solution lengths.

To map back a plan π' for T' into a plan π for T, remove all pickup and drop actions and all movements to and from portable locations, and replace all movements to and from clones by movements to and from their corresponding target location.

This clearly results in a plan which only moves along the roads of the original roadmap and satisfies the access restrictions. Moreover, it visits all target locations, because for each target location, some vehicle must move through $v_\star$ or its clone to the corresponding portable location in π' to satisfy the goals of the TRANSPORT task. To see that π also satisfies the fuel constraints, we only need to consider target locations with restricted fuel; unrestricted fuel locations and non-target locations are not critical. So let $v_\star$ be a target location with initial fuel level f in the original task T. The combined initial fuel levels of $v_\star$ and its clone in T' is then $f + 1$, so these two locations are left at most $f + 1$ times in π'. One of these movements must be towards the portable location $v_\star^p$ and is thus no longer represented in π, so that π indeed contains at most f movement actions leaving $v_\star$, satisfying the fuel constraints.

The mapping from π' to π does not introduce new actions, it only modifies actions by replacing clones with their corresponding target locations

*and removes some actions. In particular, π' must contain at least four ac-
tions for each target location which are removed by the mapping: a movement from the target location or its clone to the corresponding portable location, a movement in the opposite direction, a pickup action, and a drop
action. We thus have $m(\pi) \leq m(\pi') - c|V_\star|$. In particular, this implies
$m^*(T) \leq m^*(T') - c|V_\star|$ and hence, together with the earlier upper bound
on $m^*(T')$, $m^*(T) = m^*(T') - c|V_\star|$.*

*We can now see that the reduction is approximation-preserving. Given a
plan π' for T' with performance ratio at most $r > 1$ (thus $m(\pi') \leq rm^*(T')$),
the mapped-back solution π to T has the performance ratio*

$$\frac{m(\pi)}{m^*(T)} \leq \frac{m(\pi') - c|V_\star|}{m^*(T)} \leq \frac{rm^*(T') - c|V_\star|}{m^*(T') - c|V_\star|} = 1 + (r-1)\frac{m^*(T')}{m^*(T') - c|V_\star|}$$

$$\leq 1 + (r-1)\frac{m^*(T) + c|V_\star|}{m^*(T)} \overset{(*)}{\leq} 1 + (r-1)\frac{m^*(T) + cm^*(T)}{m^*(T)}$$

$$= 1 + (r-1)(1+c),$$

where inequality $(*)$ holds because $|V_\star| \leq m^*(T)$: Any solution to the ROUTE
*task must contain at least one movement to each target location, and each
movement has a cost of at least 1. Setting $\alpha = 1 + c$, we see that the reduction is approximation-preserving. Moreover, because we can determine the
optimal plan length for T' to be exactly $m^*(T) + c|V_\star|$, it is also optimization-
preserving. This concludes the proof.* □

The transformation could be simplified significantly for many of the ROUTE
variants. In particular, the introduction of clones is only necessary for the unit
fuel case. In the unbounded fuel case, clones could be omitted without further
adjustments, while in the variable fuel case, they could be omitted if the fuel
level of each target location is increased by one.

One disadvantage of the reduction is that it affects the shape of the
roadmap graphs. For example, if we could prove that some variant of ROUTE
planning were hard for planar graphs, we could not obtain hardness for planar graphs for the corresponding TRANSPORT variants, because the reduction
does not preserve planarity. For this reason, we provide another – much simpler – reduction which does preserve graph structures. Before we do this, we
introduce some relevant classes of graphs and formalize different notions of
"preserving graph structure". (For now, we only need the simplest kind of
preservation, but it is useful to introduce these concepts together.)

**Definition 4.2.6. Stretchable, Weight-Expandable and Reducible
Graphs**
*Let $\mathcal{G}$ be a class of (undirected, unweighted) graphs. For unweighted graphs G
and edge weight functions w, let $dist_G$ denote the graph-theoretical distance
function in G and let $dist_G^w$ denote the graph-theoretical distance function in
the graph obtained from G by weighting the edges of G according to w.*

*The class $\mathcal{G}$ is called **stretchable** if for all graphs $G = (V, E) \in \mathcal{G}$ and all numbers $K \in \mathbb{N}_1$, there exists a graph $G_K = (V_K, E_K) \in \mathcal{G}$ such that $V \subseteq V_K$ and for all $v, v' \in V$, $dist_{G_K}(v, v') = K \cdot dist_G(v, v')$.*

*The class $\mathcal{G}$ is called **weight-expandable** if for all graphs $G = \langle V, E \rangle \in \mathcal{G}$ and all weight functions $w : E \to \mathbb{N}_1$, there exists a graph $G_w = \langle V_w, E_w \rangle \in \mathcal{G}$ such that $V \subseteq V_w$ and for all $v, v' \in V$, $dist_{G_w}(v, v') = dist_G^w(v, v')$.*

*The class $\mathcal{G}$ is called **reducible** if for all graphs $G = \langle V, E \rangle \in \mathcal{G}$ and all subsets $E' \subseteq E$, the graph $G' = \langle V, E' \rangle$ is also contained in $\mathcal{G}$.*

Both for stretchability and weight-expandability, we require the mapping from G and K (or w) to G_K (or G_w) to be computable in time polynomial in the size of G and the magnitude of K (or the maximal edge weight $w(e)$).

Weight-expandability is a generalization of stretchability, because expanding a graph with a constant weight function has the same effect as stretching it. It is useful to introduce both concepts separately because we will require stretchability more often than general weight-expandability.

Examples of weight-expandable graph classes include the class of all graphs, the class of planar graphs, and the class of trees. Grid graphs are stretchable, but not weight-expandable, and complete graphs are neither.

Definition 4.2.7. *Preserving Graph Classes*
Let R be a reduction between two decision or optimization problems defined (partially or completely) in terms of graphs, and let $\mathcal{G}$ be a class of graphs.

*We say that R **preserves** $\mathcal{G}$ if graphs from $\mathcal{G}$ are always mapped to graphs from $\mathcal{G}$ by R.*

When using this definition, we will normally state that a certain reduction preserves a *family* of graph classes, rather than a single graph class. For example, instead of stating that a reduction preserves planarity, we will state that it preserves all stretchable graph classes, which implies that it preserves not just planarity, but also grid or tree shape.

In the following proof, we use this definition in its most general form by stating that the reduction preserves *all* graph classes (which is only true of reductions mapping each graph to itself).

Theorem 4.2.8. $\text{PLAN-ROUTE}_1^k\text{-}[l] \leq_{\text{OP}} \text{PLAN-TRANSPORT}_{\infty 1}^k\text{-}[l, 0]$
For all $k \in \{1, \infty, *\}$ and $l \in \{1, *\}$, we have:
$\text{PLAN-ROUTE}_j^k\text{-}[l] \leq_{\text{OP}} \text{PLAN-TRANSPORT}_{\infty j}^k\text{-}[l, 0]$.
The reduction preserves all graph classes.

Proof. Under the given restrictions (a single mobile, unbounded capacity, free pickup), a much simpler reduction than in the previous theorem maps ROUTE tasks to TRANSPORT tasks as follows:

- *Roadmap, movement costs and initial mobile location are not changed by the mapping.*

- *For each target location, there is a portable which must be moved to that location and is initially located at the initial mobile location.*
- *The mobile capacity is unbounded, and pickup is free.*

Any plan for the ROUTE *task can be transformed into a plan for the* TRANS-PORT *task which has the same cost: At the start of the plan, pick up all the portables, and whenever a target location is first visited, drop the corresponding portable. Because pickup is free, this plan has the same cost as the original plan. On the other hand, any plan for the* TRANSPORT *task can be transformed into a plan for the* ROUTE *task with the same cost by omitting all pickup and drop actions.*

It is obvious that these mappings define an optimization-preserving reduction.   □

This concludes our discussion of general complexity results for TRANSPORT and ROUTE, and we will now start examining more restricted cases.

4.3 Plan Existence

In our first foray into the complexity of the TRANSPORT and ROUTE domains, we disregard optimization altogether, asking how difficult it is to find *arbitrary* solutions. The following two theorems provide a very clear answer to this question. We start with the positive result, which shows that in the absence of fuel constraints, transportation tasks are polynomially solvable. (Recall that if we leave out the fuel parameter from our notation, this denotes the unbounded fuel case.)

Theorem 4.3.1. PLAN-TRANSPORT$_{**}$-[$*,*$] $\in$ poly-APX
The planning problem for TRANSPORT$_{**}$-[$*,*$] *is poly-approximable.*

Proof. We are given a TRANSPORT$_{**}$-[$*,*$] *task T with goal portable set P, mobiles M and locations V. For all goal portables $p \in P$, mobiles $m \in M$ and locations $u, v \in V$ connected by a road, we define the **minimal transportation cost** $c_p^m(u, v)$ for transporting p from u to v with m as the sum of the following operator costs:*

- *The minimal cost of moving m from its initial location to u, determined by computing a shortest path in the roadmap of m.*
- *The cost of picking up p with mobile m.*
- *The cost of moving m from u to v.*
- *The cost of dropping p with mobile m.*

The minimal transportation cost can be infinite if m cannot reach u from its initial location, or if it may not access the road $\{u, v\}$.

*The minimal transportation cost for transporting a given portable from u to v is defined as $c_p(u, v) = \min_{m \in M} c_p^m(u, v)$. The **delivery threshold** for a* *portable $p \in P$ is defined as the minimum number $b(p)$ for which there exists a path $l_1 \ldots l_n$ in the roadmap from the initial location of p to the goal location of p such that $c_p(l_i, l_{i+1}) \leq b(p)$ for all $i \in \{1, \ldots, n-1\}$. The **global delivery threshold** is defined as $b_g = \max_{p \in P} b(p)$. It is easy to see that the cost of any solution to T must be at least b_g: If a portable is moved along a road from u to v by some mobile m in the plan, then the plan must at least include a pickup and drop action for that portable and mobile and movements from the initial location of m via u to v. In particular, if the global delivery threshold is infinite, the task is not solvable.*

We now show how to construct a solution for solvable tasks which has a cost which is at most polynomially higher than the global delivery threshold.

For each portable $p \in P$, perform the following steps:

1. *Compute the minimal transportation cost $c_p^m(u, v)$ for all mobiles $m \in M$ and locations $u, v \in V$ connected by a road.*
2. *Construct a labelled, weighted, directed multigraph G_p with vertex set V and an arc from u to v with label m and weight w iff $c_p^m(u, v) = w < \infty$.*
3. *Compute a shortest path in G_p leading from the initial location of p to its goal location, called the **delivery path** π_p. Fail if no such path exists.*
4. *Iterating over the arcs that constitute the delivery path, let u be the source and v be the target location of the current arc, and let m be the mobile with which it is labelled. Add the following steps to the plan:*
   - *Movements of m from its initial location to u (minimizing cost).*
   - *A pickup action for picking up p with m.*
   - *A movement of m to v.*
   - *A drop action for dropping p with m.*
   - *Movements of m back to its initial location (minimizing cost).*

These computations can be performed in polynomial time and produce a solution for I iff the task is solvable. To bound the performance ratio, observe that for each arc traversed in an iteration of step 4., the cost of the actions added to the plan is bounded by twice the global delivery threshold. Because delivery paths are shortest paths, they do not revisit vertices of the graphs, and hence they consist of no more than $|V| - 1$ arcs. Because there are $|P|$ portables to deliver, we can thus bound the performance ratio of the algorithm by $2|P|(|V| - 1)$, which is polynomial in the task size. □

In the presence of fuel constraints, however, planing becomes much more difficult. Plan existence is already NP-hard for the simplest ROUTE variant involving fuel constraints, and is thus also hard for all TRANSPORT variants involving fuel constraints due to Theorem 4.2.5.

Theorem 4.3.2. PLANEX-ROUTE$_1^1$ *is* NP-*hard.*
Plan existence in the ROUTE$_1^1$ *domain is* NP-*hard, even when restricted to planar roadmaps.*

Proof. We reduce from the NP-*complete problem of deciding the existence of a Hamiltonian path with a fixed start vertex in a planar graph [43, Problem GT39].*

Let I be a HAMILTONIAN PATH *instance given by a planar graph $\langle V, E \rangle$ and start vertex $v_0 \in V$. We map this to a* ROUTE$_1^1$ *task T' with roadmap $\langle V, E \rangle$ and initial mobile location v_0 where all locations are target locations. Clearly, this mapping can be computed in polynomial time.*

If π is a Hamiltonian path in $\langle V, E \rangle$ starting at v_0, then the planning task can be solved by moving the mobile along the path π. On the other hand, if π is the trajectory of the mobile in a solution to the planning task, then we can assume that π never revisits any vertex: If the last visited location is a revisit, then the last step of the plan can be omitted, and revisits cannot happen in an earlier stage of the plan because a location cannot be left twice due to fuel constraints. Because π must visit all target locations (i. e., all locations) to be a solution and starts in v_0, this means that $\langle V, E \rangle$ has a Hamiltonian path starting at v_0.   □

Together with Theorem 4.2.4 and the reduction from ROUTE to TRANSPORT, the previous two results define a sharp boundary: All TRANSPORT and ROUTE variants with potentially bounded fuel belong to NPO \ PS (unless P = NP), and all variants with unbounded fuel belong to poly-APX. In the following sections, we will refine the latter classification result.

4.4 Hardness of Optimization

The approximation algorithm presented in Theorem 4.3.1 does not achieve very good performance ratios. We will later show that this is no surprise, and that we cannot achieve constant factor approximations in general. However, there might well be special cases in our domain families which do admit very good polynomial approximation algorithms, maybe even optimal solutions. We now show that the latter is not the case.

Theorem 4.4.1. PLAN-ROUTE$_1$-[1] $\notin$ PTAS *unless* P = NP
If P $\neq$ NP, *there is no polynomial-time approximation scheme for planning in the* ROUTE$_1$-[1] *domain.*

The bounded plan existence problem for ROUTE$_1$-[1] *is* NP-*hard.*

Proof. We provide an optimization-preserving reduction from 1-2-TSP, *the variant of* TRAVELLING SALESPERSON PROBLEM *where distances between sites are symmetric and either 1 or 2, to* ROUTE$_1$-[1]. *Because* 1-2-TSP $\notin$ PTAS *if* P $\neq$ NP *and the corresponding decision problem is* NP-*hard [7, problem ND33], this proves the theorem.*

Let I be the given 1-2-TSP *instance, defining a weighted graph $\langle V, E, w \rangle$. This is mapped to a* ROUTE$_1^1$ *task T' as follows:*

- *Compute the unweighted graph $G_w = \langle V_w, E_w \rangle$ by weight-expanding the graph (V, E) with weight function w. This can be done in polynomial time because weights are polynomially bounded (indeed, even bounded by 2).*
- *Select an arbitrary site $v_0 \in V$ and set $G'_w = \langle V_w \cup \{v'_0, v''_0\}, E_w \cup \{\{v_0, v'_0\}, \{v'_0, v''_0\}\}\rangle$, i.e., introduce a new dead-end path of length 2 accessible only from v_0.*
- *The ROUTE task is defined by roadmap graph G'_w, initial mobile location v_0, and target location set $V \cup \{v''_0\}$.*

This computation can clearly be performed in polynomial time. Any TRAVELLING SALESPERSON PROBLEM solution π can be mapped to a ROUTE solution π' with cost $m(\pi') = m(\pi)+2$ by performing the movements corresponding to the TRAVELLING SALESPERSON PROBLEM tour (starting from and ending at location v_0), then moving to v'_0, and finally to v''_0. On the other hand, any plan π' for T' can be mapped back to a TRAVELLING SALESPERSON PROBLEM solution π with $m(\pi) \leq m(\pi') - 2$ as follows:

1. *Ensure that the last action of the plan visits some target location for the first time, by omitting unnecessary actions at the end of the plan.*
2. *Ensure that the mobile ends in location v''_0. If it ends in location $v \neq v''_0$, v''_0 must be visited earlier, which requires four actions for moving to v''_0 from v_0 and back to v_0. Remove these four actions, and add actions to move from v to v''_0 to the end of the plan. This does not increase plan length, because the graph distance between any location $v \in V$ and v''_0 is at most 4.*
3. *The last location from V visited by the plan must now be v_0. Remove all actions occurring after the last visit to v. There must be at least two such actions, namely a movement from v_0 to v'_0 and a movement from v'_0 to v''_0.*
4. *Return the travelling salesperson tour in which site $v \in V$ is the i-th site of the tour iff it is the i-th distinct location from V visited by the modified plan. The cost of that tour is then no larger than the length of the modified plan.*

It is easy to see from these observations that $m^*(T') = m^*(I) + 2$. If π' is the given ROUTE solution with performance ratio at most $r > 1$ and π is the derived 1-2-TSP solution, we can bound the performance ratio of the latter by

$$\frac{m(\pi)}{m^*(I)} \leq \frac{m(\pi') - 2}{m^*(T') - 2} \leq \frac{rm^*(\pi') - 2}{m^*(T') - 2} = 1 + \frac{(r-1)m^*(T')}{m^*(T') - 2}$$

$$= 1 + (r-1)\frac{m^*(I) + 2}{m^*(I)} \overset{(*)}{\leq} 1 + (r-1)\frac{m^*(I) + m^*(I)}{m^*(I)}$$

$$= 1 + (r-1) \cdot 2,$$

where inequality $(*)$ holds because we can assume $m^*(I) \geq 2$; trivial instances not satisfying this property can be identified and special-cased in polynomial

time. Thus the reduction is approximation-preserving, and due to the equality
$m^*(T') = m^*(I) + 2$ *also optimization-preserving.*   □

Together with Theorem 4.2.5, this result provides us with a bound to approximability for all unrestricted fuel domains in the TRANSPORT and ROUTE family. Therefore, if P $\neq$ NP, none of their planning problems belongs to PO or PTAS.

One interesting question, also relevant to some of the IPC domains, is whether this result extends to restricted classes of graphs. Indeed, using very similar reductions from TRAVELLING SALESPERSON PROBLEM, we could prove NP-completeness of bounded plan existence for ROUTE$_1$-[1] also for planar roadmaps, and even for grids [57]. However, this would only prove that these planning problems do not belong to PO; in fact, they *do* admit polynomial-time approximation schemes when restricted to such graph classes, because this is the case for the corresponding TRAVELLING SALESPERSON PROBLEM variants [6].

Another graph class relevant to the competition domains are *complete* graphs. It is very easy to see that the ROUTE$_1$-[1] problem for complete graphs can be optimally solved in polynomial time. However, the same is not true for the corresponding TRANSPORT variants with unbounded capacity.

Theorem 4.4.2. PLAN-TRANSPORT$_{\infty 1}$-[1, 0/1] $\notin$ PTAS
If P $\neq$ NP, there is no polynomial-time approximation scheme for planning in the TRANSPORT$_{\infty 1}$-[1, 0] or TRANSPORT$_{\infty 1}$-[1, 1] domains, even if the set of roadmap graphs is restricted to complete graphs.

The corresponding bounded plan existence problems are NP-hard.

Proof. Papadimitriou and Yannakakis proved that the restriction of MINIMUM VERTEX COVER *to graphs where all vertices have a degree of at most 3 is not in* PTAS *unless* P = NP *[96]. In addition to the degree upper bound, we make the further restriction that the degree of each vertex must be at least 2. This does not affect the hardness result since other vertices can be trivially dealt with: As long as there are any vertices of degree 1, pick any such vertex, add its neighbour to the vertex cover, and remove the vertex and its neighbour along with any incident edges from the graph. If no vertices of degree 1 are left, remove all vertices of degree 0. All decisions made in this preprocessing step are optimal, so if there were arbitrarily good approximation results for our restricted graph class, then we could obtain an arbitrarily good approximation for arbitrary graphs with degree bound of 3 by combining the algorithm for the restricted graph class with the preprocessing technique. A Karp reduction between the related decision problems can be defined in a similar fashion.*

We now describe the reductions to TRANSPORT$_{\infty 1}$-[1, 0] *planning and* TRANSPORT$_{\infty 1}$-[1, 1] *planning. Given a* MINIMUM VERTEX COVER *instance (i. e., a graph) $G = \langle V, E \rangle$ where all vertex degrees are either 2 or 3, the corresponding* TRANSPORT *task T' is defined as follows:*

- *The set of locations is $V \cup \{v_0\}$, where v_0 is a new location that serves as the initial location of the only mobile. The roadmap graph is a complete graph over the location set.*
- *For each edge $\{u, v\} \in E$, there is one portable with initial location u and goal location v and one portable with initial location v and goal location u.*
- *Pickup is free when reducing to $\text{TRANSPORT}_{\infty 1}\text{-}[1, 0]$ and has cost 1 when reducing to $\text{TRANSPORT}_{\infty 1}\text{-}[1, 1]$.*

This is clearly a polynomial mapping, and G has a vertex cover iff T' is solvable. (In fact, every graph has a vertex cover, and every task T' generated by the reduction has a solution.)

To map a solution π' to T' back into a vertex cover U for G, we define U to include exactly those vertices from V corresponding to locations visited at least twice in the plan. To see that this is a vertex cover, assume that it were not, i. e., that there existed an edge $\{u, v\} \in E$ with $u \notin U$ and $v \notin U$. Then neither u nor v is visited twice in π', in which case it is not possible that the mobile has transported a portable from u to v and a portable from v to u.

To compare the quality of the solutions, we make the following observations:

- *For any vertex cover $U \subseteq V$, the planning task can be solved by the following three-phase plan π':*
  - *In the first phase, move to the locations U in any order, picking up all portables located there.*
  - *In the second phase, move to the locations $V \setminus U$ in any order, picking up portables located there and dropping carried portables when passing through their goal location.*
  - *In the third phase, again move to the locations U in any order, dropping carried portables when passing through their goal location.*
  To see that π' is indeed a plan for T', observe that portables which need to be delivered from $u \in U$ to $v \notin U$ are picked up in the first and dropped in the second phase, and portables which need to be delivered to locations $v \in U$ are picked up in the first or second and dropped in the third phase. The only critical portables are those that need to be moved between two locations $u, v \notin U$. However, such portables do not exist: If there is a portable to be moved from u to v, then $\{u, v\} \in E$ and the vertex cover U must include at least one of the two vertices.
- *Moreover, if U is an optimal vertex cover, then the plan π' is optimal. Each portable needs to be picked up and dropped once, each location visited at least once, so the only potential savings could be obtained by revisiting fewer locations. However, we have argued before that the set of revisited locations for any plan forms a vertex cover, and there is no smaller vertex cover than U.*
- *We thus have the relationship $m^*(T') = m^*(G) + |V| + 4c|E|$, where c is 0 if pickup is free and 1 otherwise: Each location in visited in V is visited at least once (cost $|V|$), the locations in the optimal vertex cover are revisited*

(cost $m^*(G)$), and there are $2|E|$ pickup and $2|E|$ drop actions (total cost $4c|E|$).

- *On the other hand, for the vertex cover U obtained from a solution π' to the planning task, we have $m(U) \leq m(\pi') - |V| - 4c|E|$: π' must contain at least one pickup and drop action for each portable (for a total cost of $4c|E|$) and visit each location in $|V|$ at least once (for a total cost of $|V|$). The remaining actions in the plan thus have a cost of at most $m(\pi') - |V| - 4c|E|$, and therefore this number bounds the number of locations that are revisited by π'.*
- *We have $|V| \leq |E|$, because the number of edges is half the sum of all vertex degrees and all vertices have at least degree 2. Moreover, we have $|E| \leq 3m^*(G)$ because every edge must be covered by some vertex in any vertex cover (including an optimal one), vertex degrees are bounded by 3.*

Putting the pieces together, we can estimate the performance ratio of the vertex cover U obtained from a plan π' with performance ratio at most $r > 1$:

$$\frac{m(U)}{m^*(G)} \leq \frac{m(\pi') - |V| - 4c|E|}{m^*(T') - |V| - 4c|E|} \leq \frac{rm^*(T') - |V| - 4c|E|}{m^*(T') - |V| - 4c|E|}$$

$$= 1 + (r-1)\frac{m^*(T')}{m^*(T') - |V| - 4c|E|}$$

$$= 1 + (r-1)\frac{m^*(G) + |V| + 4c|E|}{m^*(G)}$$

$$\overset{(*)}{\leq} 1 + (r-1)\frac{m^*(G) + 3m^*(G) + 4c \cdot 3m^*(G)}{m^*(G)}$$

$$= 1 + (r-1) \cdot (4 + 12c),$$

where inequality $(*)$ holds because $|V| \leq |E| \leq 3m^*(G)$. This proves that the reduction is approximation-preserving, and due to the equality $m^*(T') = m^*(G) + |V| + 4c|E|$ also optimization-preserving. $\square$

While the theorem shows that there must be *some* bound $r > 1$ for which there is no r-approximating algorithm for TRANSPORT tasks with complete roadmap graphs if $P \neq NP$, it does not provide an actual lower bound for approximability. Indeed, the result by Papadimitriou and Yannakakis does not provide a lower bound either, but it is known that the MINIMUM VERTEX COVER problem for graphs with a vertex degree bound of 5 (rather than 3) is not approximable within 1.0029 unless $P = NP$ [7]. By adjusting the reduction accordingly, we can exploit this result to show that the planning problem is not approximable within 1.000111 (with unit pickup cost) or within 1.000483 (with free pickup) unless $P = NP$. The best approximation algorithm we found for this class of TRANSPORT tasks (with pickup costs) guarantees a performance ratio of $\frac{7}{6}$. Obviously, there is still a significant gap between these two bounds.

4.5 Constant Factor Approximation

With the results from the previous section, we have now established that all ROUTE and TRANSPORT variants with unrestricted fuel belong to poly-APX, but none of them admits a polynomial-time approximation scheme. This leaves two possibilities fur such domains: Either they allow constant-factor approximations and thus belong to APX, or they do not, and belong to poly-APX \ APX. In this section, we prove results of the former kind. In summary, we can obtain constant-factor approximations whenever we have only one mobile type and mobiles either have unbounded or unit capacity. We first prove the unbounded capacity result.

Theorem 4.5.1. PLAN-TRANSPORT$_{\infty+}$-$[*, *] \in$ APX
The planning problem for TRANSPORT$_{\infty+}$-$[*, *]$ *is approximable by a constant factor.*

Proof. We are given a TRANSPORT$_{\infty+}$-$[*, *]$ *task T. We can assume that all portables are goal portables, since other portables can be ignored. Let G be the roadmap graph of T weighted with the movement cost function (note that movement costs are independent of the moving mobile for this* TRANSPORT *variant). Throughout the proof, we write $w(G')$ to denote the **weight** of a subgraph G' of G, which is defined as the sum over the weights of all roads contained in the subgraph.*

*A **delivery graph** for T is a subgraph of G in which every portable start location is connected to its corresponding goal location. (A delivery graph may have several connected components.) Computing delivery graphs of small weight is the* MINIMUM POINT-TO-POINT CONNECTION *problem, which is 2-approximable [7, problem GT52]. We can thus, in polynomial time, compute a delivery graph G_d with cost at most twice as high as the cost of a minimal weight delivery graph. The roads traversed in any solution to T must form a delivery graph (otherwise not all portables could be delivered), and thus we have $w(G_d) \leq 2m^*(T)$. We can assume that every connected component of G_d contains the initial location of a portable; connected components which do not have this property can be safely removed, which may only reduce the weight. (Of course, a reasonable approximation algorithm for* MINIMUM POINT-TO-POINT CONNECTION *would not generate such connected components in the first place.)*

*A **pickup graph** for T is a subgraph of G in which the initial location of each portable is connected to the initial location of some mobile. By extending G with a new vertex v_0 which is connected to each initial mobile location by an edge of weight 0, we obtain a modified roadmap graph $\tilde{G}$. It is easy to see that a subgraph G' of G is a pickup graph iff G' together with vertex v_0 and its incident edges is a subgraph of $\tilde{G}$ in which v_0 and all initial portable locations are in the same connected component. Subgraphs of small weight where a certain set of vertices belong to the same connected component are* *called* Steiner graphs, *or more typically* Steiner trees, *because Steiner graphs which are not trees can be pruned to obtain a tree which is also a Steiner graph. Computing Steiner trees of small weight is the* MINIMUM STEINER TREE *problem, which can also be 2-approximated [7, problem ND8]. We can thus, in polynomial time, compute a pickup graph G_p with cost at most twice as high as the cost of a minimal weight pickup graph. The roads traversed in any solution to T must form a pickup graph (otherwise not all portables could be picked up), and thus we have $w(G_p) \leq 2m^*(T)$. We can assume that every connected component of G_p contains the initial location of a portable, since other components could be safely removed.*

*Having computed a delivery graph and a pickup graph, we build the union of these two graphs and repeatedly eliminate some edge participating in a cycle until the resulting graph is a forest, which we call the **solution graph** G_s. Clearly, $w(G_s) \leq w(G_d) + w(G_p) \leq 4m^*(T)$. The solution graph is both a delivery graph and a pickup graph because it is constructed from the union of such graphs, and eliminating edges participating in cycles does not affect connectivity. Moreover, every connected component of G_s contains the initial location of a portable (because this is the case both for G_d and G_p), the goal location of all portables whose initial locations it contains (because these are connected in G_d), and the initial location of some mobile (because the initial location of any portable must be connected to such a location in G_p). We can thus solve the task as follows:*

1. *For each connected component $C = \langle V_C, E_C \rangle$ of G_s, choose a mobile m with initial location $v_m \in V_C$. Compute a tour π_C through C which starts at v_m, traverses each road in E_C twice, and ends at v_m. This is easily possible in polynomial time by performing a depth-first traversal of C starting at v_m.*
2. *Move the mobile m along the tour π_C, picking up each portable as its initial location is passed for the first time.*
3. *After returning to v_m, move the mobile m along the tour π_C a second time, dropping each portable at its goal location as it is passed for the first time on this tour.*

This clearly solves the task. To see that this is a constant factor approximation, observe that the total weight of all edge sets E_C is equal to $w(G_s)$, and that each edge in all such edge sets is traversed exactly four times by the plan, leading to total movement costs of $4w(G_s) \leq 16m^*(T)$.

Moreover the cost of all pick-up and drop operators in the generated plan is bounded by $m^*(T)$, because each portable is only picked up and dropped once, and it must be picked up and dropped at least once in any plan, including optimal ones. (Remember that pickup and drop costs are independent of the mobile involved for this TRANSPORT *variant). Thus, the generated plan is 17-approximating.* □

With some minor adaptations and a slightly more careful analysis, the approximation factor in the previous proof could be reduced to 10. Drescher proves this for the unit cost model in his Master's thesis [31].

The unit capacity case can be solved with similar techniques.

Theorem 4.5.2. PLAN-TRANSPORT$_{1+}$-[$*$, $*$] $\in$ APX
The planning problem for TRANSPORT$_{1+}$-[$*$, $*$] *is approximable by a constant factor.*

Proof. We are given a TRANSPORT$_{1+}$-[$*$, $*$] *task* T*. Again, we assume that all portables are goal portables.*

To solve T*, we first compute a pickup graph* G_p *as in the previous proof. Using the same arguments and techniques as before, we can assume that* $w(G_p) \leq 2m^*(T)$*, that the computation can be performed in polynomial time, that every connected component of* G_p *contains the initial locations of at least one portable and mobile and that the initial locations of all portables are present in* G_p*. The task can then be solved as follows:*

1. *For each connected component* $C = \langle V_C, E_C \rangle$ *of* G_p*, choose a mobile* m *with initial location* $v_m \in V_C$*. Compute a tour* π_C *through* C *which starts at* v_m*, traverses each road in* E_C *twice, and ends at* v_m*.*
2. *Move the mobile* m *along the tour* π_C*. Whenever the initial location of a portable* p *is reached, interrupt the tour, pick up* p*, move to the goal location of* p *on a shortest path, move back to the initial location of* p *on a shortest path, and continue the tour.*

*This clearly solves the task. To see that this is a constant factor approximation, we distinguish between movements of empty mobiles (**empty moves**), movements of mobiles carrying a portable (**delivery moves**), and other actions (**pick-ups and drops**):*

- *Delivery moves: For every portable, the total cost for delivery moves in the generated plan is no higher than the total cost for delivery moves of the same portable in an optimal plan, because portables are delivered on a shortest path. Moreover, the total cost of delivery in an optimal plan is equal to the sum of the delivery costs for the individual portables, because only one portable can be carried at a time. Thus, the total cost for delivery moves is bounded by* $m^*(T)$*.*
- *Empty moves: There are two kinds of empty moves: Movements along the tours in the pickup graphs, and movements from the goal location of a portable back to its initial location after it has been delivered. The total cost of movements of the first type is* $2w(G_p) \leq 4m^*(T)$*, and the total cost of movements of the second type is equal to the total cost of delivery moves, because the same roads are traversed, only in the opposite direction. Thus, we obtain a total bound of* $5m^*(T)$ *for these actions.*

- Pickups and drops: *These can again be bounded by $m^*(T)$ because each portable is only picked up and dropped once.*

  Combining the three bounds, we see that the algorithm is 7-approximating.
  □

With a slightly more careful analysis, we could see that the algorithm is actually 6-approximating. In his Master's thesis, Drescher proves a 4-approximation result for the unit cost model, using more sophisticated techniques [31].

Note that the algorithm from the previous proof could also be applied to the unbounded capacity case; after all, mobiles with unbounded capacity can do everything that mobiles with unit capacity can do. However, this would not result in a constant-factor approximation. To see this, consider the case with only one mobile and N portables, all initially located at the same location as the mobile, and all with the same goal location, which has a distance of M from the initial location. The algorithm for the unbounded capacity case would generate a solution of size $\Theta(N + M)$, while the algorithm for the unit capacity case would generate a solution of size $\Theta(NM)$.

Summarizing the results of this section, we have shown that constant-factor approximations for TRANSPORT tasks with unbounded fuel can be provided if we do not allow mobiles to have arbitrary capacity and if we do not allow mobiles to have different roadmaps. In the following section, we complete our analysis by proving that both restrictions are necessary.

4.6 Hardness of Constant Factor Approximation

In our earlier analysis of transportation planning [54, 57], we showed that whether or not there are mobiles of different types does not affect the complexity of the plan existence and bounded plan existence problems for the family of transportation problems we then studied. Indeed, the results we presented earlier in this chapter imply that the same is true for the richer family of transportation and route planning considered here. However, we will now show that the presence of mobiles of different types has a marked impact on the achievable approximation quality: Constant factor approximations are no longer possible, even for simple unit cost route planning problems.

Theorem 4.6.1. PLAN-ROUTE$_*$-[1] $\notin$ APX
If P $\neq$ NP, there is no constant-factor approximation algorithm for planning in the ROUTE$_*$-[1] domain, even in the restricted case where the roadmap graphs of all mobiles are trees and all mobiles start at the same location.

Moreover, this problem variant is NP-hard.

Proof. Note that NP-hardness is not implied by our earlier results due to the restriction to tree roadmaps. We prove both claims by providing an optimization-preserving reduction from the MINIMUM SET COVER problem, which is not in APX unless P = NP, and which has an NP-hard decision problem.

Let $I = \langle S, C \rangle$ be the given set cover instance, and let $r > 1$ be the given performance ratio. We define $M = \left\lceil \frac{2|S|}{r-1} + 2|S| \right\rceil$. The corresponding ROUTE *task T' has the following components:*

- *For each element $s \in S$, there is a location v_s (called the **element location** for s), for each subset $\widehat{S} \in C$, there is a location $v_{\widehat{S}}$ (called the **subset location** for $\widehat{S}$), and there are $M + 1$ **start-up locations** $v^0, \ldots, v^M$. The set of target locations consists of the element locations.*
- *There are roads connecting each start-up location v^i with $i < M$ to the next start-up location v^{i+1}, roads connecting v^M to each subset location $v_{\widehat{S}}$ ($\widehat{S} \in C$), and roads connecting each subset location $v_{\widehat{S}}$ ($\widehat{S} \in C$) to those element locations v_s with $s \in \widehat{S}$.*
- *There is one mobile $m_{\widehat{S}}$ for each subset $\widehat{S} \in C$. All mobiles are initially located at v^0, and all may access all roads except for those incident to a subset location $v_{\widehat{S}}$, which may only be accessed by the corresponding mobile $m_{\widehat{S}}$.*

Plans π' for T' are mapped back to a collection of subsets by choosing all those subsets $\widehat{C} \subseteq C$ for which the corresponding subset locations $v_{\widehat{S}}$ are visited in the plan. This is indeed a set cover: If some element $s \in S$ were included in no subset from $\widehat{C}$, then no subset location $v_{\widehat{S}}$ with $s \in \widehat{S}$ would be visited in π'. These are the only locations from which the target element location v_s can be reached, so π' would not solve T'.

To compare the quality of the solutions, we make the following observations:

- *For any set cover $\widehat{C} \subseteq C$, the planning task can be solved by a plan of length $M|\widehat{C}| + 2|S|$: First, move all mobiles corresponding to subsets $\widehat{S} \in \widehat{C}$ from v^0 to v^M. Second, for each element $s \in S$, select a subset $\widehat{S} \in \widehat{C}$ with $s \in \widehat{S}$ and move $m_{\widehat{S}}$ from its current location to the element location m_s. In the first phase, each of the $|\widehat{C}|$ mobiles performs M movements, for a total cost of $M|\widehat{C}|$. In the second phase, two movements are necessary for each element of S, for a total cost of $2|S|$. (Note that both moving from v^M to an element location and moving from one element location to the next requires two movement actions.)*
- *Due to the preceding observation, the optimal plan length for T' is at most $Mm^*(I) + 2|S|$ and thus $m^*(I) \geq \frac{1}{M}(m^*(T') - 2|S|)$.*
- *Set covers generated from a given solution π' to T' have size at most $\frac{1}{M}m(\pi')$, because each mobile can only reach a single subset location, and this incurs a cost greater than M for each mobile.*

We can thus bound the performance ratio of the set cover $\widehat{C}$ obtained from a plan π' with performance ratio at most $r > 1$ as follows:

$$\frac{m(\widehat{C})}{m^*(I)} \leq \frac{\frac{1}{M}m(\pi')}{\frac{1}{M}(m^*(T') - 2|S|)} \leq \frac{rm^*(T')}{m^*(T') - 2|S|}$$

$$= 1 + \frac{(r-1)m^*(T') + 2|S|}{m^*(T') - 2|S|} = 1 + (r-1)\frac{m^*(T') + \frac{2|S|}{r-1}}{m^*(T') - 2|S|}$$

$$= 1 + (r-1)\left(1 + \frac{\frac{2|S|}{r-1} + 2|S|}{m^*(T') - 2|S|}\right)$$

$$\leq 1 + (r-1)\left(1 + \frac{M}{m^*(T') - M}\right)$$

$$\overset{(*)}{\leq} 1 + (r-1)\left(1 + \frac{M}{\frac{3}{2}M - M}\right) = 1 + (r-1)\cdot 3$$

where inequality $(*)$ *holds because we can assume that* $m^*(T') \geq \frac{3}{2}M$. *If this were not the case, then there would be a set cover of cardinality at most* $\frac{3}{2}$, *and thus (because the cardinality cannot be fractional) of cardinality at most 1. Such trivial instances can be special-cased and solved directly. The reduction is thus approximation-preserving.*

To see that it is optimality-preserving, notice that $m^*(I) \leq K$ *iff* $m^*(T') \leq KM + 2|S|$: *If* $m^*(I) \leq K$, *then an optimal set cover can be transformed into a plan of cost at most* $KM + 2|S|$. *On the other hand, if* $m^*(I) > K$ *and thus* $m^*(I) \geq K + 1$, *then any solution to* T' *must have cost at least* $(K+1)M > KM + 2|S|$. $\qquad\square$

The only remaining question is whether or not we can find constant factor approximations for TRANSPORT variants with unbounded fuel, a single mobile type and variable capacity. We will show that this is likely not the case; however, for these results, we need a somewhat stronger assumption than $P \neq NP$. In particular, we can prove that none of the variable capacity domains belongs to APX unless each problem in NP belongs to $ZPTIME(n^{poly \log n})$, the class of problems which can be decided by *randomized* algorithms with *expected* running time $O(n^{\log n^k})$ for some problem-dependent parameter $k \in \mathbb{N}_0$. Clearly, $P = NP$ implies $NP \subseteq ZPTIME(n^{poly \log n})$, but we cannot prove the converse. Thus, the assumption is stronger than $P \neq NP$.

To complete our analysis, we must prove hardness results for two TRANSPORT variants, namely TRANSPORT$_{*1}$-$[1, 0]$ and TRANSPORT$_{*1}$-$[1, 1]$. However, instead of doing this directly, we start with a proof for the weighted movement-cost variant TRANSPORT$_{*1}$-$[*, 0]$, and later reduce this to the "easier" cases.

Theorem 4.6.2. PLAN-TRANSPORT$_{*1}$-$[*, 0] \notin$ APX
If NP $\not\subseteq$ ZPTIME$(n^{poly \log n})$, *there is no constant-factor approximation algorithm for* TRANSPORT$_{*1}$-$[*, 0]$ *planning.*

Proof. The MINIMUM PREEMPTIVE CAPACITATED DIAL-A-RIDE *problem is the variant of* TRANSPORT$_{*1}$-$[*, 0]$ *where the mobile must return to its initial location at the end of the plan. Gørtz has shown that this problem cannot be*

approximated within a constant factor unless $\mathsf{NP} \subseteq \mathsf{ZPTIME}(n^{poly\log n})$ *[51].*
From a c-approximation algorithm for PLAN-TRANSPORT$_{*1}$-$[*, 0]$, *we could*
generate a constant-factor approximation for MINIMUM PREEMPTIVE CA-
PACITATED DIAL-A-RIDE *by extending the* TRANSPORT *plan by movements*
of the mobile back to its original location. This increases plan length by a factor
of 2 at most, and the optimal solution length for the MINIMUM PREEMPTIVE
CAPACITATED DIAL-A-RIDE *instance is at least as large as the optimal solu-*
tion length for the TRANSPORT *task. Therefore, the performance ratio of the*
MINIMUM PREEMPTIVE CAPACITATED DIAL-A-RIDE *solution thus obtained*
would be bounded by 2c, a constant. □

It would be nice if the result could be strengthened by showing that it already
holds under the standard $\mathsf{P} \neq \mathsf{NP}$ assumption. However, this does not appear
to be easily achievable. Before Gørtz's proof, hardness of approximation for the
MINIMUM PREEMPTIVE CAPACITATED DIAL-A-RIDE problem has been an
open question for a long time. Her proof is quite complex, spanning 23 pages,
and builds on earlier results by Andrews and Zhang for the related MINIMUM
BUY-AT-BULK NETWORK DESIGN and FIBRE MINIMIZATION IN OPTICAL
NETWORKS problems [3–5]. In fact, one special case of MINIMUM BUY-AT-
BULK NETWORK DESIGN which Andrews and Zhang prove to be hard to
approximate is essentially a variant of TRANSPORT$_{*1}$-$[*, 0]$ where movements
of the mobile are free when it is empty. Andrews suggests some ideas for
a possible derandomization of his construction in the conclusions of his paper [3], but notes that so far, such derandomization attempts have borne no
fruit.

    We next show how arbitrary movement costs can be "compiled away" by
an approximation-preserving reduction, which extends the previous theorem
to TRANSPORT$_{*1}$-$[1, 0]$. Rather than presenting the result just for the TRANS-
PORT variant to which we will apply it, we prove a more general compilation,
which might also be useful for proving results for other members of the TRANS-
PORT family (for example with restricted roadmap graph classes).

Theorem 4.6.3. PLAN-TRANSPORT$_{ij}$-$[*, 0] \leq_{\mathbf{AP}}$ PLAN-TRANSPORT$_{ij}$-$[1, 0]$
For all $i \in \{1, \infty, *\}$ and $j \in \{1, +, *\}$, we have:
    PLAN-TRANSPORT$_{ij}$-$[*, 0] \leq_{\mathbf{AP}}$ PLAN-TRANSPORT$_{ij}$-$[1, 0]$.
The reduction preserves weight-expandable graph classes.

Proof. Let T be the given TRANSPORT *task with variable movement costs, and
let r > 1 be the desired performance ratio. Note that we cannot simply expand
the roadmap graph, because edge weights can be exponential in the task size.*
    *First, we check if T is solvable; if not, we map it to an unsolvable task. If
T is solvable, it is mapped to the task T′ with unit movement costs as follows:*

1. *Compute an upper bound M for the* length *(not cost) of any reasonable
   plan for T. This bound is polynomial in the task size (Theorem 4.2.3).*

2. *Compute the minimal number B such that T is still solvable if only roads with movement cost at most B may be used. This can be done in polynomial time by sorting movement costs, then repeatedly deleting the most expensive roads until the task becomes unsolvable.*

   Observe that we must have $m^*(T) \geq B$ because the task cannot be solved if restricted to roads of cost at most $B - 1$, which implies that at least one more expensive road must be traversed in any plan. Moreover, we must have $m^*(T) \leq BM$, because T has a solution where no action costs more than B (recall that pickup is free), and this solution can be converted to a reasonable solution, which has length at most M.

3. *Prune the task by eliminating all roads with cost greater than BM. Clearly, such roads cannot be traversed in an optimal plan, so this does not affect the optimal measure.*

4. *If $B \leq 2M$ or $B \leq \frac{M}{r-1}$, then the most expensive roads of the (pruned) task have cost at most $2M^2$ or at most $\frac{M^2}{r-1}$. Both values are polynomial in the task size for fixed r, and thus we can translate the task to an equivalent task with unit action costs in polynomial time by weight-expansion. This is clearly approximation-preserving (even optimization-preserving, as the resulting task has the same optimal measure as the original one). We can thus assume $B \geq \max\{2M, \frac{M}{r-1}\}$ in the following.*

5. *Set $\alpha = \frac{M}{B(r-1)}$. For each road, change the movement cost to $\lceil \alpha w \rceil$, where w is the original movement cost. The most expensive action now has cost $\lceil \alpha B \rceil = \lceil \frac{M}{r-1} \rceil$, which is polynomial in the task size for fixed r.*

6. *All movement costs are now polynomial, so the task can be converted to a task T' with unit movement cost in polynomial time by weight-expansion.*

Solutions π' to the unit-cost task T' are mapped back to solutions of π in the obvious way, by removing the intermediate locations introduced by weight-expansion from the trajectories of the mobiles. The plan costs satisfy the inequality $m(\pi) \leq \frac{1}{\alpha} m(\pi')$, because the cost for every movement between two locations of T changes from $\lceil \alpha w \rceil$ to w, and $\frac{\lceil \alpha w \rceil}{\alpha} \geq w$.

Moreover, $m^*(T') \leq \alpha m^*(T) + M$: Every plan π for T induces a plan π' for T' where every movement action in π with cost w corresponds to $\lceil \alpha w \rceil \leq \alpha w + 1$ unit cost movement actions in π'. The sum, over all such movements, over the terms αw adds up to $\alpha m^*(T)$, and the sum over the 1 terms is bounded by M because π cannot contain more than M movements by virtue of being reasonable. Solving this inequality for $m^*(T)$, we obtain $m^*(T) \geq \frac{1}{\alpha}(m^*(T') - M)$.

This allows us to bound the performance ratio of a solution π to T obtained from a solution π' to T' as follows:

$$\frac{m(\pi)}{m^*(T)} \le \frac{\frac{1}{\alpha}m(\pi')}{\frac{1}{\alpha}(m^*(T') - M)} = \frac{m(\pi')}{m^*(T') - M} \le \frac{rm^*(T')}{m^*(T') - M}$$

$$= 1 + \frac{(r-1)m^*(T') + M}{m^*(T') - M} = 1 + (r-1)\frac{m^*(T') + \frac{M}{r-1}}{m^*(T') - M}$$

$$\le 1 + (r-1)\frac{m^*(T') + B}{m^*(T') - \frac{1}{2}B} = 1 + (r-1)\left(1 + \frac{\frac{3}{2}B}{m^*(T') - \frac{1}{2}B}\right)$$

$$\le 1 + (r-1)\left(1 + \frac{\frac{3}{2}B}{B - \frac{1}{2}B}\right) = 1 + (r-1) \cdot 4.$$

The reduction is thus approximation-preserving, which concludes the proof.

□

Note that this one of the few reductions we present which is not shown to be optimization-preserving: The necessary rounding step occurring in the reweighting from w to $\lceil \alpha w \rceil$ loses precision, so that we cannot accurately map solution bounds for the weighted task to solution bounds for the unit-weight task. Of course, having established NP-hardness for bounded plan existence already, we do not need an optimization-preserving reduction anyway.

We remark that the reduction could be easily extended to the case where pickup and drop actions have unit cost, rather than being free. However, providing a similar reduction for *variable* pickup costs seems much more challenging, because the introduction of variable pickup costs leads to a somewhat complicated trade-off between delivering portables along short routes and delivering them with a small number of pickup and drop actions.

Being able to compile away weighted movement costs, we can now extend the hardness result for TRANSPORT$_{*1}$-$[*, 0]$ to TRANSPORT$_{*1}$-$[0, 0]$. All that remains to show is that the problem with unit pickup costs is no easier. Again, we present a reduction which is more general than required.

Theorem 4.6.4. PLAN-TRANSPORT$_{ij}$-$[l, 0] \le_{\mathbf{OP}}$ PLAN-TRANSPORT$_{ij}$-$[l, 1]$
For all $i \in \{1, \infty, *\}$, $j \in \{1, +, *\}$ and $l \in \{1, *\}$, we have:
    PLAN-TRANSPORT$_{ij}$-$[l, 0] \le_{\mathbf{OP}}$ PLAN-TRANSPORT$_{ij}$-$[l, 1]$.
The reduction preserves stretchable graph classes.

Proof. Let T be the given TRANSPORT *task with free pickup, and let $r > 1$ be the desired performance ratio.*

We first compute a polynomial bound M on reasonable plan length for T and define $\alpha = \lceil \frac{M}{r-1} + M \rceil$, which is polynomial in the size of T for fixed r. The TRANSPORT$_{ij}$-$[l, 1]$ *task T' generated by the reduction is identical to T except that the roadmap graph is stretched by α. Solutions π' to T' are mapped back to solutions π to T in the obvious way, by removing actions pertaining to auxiliary locations introduced through stretching.*

Clearly, we have $m(\pi) \le \frac{1}{\alpha}m(\pi')$, because each movement in π corresponds to (at least) α movements in π' and non-movement actions do not contribute to the cost of π.

On the other hand, for every reasonable solution π to T, there exists a solution π' to T' with $m(\pi') \leq \alpha m(\pi) + M$, where every movement action has been replaced with α movement actions: The movement cost in π' is α times the movement cost in π, which is $m(\pi)$, and the cost for pickup and drop actions in a reasonable plan is at most M. In particular, this implies that $m^*(T') \leq \alpha m^*(T) + M$ and hence $m^*(T) \geq \frac{1}{\alpha}(m^*(T') - M)$.

This leads to the following bound on the performance ratio of the mapped-back solution π:

$$\frac{m(\pi)}{m^*(T)} \leq \frac{\frac{1}{\alpha}m(\pi')}{\frac{1}{\alpha}(m^*(T') - M)} = \frac{m(\pi')}{m^*(T') - M} \leq \frac{rm^*(T')}{m^*(T') - M}$$

$$= 1 + \frac{(r-1)m^*(T') + M}{m^*(T') - M} = 1 + (r-1)\frac{m^*(T') + \frac{M}{r-1}}{m^*(T') - M}$$

$$= 1 + (r-1)\left(1 + \frac{\frac{M}{r-1} + M}{m^*(T')}\right) \leq 1 + (r-1)\left(1 + \frac{\alpha}{m^*(T')}\right)$$

$$\overset{(*)}{\leq} 1 + (r-1)(1 + \frac{\alpha}{\alpha}) = 1 + (r-1) \cdot 2,$$

where inequality (∗) holds because we can assume that $m^*(T') \geq \alpha$, because otherwise T' (and consequently T) is solvable by an empty plan, a situation which we can special-case. The reduction is thus approximation-preserving.

It is also optimization-preserving because $m^*(T) \leq K$ iff $m^*(T') \leq \alpha K + M$: We have already argued that $m^*(T') \leq \alpha m^*(T) + M$, which shows the "only if" part, and to show the "if" part, note that $m^*(T) \geq K + 1$ implies $m^*(T') \geq \alpha(K+1) > \alpha K + M$. $\square$

This concludes our presentation of complexity results for TRANSPORT and ROUTE, and we can turn to discussion.

4.7 Summary

We have introduced domain families for transportation and route planning which are parameterized along five dimensions:

- Mobiles can have unit capacity, unbounded capacity, or arbitrary capacity. (This only applies to the TRANSPORT family.)
- There can be one mobile, several mobiles with identical movement capabilities, or several mobiles with different movement capabilities.
- Locations can have unbounded fuel, one fuel unit each, or arbitrary amounts of fuel.
- Movement can incur unit costs, or varying costs depending on the road and mobile.

– Pickup and drop actions can incur no cost, unit cost, or varying costs depending on the portable and mobile. (This only applies to the TRANS-PORT family.)

We have then analysed the decision complexity of planning in these domain variants by considering the respective *plan existence* and *bounded plan existence* problems, and we have analysed the approximation complexity of the respective planning problems with respect to the classes PO, FPTAS, PTAS, APX, poly-APX, exp-APX, PS, NPO, NPS, EXPO and EXPS. For the decision problems, we get a very simple classification:

– Plan existence is a polynomial problem if fuel is unrestricted, and NP-complete otherwise.
– Bounded plan existence is always NP-complete.

These results might lead one to suppose that making distinctions based on the number, movement capabilities or carrying capacity of mobiles has no impact on the planning problem at all. However, taking a closer look at *approximability* properties, we see that this is not true. Here, we obtained the following results:

– For domain variants with restricted fuel, the planning problem is in NPO \ PS, unless P = NP.
  In other words, for these domains we cannot generate plans in polynomial time.
– For domain variants with unrestricted fuel and mobiles with different movement capabilities, the planning problem is in poly-APX \ APX, unless P = NP. The same result holds for domain variants with unrestricted fuel and mobiles with arbitrary capacity. However, this result is subject to the assumption $\mathsf{ZPTIME}(n^{\mathrm{poly}\log n}) \not\subseteq \mathsf{NP}$, which is a stronger statement than P ≠ NP.
  In other words, for these domains we can generate plans in polynomial time, but we cannot guarantee that the generated plans are at most a constant factor more expensive than optimal ones.
– For all other domain variants, the planning problem is in APX \ PTAS, unless P = NP.
  In other words, for these domains we can generate plans in polynomial time, and we can guarantee that the generated plans are at most a constant factor more expensive than optimal ones. However, this factor cannot be reduced arbitrarily close to 1.

It is interesting to note that the only classification parameters that make *no* difference to either decision or approximation complexity are those which determine the costs of operators. For propositional planning systems, this can be considered encouraging news, because it shows that abstracting away action costs, as is commonly done in PDDL-style planning, does not significantly affect the computational properties of the problem at least for the large class

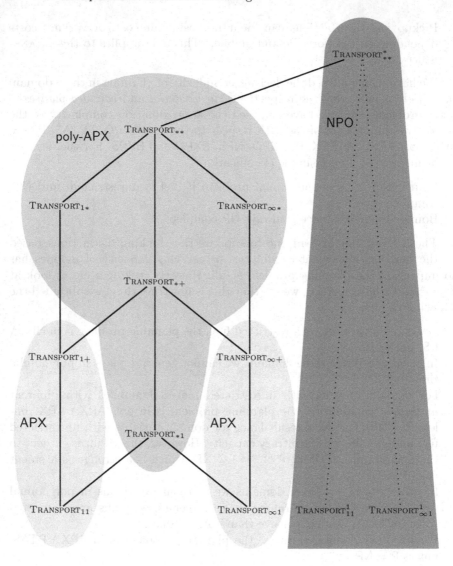

Fig. 4.4. Overview of results for TRANSPORT^k_{ij} (i: capacity constraints, j: number of mobiles/mobile types, k: fuel constraints). Dotted lines indicated omitted variants with restricted fuel

of transportation and route planning benchmarks. Thus, there is hope that established techniques for propositional planning could be applicable to more realistic settings without dramatic changes.

To conclude the summary of results, Figs. 4.4 and 4.5 give a graphical representation of the boundaries of approximability in the TRANSPORT and ROUTE domain families. Due to the large number of domain variants, most

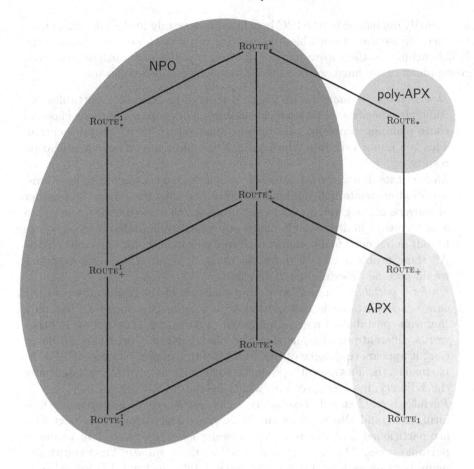

Fig. 4.5. Overview of results for ROUTE_j^k (j: number of mobiles/mobile types, k: fuel constraints)

TRANSPORT variants with restricted fuel are omitted – all of these belong to NPO \ PS. Moreover, the parameters defining action costs are omitted in the figures since we could show that they do not affect approximation behaviour. Each entry in Fig. 4.4 thus actually represents six different TRANSPORT variants, and each entry in Fig. 4.5 represents two different ROUTE variants.

4.8 Beyond TRANSPORT and ROUTE

Altogether, we have classified the approximation complexity of 162 transportation domains and 18 route planning domains. Our choice of domain variants was heavily influenced by the IPC benchmarks we study in the following chapter, because we want to be able to apply our result to those. Disregarding the IPC benchmarks, there are a number of other variations of transportation and route planning we have not considered, but which appear relevant:

- *Destinations for mobiles:* In addition to goal locations for portables, we could also specify goal locations for mobiles. For example, in many classical route planning problems, the mobile must return to its initial location. This is a minor variation which should not affect any of our classification results.

- *Mobiles with limited fuel:* Instead of associating fuel bounds with locations, we could associate fuel bounds with mobiles, thus bounding the number of movements a given mobile can make. In the one-mobile cases, plan existence with limited mobile fuel is equivalent to bounded plan existence for our unbounded fuel domains with free pickup and unit movement costs. All these problems are NP-hard, so under this variation, plan existence would become NP-complete in all cases.

- *Uniform portables:* In another classic variation of the problem, we do not care *which* portable is transported to a certain goal location, but only *that* some portable, or a certain number of portables, is delivered. In other words, different portables need not be distinguished. Under this modification, it appears that some of the TRANSPORT variants become simpler. In particular, the TRANSPORT$_{*1}$ domain with uniform portables is essentially the k-DELIVERY-TSP problem, which belongs to APX [22].

- *Portable types:* Instead of considering the case where *all* portables are uniform, one could also consider the more general situation where portables are partitioned into different types, and goals are specified in terms of portable types. This is at least as hard as the standard TRANSPORT domain, because each portable might have a different type. We hypothesize that this variation no longer admits constant-factor approximations for TRANSPORT$_{\infty 1}$, but still does for TRANSPORT$_{1+}$.

- *Directed roadmaps:* The domains could be easily generalized to directed roadmaps in place of undirected ones. Of course, this can quickly lead to dead-end situations where no further movement is possible, even with unlimited fuel. We hypothesize that some cases, such as one mobile of unbounded or unit capacity, remain polynomially solvable with directed roadmaps, but many others, such as multiple mobiles of unbounded capacity, no longer admit polynomial solution algorithms unless P = NP.

- *Parallel plans:* Instead of considering the sum of individual action costs, parallel plan length or *makespan* could be used as a solution measure. In this setting, multiple compatible actions, such as movements of different mobiles, could occur in parallel, and the cost of a plan would be the time at which it finishes execution. There are two common notions of parallelism for transportation problems in planning benchmarks, depending on whether it is possible to pick up and drop several portables with the same mobile simultaneously (as in LOGISTICS or GRIPPER) or not (as in MYSTERY).

In the case where such simultaneous pick up and drop activities are disallowed, our hardness results for TRANSPORT with a single mobile and the unit pickup cost model apply. In the case where simultaneous pickups and drops are allowed, our hardness results for TRANSPORT with a single mobile and free pickup apply with a few simple changes. Thus, approximating parallel plans is no easier than the problems we studied. Determining whether or not it is harder would require some deeper study.

— *Restricted roadmap graphs:* We have shown that some of our hardness results extend to restricted graph classes such as trees, grids, planar graphs or complete graphs. This analysis could be extended to systematically study all variants of TRANSPORT and ROUTE and a number of relevant graph classes. Such an undertaking would not just constitute an enormous amount of work, but also require answers to some exceedingly difficult complexity questions. In particular, some TRANSPORT variants for planar or grid graphs scrape the border between PTAS and APX, and we believe that it would be difficult to prove either membership or non-membership in PTAS.

This concludes our discussion of general transportation and route planning. In the next chapter, we reap the fruits of our analysis by applying the results of this chapter to the IPC domains.

IPC Domains: Transportation and Route Planning

This chapter discusses the approximation properties of the IPC benchmarks related to transportation planning or route planning. We easily obtain results for the GRIPPER (Sect. 5.1), MYSTERY and MYSTERYPRIME (Sect. 5.2), LOGISTICS (Sect. 5.3), ZENOTRAVEL (Sect. 5.4), DEPOTS (Sect. 5.5), two variants of MICONIC-10 (Sect. 5.6) and ROVERS (Sect. 5.7). The third variant of MICONIC-10 (also Sect. 5.6) as well as the GRID (Sect. 5.8), DRIVERLOG (Sect. 5.9) and AIRPORT (Sect. 5.10) domains differ significantly from the TRANSPORT and ROUTE families. For them, we provide proofs from first principles. The chapter ends with a brief summary of results (Sect. 5.11).

Throughout the chapter, our proofs typically assume that all portables are goal portables and the goal location of each portable differs from its initial location, which is a valid simplification because non-goal portables can be ignored. We only make an exception for the DEPOTS and GRID domains, where portables have other relevant properties in addition to being carried around. (The restriction does not apply to ROVERS and AIRPORT, which are not transportation domains, and in some domains such as MICONIC-10, all portables are required to be goal portables anyway.)

5.1 GRIPPER

The GRIPPER domain is a transportation domain with a single mobile (a robot) with a fixed capacity of two, modelling two *grippers* for holding objects which give the domain its name. Despite the fact that there is in general no polynomial-time approximation scheme for the subproblem of TRANSPORT planning defined by these constraints, GRIPPER planning is actually very easy due to another, very significant restriction: There are only two locations in GRIPPER tasks.

Definition 5.1.1. GRIPPER *Tasks and Domain*
A GRIPPER *task is a* TRANSPORT$_{*1}$ *task with the following properties:*

– *There are exactly two locations, called A and B, which are connected.*
– *The mobile has a capacity of 2 and starts at location A.*
– *All portables start at location A and must be moved to location B.*

The GRIPPER **domain** *is the* TRANSPORT *domain restricted to* GRIPPER *tasks.*

It is not difficult to see that GRIPPER planning is an easy problem.

Theorem 5.1.2. PLAN-GRIPPER ∈ PO
Optimal plans for GRIPPER *tasks can be generated in polynomial time.*

Proof. An optimal plan can clearly be generated by repeating the following steps until the task is solved:

– *Pick up any two portables at A, or one if only one is located at A.*
– *Move to B.*
– *Drop all carried portables.*
– *If the goal is not yet satisfied, move to A.*

□

This easiness of GRIPPER planning – which easily extends, with the same algorithm, to parallel planning – is mostly due to the fact that the roadmap graph is fixed. Interestingly, the restriction to the fixed roadmap and fixed carrying capacity are *not* inherent in the PDDL specification of the GRIPPER domain, which allows for varying capacities and more locations (although the roadmap graph must always be fully connected). In this more general form, the GRIPPER domain would belong to APX \ PTAS unless P = NP. However, given that all IPC GRIPPER tasks are of the simple form demanded by our definition, such a general definition of the GRIPPER domain would be inappropriate.

5.2 MYSTERY and MYSTERYPRIME

The MYSTERY domain earns its name from the curious terminology used in its PDDL specification. It is a prototypical transportation domain, but this is obscured by the fact that locations are called *foods*, mobiles are called *pleasures*, portables are called *pains* and actions are named along similar lines. Both MYSTERY and MYSTERYPRIME are fuel-restricted domains.

Definition 5.2.1. MYSTERY **Tasks and Domain**
A MYSTERY **task** *is a* TRANSPORT$^*_{*+}$ *task with a planar roadmap and finite fuel at all locations.*

The MYSTERY **domain** *is the* TRANSPORT *domain restricted to* MYSTERY *tasks.*

The MYSTERYPRIME differs from MYSTERY only by having an additional type of actions, which allows moving fuel between locations.

Definition 5.2.2. MYSTERYPRIME *Tasks and Domain*
A MYSTERYPRIME *task is a* TRANSPORT$^*_{*+}$ *task with a planar roadmap and finite fuel at all locations.*

The MYSTERYPRIME *domain is the* TRANSPORT *domain restricted to* MYSTERYPRIME *tasks, with an additional operator type: Fuel may be* **squirted** *from a location v to another location v' in state $\langle l, fuel \rangle$ iff $fuel(v) \geq 2$. This action reduces $fuel(v)$ by 1 and increases $fuel(v')$ by 1.*

This definition diverges from the PDDL specification of MYSTERYPRIME in two respects. First, in the PDDL domain there is a (task-specific) upper bound to the amount of fuel that can be stored at any location, due to the way numbers are encoded in propositional PDDL. Second, the PDDL domain contains an error which allows producing fuel by squirting it from a location v to itself: If location v has $k \geq 2$ units of fuel, then squirting fuel from v to v leads to two separate fuel depots of $k+1$ and $k-1$ fuel units at v, effectively doubling the fuel reserve for the location. In both cases, we chose not to model these aspects of the domain in our definition because they are somewhat accidental aspects of the PDDL definition. (In any case, neither of the differences amounts to a difference in complexity.)

Due to the fuel constraints, MYSTERY and MYSTERYPRIME planning is hard.

Theorem 5.2.3. PLAN-MYSTERY, PLAN-MYSTERYPRIME $\in$ NPO $\setminus$ PS
MYSTERY *and* MYSTERYPRIME *planning is in* NPO, *but not in* PS *unless* P $=$ NP. *Hardness already holds if there is only one mobile, which has unbounded capacity.*

Proof. For non-membership in PS, *we show that the plan existence problem for these domains is* NP-*hard. This follows immediately from Theorem 4.3.2, which shows hardness for* ROUTE$^1_1$ *with planar roadmaps, and Theorem 4.2.8, which shows that this problem is reducible to* TRANSPORT$^1_{\infty 1}$ *preserving graph classes. Note that both* MYSTERY *and* MYSTERYPRIME *generalize* TRANSPORT$^1_{\infty 1}$; *in particular, no squirt actions are possible for tasks with unit fuel, because squirting requires two fuel units at the origin location. (This is one of the reasons why we chose to consider unit fuel domains in the first place.)*

Membership in NPO *for* MYSTERY *follows from the general* TRANSPORT *result. The* MYSTERYPRIME *domain also belongs to* NPO *because it admits short optimal plans: We can bound the number of non-squirt actions in any plan by the bound for reasonable* TRANSPORT *solutions, and an optimal solution never contains more squirt actions than movements, because the only use for squirting is to enable movements.* □

5.3 LOGISTICS

We now turn to the most commonly used planning benchmark of the last decade, which has become even more ubiquitous than the BLOCKSWORLD domain.

Definition 5.3.1. LOGISTICS *Tasks and Domain*
A LOGISTICS *task is a* TRANSPORT$_{\infty*}$ *task with the following properties:*

- *Locations are partitioned into sets called **cities**, and each city has a dedicated location called an **airport**.*
- *There are two types of mobiles: **trucks** move between locations of the same city along a fully connected road network, and **airplanes** move between airports, along a fully connected network of airways.*

A LOGISTICS *task is called **simple** if there is only one truck, only one city and no airplane, or if there is only one airplane, all locations are airports, and there is no truck.*

The LOGISTICS ***domain*** *is the* TRANSPORT *domain restricted to* LOGISTICS *tasks. The* SIMPLELOGISTICS ***domain*** *is the* LOGISTICS *domain restricted to simple tasks.*

The LOGISTICS domain is one of the few domains with different mobile types, which is generally an indication of difficulty of constant-factor approximations. However, because there are only two different types of mobiles and the diameter of their roadmaps is small, it is actually fairly easy to prove that LOGISTICS belongs to APX. In fact, a greedy algorithm which delivers one portable at a time in the obvious way is already 2-approximating, because it contains as few non-movement actions as possible and no more movements than non-movement actions.

However, due to its significance as a commonly used benchmark, we expend a bit more effort and present some tighter approximations for LOGISTICS. For this purpose, we first define a minimization problem which is closely related to the SIMPLELOGISTICS domain.

Problem 5.3.2. MINIMUM FEEDBACK VERTEX SET
The MINIMUM FEEDBACK VERTEX SET problem is defined as follows:

INSTANCE: A directed graph $\langle V, A \rangle$.
SOLUTION: A **feedback set**, i.e., a subset of vertices $U \subseteq V$ such that the subgraph induced by $V \setminus U$ is acyclic.
MEASURE: $|U|$.

If $P \neq NP$, then the MINIMUM FEEDBACK VERTEX SET problem does not admit a polynomial-time approximation scheme. Whether or not it is in APX is an open question [7, problem GT 8]. However, here we are not interested in its approximation properties, but in its relationship to SIMPLELOGISTICS planning, shown by the following theorem.

Theorem 5.3.3. SIMPLELOGISTICS *vs.* MINIMUM FEEDBACK VERTEX SET
Let T be a SIMPLELOGISTICS *task, let P the set of portables in T, and let*
$\widehat{P} \subseteq P$ *be the set of those portables whose initial location is different from the
initial location of the mobile.*

The **delivery graph** *of T is the directed graph $G' = \langle V', A' \rangle$ where V'
consists of all locations which are initial or goal location of portables in $\widehat{P}$,
and A' contains an arc from u to v iff there is some portable in $\widehat{P}$ with initial
location u and goal location v.*

Then $m^*(T) = 2|P| + |V'| + m^*(G')$, where $m^*(G')$ is with respect to the
MINIMUM FEEDBACK VERTEX SET *problem.*

*Moreover, for each feedback vertex set U' of G', a plan for T of length
$m^*(T) = 2|P| + |V'| + m(U')$ can be efficiently constructed.*

*Proof. We first prove the second property. Given a feedback vertex set $U' \subseteq V'$,
we generate the following plan:*

1. *Pick up all portables at the initial mobile location, if any.*
2. *Move to all locations in U' in any order, picking up all portables located
   there.*
3. *Move to all locations in $V' \setminus U'$ in an order which is consistent with the
   arcs in the delivery graph, picking up and dropping portables as required.*
4. *Move to all locations in U' in any order, dropping all portables at their
   goal locations.*

*Each portable is picked up and dropped exactly once, accounting for $2|P|$ ac-
tions. Each location in V' is moved to exactly once in the second and third
step, accounting for $|V'|$ actions. Finally, there are $|U'| = m(U')$ more actions
generated in the fourth step.*

*The construction is very similar to that used in the proof to Theorem 4.4.2,
so that we refrain from a further discussion.*

*To prove the first property, note that if the feedback vertex set U' considered
for the second property is optimal, then a plan of length $2|P| + |V'| + m^*(G')$
is obtained. Moreover, there cannot be a shorter plan: The number of pickup
and drop actions cannot be reduced, each location in V' must be visited at least
once, and the set of locations visited at least twice must form a feedback vertex
set. Again, consult the proof to Theorem 4.4.2 for details.* □

We now describe an approximation algorithm for MINIMUM FEEDBACK VER-
TEX SET which might generate quite bad solutions in terms of their performance ratio, but exceeds the optimal measure only by a term growing slowly
in the number of *arcs* in the given directed graph. Because each arc is related
to a portable in the LOGISTICS task, and each portable requires at least one
pickup and drop action, this extra cost is quite small compared to the optimal
plan length.

Theorem 5.3.4. MINIMUM FEEDBACK VERTEX SET *approximation algorithm*
Given a directed graph $G = \langle V, A \rangle$, a feedback vertex set of size at most $\frac{1}{3}|A| + m^*(G)$ can be computed in polynomial time.

Proof. The algorithm runs in two stages. In the first stage, the algorithm picks any vertex with degree at least 3, removes it from the input graph along with all incident arcs, and inserts it into the result set. The stage ends as soon as no such vertex exists any more.

In the second stage, we compute an optimal feedback vertex set for the remaining graph and add it to the result set. This is easily possible: Because no vertex has a degree greater than 2, any cycle in the graph must be disconnected from the rest of the graph. (The arcs in the cycle already account for a degree of 2 of all vertices involved, so none of these vertices can have further neighbours.) Thus, an optimal feedback vertex set for the remaining graph simply contains any one vertex from each of the disconnected cycles.

In the first stage, we add at most $\frac{1}{3}|A|$ vertices to the result set: In each step, at least 3 arcs are eliminated, and this can be done at most $\lfloor \frac{1}{3}|A| \rfloor$ times. In the second stage, we add at most $m^*(G)$ vertices to the result set: The vertices picked in this stage form an optimal feedback vertex set for the reduced graph, and since this is a subgraph of the original graph, feedback vertex sets for the original graph cannot be smaller. □

Combining the previous two results, we get the following classification of SIMPLELOGISTICS.

Theorem 5.3.5. PLAN-SIMPLELOGISTICS $\in$ APX \ PTAS
There exists a polynomial $\frac{7}{6}$-approximation algorithm for SIMPLELOGISTICS planning. However, the problem does not admit a PTAS unless P = NP.

Proof. SIMPLELOGISTICS is identical to TRANSPORT$_{\infty}1$ with complete graphs, for which non-membership in PTAS was shown in Theorem 4.4.2.

For the $\frac{7}{6}$-approximation, we combine the MINIMUM FEEDBACK VERTEX SET approximation algorithm with the result from Theorem 5.3.3. Using the same notation as in the proof of that theorem, the optimal plan length is $m^*(T) = 2|P| + |V'| + m^*(G')$, while the length of the generated plan is $m(T) = 2|P| + |V'| + m(U')$ for $m(U') \leq \frac{1}{3}|A'| + m^*(G')$. With $|A'| \leq |P|$, we can thus bound the performance ratio by

$$\frac{m(T)}{m^*(T)} \leq \frac{2|P| + |V'| + \frac{1}{3}|P| + m^*(G')}{2|P| + |V'| + m^*(G')} = 1 + \frac{\frac{1}{3}|P|}{2|P| + |V'| + m^*(G')}$$

$$\leq 1 + \frac{\frac{1}{3}|P|}{2|P|} = \frac{7}{6}.$$

□

Unfortunately, this result does not easily extend to the general LOGISTICS domain. General LOGISTICS tasks do not decompose into simple ones because of the interactions between the subtasks for the different cities. If a portable must be moved from city A to city B and another portable must be moved from city B to city A, then the plans for these two cities must be interleaved.

One way of limiting such interactions is to partition the overall planning problem into three *phases*: In the first phase, called the *outbound phase*, portables are picked up and those which need to be transported to a different city are brought to the airport. In the second phase, called the *airplane phase*, portables are transported between the airports of different cities. Finally, in the third phase, called the *inbound phase*, portables are moved to their goal locations within cities.

The subtasks which need to be solved in each phase are essentially SIMPLELOGISTICS tasks, apart from the fact that multiple mobiles may be available. We could thus use the SIMPLELOGISTICS planning algorithm for these subtasks in hopes of generating a similarly tight approximation for the general domain. However, such an analysis would only allow us to compare the solution quality to that of the *best such three-phase solution*, which is not very useful on its own. We would require an additional analysis of how bad three-phase solutions can get compared to optimal ones. To avoid these complications, we instead present a much simpler direct algorithm (and analysis) for the general LOGISTICS planning problem which does not rely on our results for SIMPLELOGISTICS, but still leads to a reasonable approximation.

Theorem 5.3.6. PLAN-LOGISTICS $\in$ APX $\setminus$ PTAS
There exists a polynomial $\frac{4}{3}$-approximation algorithm for LOGISTICS planning. However, the problem does not admit a PTAS unless P = NP.

Proof. Non-membership in PTAS follows from the SIMPLELOGISTICS result. We now describe the general approximation algorithm. We can assume that the given task T is solvable – solvability can easily be decided in polynomial time by the TRANSPORT$_{**}$ algorithm.

*An **intra-city portable** is a portable where initial and goal location belong to the same city; other portables are called **inter-city portables**.*

Let L_T^* be the minimal number of truck movements over all plans, let L_A^* be the minimal number of airplane movements over all plans, and let L_O^* (other) be the minimal number of non-movement actions over all plans.

We define the following location sets:

- *The set of locations V_T^- contains all goal locations of intra-city portables, all non-airport goal locations of inter-city portables, and all airports of cities where a non-airport location is the initial location of an inter-city portable. We have $|V_T^-| \leq L_T^*$ because each location in V_T^- must be entered by some truck in any plan.*
- *The set of locations V_T^+ contains all initial locations of intra-city portables, all non-airport initial locations of inter-city portables, and all airports of*

cities where a non-airport location is the goal location of an inter-city portable. We have $|V_T^+| \leq L_T^*$ because each location in V_T^+ must be left by some truck in any plan.

- The set of locations V_A^- contains all airports of cities containing the goal location of an inter-city portable. We have $|V_A^-| \leq L_A^*$ because each location in V_A^- must be entered by some airplane in any plan.
- The set of locations V_A^+ contains all airports of cities containing the initial location of an inter-city portable. We have $|V_A^+| \leq L_A^*$ because each location in V_A^+ must be left by some airplane in any plan.

Using these sets, we generate a plan as follows:

1. **Outbound phase:** In each city with at least one location in $V_T^+ \cup V_T^-$, choose some truck called the **city truck** for that city. (Such a truck must exist if the task is solvable.) Move the city truck to all non-airport city locations in V_T^+ in any order, picking up all portables initially located there. Then move it to the airport if it is in $V_T^+ \cup V_T^-$ and drop all carried inter-city portables. All inter-city portables are now at airports.
2. **Airplane phase:** If $V_A^+ \cup V_A^- \neq \emptyset$, some airplane must exist if the task is solvable. Choose one, and move it to all airports in V_A^+ in any order, picking up all inter-city portables located there. Then move the airplane to all airports in V_A^- in any order, dropping all portables at their respective city and picking them up with the city truck if their goal location is not the airport. All portables are now in their correct city and either carried by the city truck or already at their goal location.
3. **Inbound phase:** In each city where the city truck is carrying portables, move to the non-airport locations in V_T^- in any order, dropping portables at their goal location. (If the city airport is in V_T^-, the truck is already located there, so it need not be moved there.)

This indeed generates a solution to the planning task. (If this is not immediately clear, verify that it works for all seven qualitatively different cases: portables can be intra- or inter-city, initial locations and goal locations can be airports or non-airports, and the combination intra-city/airport/airport is impossible.)

To estimate the performance ratio of the solution, we consider the number of truck movements L_T, the number of airplane movements L_A and the number of non-movement actions L_O in the generated plan π. Clearly, $m(\pi) = L_T + L_A + L_O$.

First observe that the number of pickup and drop actions in the plan is minimal, i.e., $L_O = L_O^*$. In the outbound and inbound phases, the algorithm generates at most a total of $|V_T^+| + |V_T^-| \leq 2L_T^*$ movement actions, moving to each non-airport in V_T^+ once in the outbound phase, to each non-airport in V_T^- once in the inbound phase, and moving to an airport once in the outbound phase only if it is in $V_T^+ \cup V_T^-$. Thus, the total number of truck movements M_T is bounded as $L_T \leq 2L_T^*$. Similarly, we obtain the bound $L_A \leq 2L_A^*$ on the number of airplane movements.

Moreover, there are no more movements than non-movements in the plan, because no mobile moves twice without picking up or dropping a portable in between, and no mobile moves after its last pickup or drop. Thus, we have $L_T + L_A \leq L_O$.

Finally, we must have $m^*(T) \geq L_T^* + L_A^* + L_O^*$ actions (with equality holding only if some optimal plan minimizes all three types of actions simultaneously).

We can thus bound the performance ratio as

$$\frac{m(\pi)}{m^*(T)} \leq \frac{L_T + L_A + L_O}{L_T^* + L_A^* + L_O^*} \overset{(*)}{\leq} \frac{L_T + L_A + L_O}{\frac{1}{2}L_T + \frac{1}{2}L_A + L_O} = 1 + \frac{\frac{1}{2}L_T + \frac{1}{2}L_A}{\frac{1}{2}L_T + \frac{1}{2}L_A + L_O}$$

$$\leq 1 + \frac{\frac{1}{2}L_T + \frac{1}{2}L_A}{\frac{1}{2}L_T + \frac{1}{2}L_A + L_T + L_A} = 1 + \frac{\frac{1}{2}}{\frac{3}{2}} = \frac{4}{3},$$

where inequality $(*)$ *holds because $L_T + L_A \leq 2L_T^* + L_A^*$ and hence $L_T^* + L_A^* \geq \frac{1}{2}L_T + \frac{1}{2}L_A$.* □

5.4 ZENOTRAVEL

We continue with a domain to which the LOGISTICS results can be applied fairly straightforwardly.

Definition 5.4.1. ZENOTRAVEL **Tasks and Domain**
A ZENOTRAVEL **task** *is given by a* TRANSPORT$_{\infty+}$ *task with a complete graph roadmap and an* **initial airplane fuel function** $fuel_0 : M \to \mathbb{N}_0$, *where M is the set of mobiles of the task. The mobiles of a* ZENOTRAVEL *task are called* **airplanes**, *the portables are called* **passengers**.

The ZENOTRAVEL **domain** *maps* ZENOTRAVEL *tasks with locations V, airplanes M and passengers P to state spaces as follows:*

STATES:　　　　*Pairs $\langle l, fuel \rangle$, where $l : M \cup P \to V \cup M$ is the **location** function and fuel $: M \to \mathbb{N}_0$ is the **airplane fuel reserve** function. Only passengers may have airplanes as their location.*

INITIAL STATE:　*$\langle l_0, fuel_0 \rangle$, where l_0 is the initial location function and $fuel_0$ is the initial airplane fuel function of the task.*

GOAL STATES:　*Any state $\langle l, fuel \rangle$ where $l(p) = l_*(p)$ for all goal passengers p and the goal location function l_*.*

OPERATORS:　　*Move actions are defined as in the TRANSPORT domain, except that movements require a non-zero fuel reserve of the moving airplane and reduce it by 1.*

　　　　　　　Zoom actions are like move actions, but require a fuel reserve of at least 2 and reduce it by 2.

　　　　　　　Refuel actions for a given airplane are always applicable and increase its fuel reserve by 1.

　　　　　　　Pick up and drop actions are defined as in the TRANSPORT domain.

Zoom actions have no practical use – this is different in non-propositional variants of ZENOTRAVEL, where they require less time to execute than regular movements. The only slight difference between our definition of the domain and the PDDL definition is that the latter imposes an upper bound on mobile fuel. This makes no difference for planning, because it is sensible to only refuel when and if additional fuel is needed for a movement, i. e., when the fuel reserve of an airplane is zero. Strictly speaking, our definition leads to an infinite state space, but it is apparent that this does not cause any practical problems.

Using our earlier results, the ZENOTRAVEL domain is easy to classify.

Theorem 5.4.2. PLAN-ZENOTRAVEL $\in$ APX $\setminus$ PTAS
The planning problem for ZENOTRAVEL *is 2-approximable, but it does not admit a polynomial-time approximation scheme unless* P $=$ NP. *The latter result holds even in the restricted case where there is only one airplane, which has sufficient fuel to never require refuelling in a reasonable plan.*

Proof. Non-membership in PTAS *follows from the* SIMPLELOGISTICS *result, because* ZENOTRAVEL *generalizes this domain in the case where fuel constraints do not matter, i. e., where all airplanes have more fuel than required in any reasonable plan.*

A given task T is solvable if there is at least one airplane or if it is solved by the empty plan. For a solvable task not solved by the empty plan, we generate the following two-phase plan π, where m is an an arbitrary airplane:

1. **Pickup phase:** *Move m to the initial locations of all passengers in any order, refuelling when necessary. Pick up any passenger upon arriving at their initial location.*
2. **Delivery phase:** *Move m to the goal locations of all passengers in any order, refuelling when necessary. Drop any passengers upon arriving at their initial location.*

Let L_M^*, L_R^* and L_O^* be the total number of movement, refuelling and other operators in some optimal plan, and let L_M, L_R and L_O be the corresponding numbers in π. Clearly, $m(\pi) = L_M + L_R + L_O$ and $m^*(T) = L_M^* + L_R^* + L_O^* \geq L_M^* + L_O^*$.

All locations moved to in the pickup phase must be left by some airplane at least once in an optimal plan. Thus the number of movements in that phase is bounded by L_M^*. Similarly, every location moved to in the delivery phase must be moved to at least once in some optimal plan, so the number of movements in that phase is bounded by L_M^* as well, for a total bound $L_M \leq 2L_M^*$.

The total number of refuellings is bounded by the number of movements, so $L_R \leq L_M$. Moreover, $L_O \leq L_O^*$ because the generated plan contains a minimal number of pickup and drop actions. Finally, $L_M \leq L_O$ because every movement is followed by at least one pickup or drop action.

We can thus bound the performance ratio as

$$\frac{L_M + L_R + L_O}{L_M^* + L_R^* + L_O^*} \leq \frac{2L_M + L_O}{L_M^* + L_O^*} \overset{(*)}{\leq} \frac{2L_M + L_O}{\frac{1}{2}L_M + L_O} = 1 + \frac{\frac{3}{2}L_M}{\frac{1}{2}L_M + L_O}$$

$$\leq 1 + \frac{\frac{3}{2}L_M}{\frac{1}{2}L_M + L_M} = 1 + \frac{\frac{3}{2}}{\frac{3}{2}} = 2,$$

where inequality (*) *holds because* $L_M \leq 2L_M^*$ *and hence* $L_M^* \geq \frac{1}{2}L_M$. □

The performance ratio of the solution could certainly be improved by using a better approximation algorithm akin to the one for SIMPLELOGISTICS. Note, however, that the analysis of the SIMPLELOGISTICS algorithm is not immediately applicable to ZENOTRAVEL tasks with multiple airplanes.

5.5 DEPOTS

The DEPOTS domain is a cross between BLOCKSWORLD and SIMPLELOGISTICS. It features trucks that transport objects between locations as in SIMPLELOGISTICS, but also requires the objects at the individual locations to be arranged into towers as in BLOCKSWORLD, with the particular variant of BLOCKSWORLD being one with multiple arms (called *hoists*) and limited and named table positions (called *pallets*).

Definition 5.5.1. DEPOTS *Tasks*
A DEPOTS **task** *is defined by an 8-tuple* $\langle V, T, B, P, H, l, l_0, p_\star \rangle$ *with the following components:*

- *V is a finite set of **locations**. The **roadmap** of the task is the complete graph over V.*
- *T is a finite, non-empty set of **trucks**.*
- *B is a finite set of **blocks**.*
- *P is a finite set of **pallets**.*
- *H is a finite set of **hoists**.*
  *A function p from some set of blocks to $T \cup B \cup P \cup H$ is called a **partial block position function** iff no two blocks are mapped to the same block, pallet or hoist, and the directed directed graph containing an arc (b, b') iff $p(b) = b'$ is acyclic.*
  *If p is defined for all blocks, it is called a **block position function**.*
- *$l : (P \cup H) \to V$ is the **location** function. For each location $v \in V$, there must be a pallet p and hoist h with $l(p) = v$ and $l(h) = v$.*
- *$l_0 : T \to V$ is the **initial location** function.*
- *p_0 is a block position function called the **initial block position function**.*
- *$p_\star$ is a partial block position function called the **goal block position function**, and the blocks for which it is defined are called **goal blocks**. For any goal block b, $p_\star(b)$ must be a block or a pallet, and if it is a block, then it must also be in the domain of $p_\star$.*

The sets T, B, P and H are required to be disjoint.

The definition should be intuitive enough to require little additional explanation. The acyclicity requirement for block position functions should be clear. The requirement on the goal block position functions is somewhat arbitrary, but satisfied for all DEPOTS benchmark tasks, so that we consider it part of the domain definition. Informally, it demands that if a goal position is given for a block, then a goal position must also be given for all blocks below it, including the bottom-most block in its tower, for which the pallet it rests on must be defined. It is not possible to require that a block be inside a certain truck or held by a certain hoist in a goal state, but the goal position of a block may be left unspecified. (The restrictions on the goal may be lifted or strengthened to require a completely specified goal state without affecting our complexity results.)

The requirements to have at least one hoist and pallet at each location are also part of the benchmark suite. The same is true for the requirement that there must be at least one truck. Note that this last restriction, unlike the other ones, is critical to our approximation results.

We can now define the DEPOTS domain.

Definition 5.5.2. DEPOTS *Domain*
The DEPOTS *domain maps* DEPOTS *tasks with locations V and trucks T to state spaces as follows:*

STATES: *Pairs $\langle l, p \rangle$, where $l : T \to V$ is the **truck location** function, and p is a block position function.*
*We say that hoist h **holds** block b if $p(b) = h$ and that truck t **carries** block b if $p(b) = t$. We say that block b is **on top of** block or pallet b' if $p(b) = b'$. We say that a block is **clear** if it is on top of something and nothing is on top of it, and that a pallet is **clear** if nothing is on top of it.*
*The **location** of a block that is on top of something is defined as the location of the thing it is on top of (a recursive definition, where the base case is given by the location of the pallet at the bottom). A hoist is **near to** a truck, block or pallet (and vice versa) if their location are the same.*

INITIAL STATE: *$\langle l_0, p_0 \rangle$, where l_0 is the initial location function and p_0 is the initial block position function.*

GOAL STATES: *Any state $\langle l, p \rangle$ with $p(b) = p_\star(b)$ for all goal blocks b, where $p_\star$ is the goal block position function.*

OPERATORS: *A truck can **move** from any location to any other location v, which changes its location to v.*
*A hoist can **load** a block into a truck if it holds the block and is near to the truck. This results in the truck carrying the block.*
*A hoist can **unload** a block from a truck if the truck carries the block, is near to the hoist, and the hoist does not hold anything. This results in the hoist holding the block.*

> *A hoist can **drop** a block b onto a block or pallet b' if it is holding b, it is near to b', and b' is clear. This results in b being on top of b'.*
>
> *A hoist can **pick up** a clear block b if it is near to b and it is not holding any block. This results in the hoist holding b.*

Due to its BLOCKSWORLD subproblem, it is a bit of a stretch to call DEPOTS a "transportation domain". However, the important point for our analysis is that it contains enough of a transportation subproblem to apply the approximation complexity results for the TRANSPORT family. This is indeed the case.

Theorem 5.5.3. PLAN-DEPOTS $\in$ APX \ PTAS
The planning problem for DEPOTS is 3-approximable, but it does not admit a polynomial-time approximation scheme unless P = NP. The latter result holds even in the restricted case where there is only one truck, all blocks are goal blocks, all blocks are clear in the initial and goal position and all pallets used in the goal position are initially empty, and in the restricted case where all blocks are goal blocks and there is only one hoist, one truck and one location.

Proof. Non-membership in PTAS under the first restriction follows from the SIMPLELOGISTICS result, because in this case, the domains are essentially identical. The main difference is that moving a block from a pallet into a truck and from a truck onto a pallet requires two actions instead of one, but this does not affect the proof significantly. (Putting blocks on top of other blocks is still possible, but useless under this restriction.)

Non-membership in PTAS under the second restriction follows from Selman's result for the BLOCKSWORLD domain [107]. His result applies to a slightly different variant of BLOCKSWORLD with unlabelled and unlimited table positions, but it is easy to adapt the proof to DEPOTS.

For membership, we treat the DEPOTS tasks as a single BLOCKSWORLD task, as if all pallets were present at the same location, and follow the standard BLOCKSWORLD "unstack-stack" approximation algorithm [108]. For classical BLOCKSWORLD, the algorithm works as follows [108]:

1. *Move all blocks which need to be moved onto the table.*
2. *Move all blocks moved onto the table in the previous step to their final position in the right order.*

To adapt this algorithm to DEPOTS, we use a single truck t to replace the table: Whenever a block b shall be "moved to the table" in the first phase, we move the truck to the location of b, pick up b with some hoist, and load b into the truck. Whenever a block b shall be "moved to its final position" in the second phase, we move the truck t to the location of the block or pallet b' onto which b must be dropped, unload b with some hoist, and drop it on top of b'.

It is easy to see that an optimal solution for the derived BLOCKSWORLD task of a DEPOTS task cannot be shorter than an optimal solution to the

DEPOTS *task, L**. *Because the* BLOCKSWORLD *approximation algorithm is 2-approximating, this bounds the number of non-movement actions in the generated solution by* $2L^*$.

Finally, one movement action is generated for every two other actions, bounding the number of movements by L^*, *for a total plan length of at most* $3L^*$, *proving the approximation factor of 3.*   □

We remark that the "unstack-stack" algorithm is indeed 2-approximating in BLOCKSWORLD variants with multiple arms, even though it is usually introduced for the single-arm case. To see this, observe that the approximation algorithm only uses a single arm, so the measure of the generated plan does not depend on the number of available arms. Since increasing the number of arms can never increase the optimal measure, it is sufficient to consider the performance ratio for tasks with an "unbounded" number of arms (e. g., having as many arms as blocks). In this case, it is not hard to see that the plans generated by "unstack-stack" are always *exactly* twice as long as optimal ones.

5.6 MICONIC-10

We now turn to the MICONIC-10 domain, introduced by Koehler and Schuster [82]. It requires a somewhat more detailed discussion because it is actually a family of three domains called MICONIC-10-STRIPS, MICONIC-10-SIMPLEADL and MICONIC-10-FULLADL.

In all three domains, the objective is to transport a number of passengers between different floors of a building by means of an elevator. This is, of course, a transportation planning task, and it features a single mobile of unbounded capacity, unrestricted fuel, and a complete roadmap. Indeed, the MICONIC-10-STRIPS domain is already characterized by this description almost completely.

Definition 5.6.1. MICONIC-10-STRIPS
The MICONIC-10-STRIPS *domain is a variant of* TRANSPORT$_{\infty}1$ *with a complete roadmap where the mobile is called an **elevator**, the locations are called **floors**, and the portables are called **passengers**.*

*Instead of saying that the elevator picks up or drops a passenger, we say that the passenger **enters** or **leaves** the elevator.*

The only differences between MICONIC-10-STRIPS *and* TRANSPORT$_{\infty}1$ *with complete roadmaps are that all passengers must be goal passengers and that passengers may only leave the elevator at their goal floor in* MICONIC-10-STRIPS.

This definition is not completely in accordance with the PDDL definition, in which it is possible for a passenger to enter the elevator multiple times from his initial floor, even if he is already inside the elevator or at his goal floor.

In the latter case, he can also leave at the goal floor another time. This is a harmless modelling flaw, which does not affect complexity.

We have already considered the almost identical SIMPLELOGISTICS domain, so we can keep the discussion of this MICONIC-10 variant brief.

Theorem 5.6.2. PLAN-MICONIC-10-STRIPS $\in$ APX \ PTAS
The planning problem for MICONIC-10-STRIPS *is* $\frac{7}{6}$-*approximable.*
However, the problem does not admit a PTAS *unless* P = NP.

Proof. The SIMPLELOGISTICS *proof (Theorem 5.3.5) applies. The two differences (all passengers are goal passengers, passengers may only leave at their goal location) do not affect the proof.*   □

We now introduce the two other MICONIC-10 variants. One of them, MICONIC-10-SIMPLEADL is not much more complicated than MICONIC-10-STRIPS, the difference basically amounting to a different action cost model. The MICONIC-10-FULLADL domain, however, is somewhat more complicated, introducing constraints on elevator movement caused by special service requirements. We refer to Koehler and Schuster's work for a motivation of the different features [82].

Definition 5.6.3. MICONIC-10 *Tasks*
A MICONIC-10 *task is defined by a 6-tuple* $\langle N, P, A, C, l_0, l_\star \rangle$ *with the following components:*

– $N \in \mathbb{N}_1$ *is the* **number of floors**. *The set* $F = \{1, \dots, N\}$ *is called the set of* **floors**. *We say that* $f' \in F$ *is* **between** $f \in F$ *and* $f'' \in F$ *iff* $f \leq f' \leq f''$ *or* $f'' \leq f' \leq f$.
– P *is a finite set of* **passengers**.
– $A \subseteq P \times F$ *is the* **access relation**. *We say that passenger* $p \in P$ **has access to floor** $f \in F$ *iff* $(p, f) \in A$.
– $C : P \to 2^c$ *is the* **special constraint** *function, where* $c = \{$direct, non-stop, vip, supervised, attendant, group-1, group-2$\}$ *is the set of possible constraints. We say that* $p \in P$ **is a direct-travel passenger** *if* direct $\in C(p)$, *and similarly for the other constraints.*
– $l_0 : P \to F$ *is the* **initial floor** *function.*
– $l_\star : P \to F$ *is the* **goal floor** *function, which satisfies* $l_\star(p) \neq l_0(p)$ *for all passengers* $p \in P$.

A MICONIC-10 *task is called* **simple** *iff all passengers have access to all floors and no passenger has special constraints, i. e.,* $C(p) = \emptyset$ *for all passengers* p.

We specify the set of floors by a number rather than defining it as an immediate component of the task (as for the other transportation domains), because using numbers is a convenient way of obtaining a total order on floors, which is needed for the direct constraint. Compared to MICONIC-10-SIMPLEADL, full MICONIC-10 tasks impose a number of restrictions on elevator movement. An elevator may only stop at a floor if all boarded passengers have access to that floor, VIP passengers must be served before others, non-stop passengers must be moved from their initial floor to their goal floor without intermediate stops, direct-travel passengers may never be transported downwards (upwards) if their goal location is above (below) their initial floor, supervised passengers may only be inside the elevator if an attendant is also present, and group 1 and 2 passengers may not be inside the elevator at the same time.

In addition to these constraints, which are not relevant to MICONIC-10-SIMPLEADL, the other change compared to MICONIC-10-STRIPS is that there are no individual *enter* or *leave* actions. Instead, there is a single *stop* action which causes all passengers who have arrived at their destination to leave the elevator, and causes all passengers waiting to be served to enter. This models the intuition that activities of the passengers are not controllable by the agent controlling the elevator. We will now formalize the domain.

Definition 5.6.4. MICONIC-10-FULLADL *and* MICONIC-10-SIMPLEADL *Domains*

The MICONIC-10-FULLADL *domain maps* MICONIC-10 *tasks with floors F and passengers P to state spaces as follows:*

STATES: *4-tuples* $\langle f, W, B, S \rangle$, *where* $f \in F$ *is the* **elevator floor** *and* $W \subseteq P$, $B \subseteq P$ *and* $S \subseteq P$ *form a partition of the passenger set into* **waiting**, **boarded** *and* **served** *passengers.*

A state is **illegal** *iff there is a boarded group 1 passenger and a boarded group 2 passenger, or if there is a boarded supervised passenger but no boarded attendant.*

INITIAL STATE: $\langle 1, P, \emptyset, \emptyset \rangle$.

GOAL STATES: *Any state* $\langle f, \emptyset, \emptyset, P \rangle$.

OPERATORS: *The elevator can* **move** *to a floor* f' *different from the current elevator floor* f *iff for all boarded direct-travel passengers* p, *the new floor* f' *is between* f *and the goal floor of* p. *The action changes the elevator floor to* f'.

The elevator can **stop** *at its current floor* f *iff the following conditions are satisfied:*

– *All boarded passengers have access to* f.
– *If any non-stop passenger is boarded,* f *must be the goal location of one of them.*
– *All VIPs are served, or there is a waiting VIP with initial floor* f, *or there is a boarded VIP with destination floor* f.
– *The resulting state is not illegal.*

In the resulting state, all boarded passengers with goal floor f *are served and all waiting passengers with initial floor* f *are boarded.*

The MICONIC-10-SIMPLEADL *domain is the restriction of* MICONIC-10-FULLADL *to simple* MICONIC-10 *tasks.*

Our domain definition is in accordance with the original (informal) description of the Miconic-10 domain by Koehler and Schuster. It repairs two flaws of the PDDL definition, which does not implement the VIP service and non-stop travel constraints correctly. In the PDDL definition, the elevator is allowed to stop at the initial or goal floor of a VIP passenger even if the passenger is already boarded or served and other VIP passengers should take priority. Additionally, the elevator cannot stop at any floor if several non-stop travel passengers with conflicting destinations have boarded. Both differences to the PDDL specification are harmless regarding our results: We will not make use of nonstop passengers in our proofs, and there will only be one VIP passenger per task, in which case no complications arise.

Note that the special constraints of Miconic-10-FullADL tasks can easily render the task unsolvable. For example, a task is trivially unsolvable if there are supervised passengers but no attendants. Before analyzing the impact of those constraints on the complexity of planning, we briefly discuss the simple variant.

Theorem 5.6.5. Plan-Miconic-10-SimpleADL $\in$ APX $\setminus$ PTAS
The planning problem for Miconic-10-SimpleADL *is 2-approximable.*
However, the problem does not admit a PTAS *unless* P $=$ NP.

Proof. The domain is very similar to Transport$_{\infty 1}$-$[1, 0]$ *planning with complete graph roadmaps, i. e., the variant of* SimpleLogistics *where pickup is free. Non-membership in* PTAS *for this domain was shown in Theorem 4.4.2.*

In Miconic-10-SimpleADL, *pickup is not actually free. However, we can mandate that every movement in a plan should be followed by a stop action, so that we can assume that stopping is part of moving and all movement actions have a cost of 2. Scaling all action costs by a constant does not change approximation properties. The only slight complication is that a solution might or might not start with a single stop action at the start of the plan. However, this only affects plan length by an additive constant of 1, which can be ignored for reasonably large plans. (Tasks with short plans can be special-cased.)*

For proving the approximation result, we use a very simple approximation algorithm: In the first phase, move to and stop at all initial floors (beginning with floor 1 if it is an initial floor). In the second phase, move to and stop at all goal floors. This solution has a performance ratio of at most 2 because all movements in the first phase must occur in an optimal plan and all movements in the second phase must occur in an optimal plan. $\square$

Note that the hardness result critically relies on the fact that the roadmap of a Miconic-10 task is a complete graph, which is a somewhat questionable aspect of the domain. In order to take into account the real costs of moving elevators between distant floors, it would also make sense to investigate a domain variant where floors are only connected to the floors *directly* above and below (or, equivalently, movements between f and f' incur a cost of $|f - f'|$).

For MICONIC-10-FULLADL, these different ways of modelling do not make a difference, because here, plan existence is already hard.

Theorem 5.6.6. PLAN-MICONIC-10-FULLADL $\in$ NPO $\setminus$ PS
MICONIC-10-FULLADL *planning belongs to* NPO, *but not to* PS *unless* P $=$ NP.

Proof. To show membership in NPO, we prove that the domain admits short optimal plans. An optimal plan never contains more than $2|P|$ stop actions for passenger set P, because stop actions have no effect if they do not lead to a passenger boarding or being served. Moreover, it never contains more movements than stop actions. Together, this provides a polynomial bound on optimal plan length.

To show non-membership in PS, we show NP-hardness of plan existence by reducing from the problem of finding a Hamiltonian path with a fixed start vertex v_1 in a digraph $\langle V, A \rangle$ [43, Problem GT39].

The corresponding MICONIC-10 task has no direct-travel passengers, so we can describe the set of floors without referring to their numbers. It consists of

- the initial elevator floor f_0,
- a **final floor** f_∞,
- for each vertex $v \in V$, a **vertex start floor** f_v and **vertex end floor** f_v^*, and
- for each arc $\langle u, v \rangle \in A$, an **arc floor** $f_{u,v}$.

For each vertex $v \in V$, F_v is the set containing f_v, f_v^* and the arc floors for all outgoing arcs of v. The set of all floors is denoted by F.

The passengers and their initial and goal floors and constraints are defined as follows:

Passenger	From	To	Access to...	Constraints
p_0	f_0	f_{v_1}	$\{f_0, f_{v_1}\}$	vip, attendant
$p_v \ (v \in V)$	f_0	f_v	$F \setminus \{f_\infty\}$	supervised
$p_v^* \ (v \in V)$	f_v	f_v^*	$F_v \cup \{f_\infty\}$	attendant
$p_v^\infty \ (v \in V)$	f_v^*	f_∞	$F \setminus \{f_v\}$	none
$p_{u,v} \ (\langle u,v \rangle \in A)$	$f_{u,v}$	f_v	$\{f_{u,v}, f_u^*, f_v\}$	attendant

Assume that it is possible to solve the task. Because p_0 is a VIP, the first stops must be at f_0 and f_{v_1}, where all supervised passengers and $p_{v_1}^*$ board. Because of the access restrictions of that passenger, the journey can only proceed to floors from F_{v_1}, and $f_{v_1}^*$ is not an option because stopping there would lead to the only attendant leaving. Thus, the elevator must move to f_{v_1,v_2} (for some vertex v_2 that is adjacent to v_1) and can then only proceed to $f_{v_1}^*$ and then f_{v_2}, where $p_{v_1}^\infty$ and $p_{v_2}^*$ board.

We are now in a similar situation as upon arrival at f_{v_1}, and again, the elevator will eventually move to some floor f_{v_3}, then f_{v_4}, following the arcs of the digraph $\langle V, A \rangle$ in a path $v_1 \ldots v_n$ until all vertices have been visited once. No vertex can be visited twice because of the access restrictions for passengers of type p_u^∞. *So plan existence implies the existence of a Hamiltonian path starting at v_1 in $\langle V, A \rangle$.*

On the other hand, the previous discussion shows that if a Hamiltonian path exists, then there exists a sequence of actions leading to a state where all supervised passengers have been served and the elevator is located at some floor f_v for $v \in V$. No longer requiring attendants, it can then immediately proceed to f_v^*, then f_∞ and finally serve the remaining passengers of type $f_{u,v}$ (for arcs $\langle u, v \rangle$ not part of the Hamiltonian path), one after the other, completing the plan. □

5.7 ROVERS

We now consider the ROVERS domain, which is not a *transportation domain* as such (although certain objects may be carried, they do not need to be transported anywhere), but is closely related to ROUTE.

The ROVERS domain is motivated by space applications and models the exploration of a planetary surface by a set of rovers of different capabilities. It has a quite complex definition.

Definition 5.7.1. ROVERS *Tasks*
The set of **camera modes** *is defined as* $M = \{\texttt{low-res}, \texttt{high-res}, \texttt{colour}\}$.

A ROVERS *task is defined by an 18-tuple* $\langle G, v_l, R, R^s, R^r, road, O, vis, C, mode, cal, r_C, l_0, V_0^s, V_0^r, V_\star^s, V_\star^r, I_\star \rangle$ *with the following components:*

- $G = \langle V, E \rangle$ *is a graph whose vertices are called* **waypoints** *and whose edges are called* **routes**.
- *Waypoint* $v_l \in V$ *is called the* **lander location**.
- R *is a finite set of* **rovers**.
- $R^s \subseteq R$ *is the set of* **rovers equipped for soil analysis**.
- $R^r \subseteq R$ *is the set of* **rovers equipped for rock analysis**.
- $road \subseteq R \times E$ *is the* **access relation**. *The* **roadmap** *of a rover* $r \in R$ *is the graph* $G_r = \langle V_r, E_r \rangle$ *where* E_r *contains all routes reachable from the lander location using only routes* e *with* $\langle r, e \rangle \in road$, *and* V_r *consists of the waypoints incident to such routes. We require that* G_r *is a tree for each rover.*
- O *is a finite set of* **image objectives**.
- $vis \subseteq V \times (V \cup O)$ *is the* **visibility relation** *between waypoints and waypoints or image objectives. If* $\langle u, v \rangle \in vis$, *we say that* u **can see** v *and that* v **can be seen from** u. *We require that* vis *restricted to* $V \times V$ *is symmetric and reflexive and that* $\langle u, v \rangle \in vis$ *for all routes* $\{u, v\} \in E$.
- C *is a finite set of* **cameras**.
- $mode : C \to 2^M$ *is the* **camera mode** *function. If* $m \in mode(c)$ *for a camera* $c \in C$, *we say that the camera* **supports** m.
- $cal : C \to O$ *is the* **calibration target** *function.*

- $r_C : C \to R$ is the **camera location** function. If $r_C(v) = r$, we say that camera c **is installed in** rover r.
- $l_0 : R \to V$ is the **initial rover location** function. We require that $l_0(r)$ is contained in the roadmap of rover r.
- $V_0^s \subseteq V$ is the set of **initial soil sample waypoints**.
- $V_0^r \subseteq V$ is the set of **initial rock sample waypoints**.
- $V_\star^s \subseteq V_0^s$ is the set of **soil sample goals**.
- $V_\star^r \subseteq V_0^r$ is the set of **rock sample goals**.
- $I_\star \subseteq O \times M$ is the set of **image goals**.

A ROVERS tasks is called **simple** if there are no rovers equipped for rock analysis, no image objectives, no cameras, no initial rock sample waypoints, no rock sample goals and no images goals, if the initial location of all rovers is the lander location, all rovers are equipped for soil analysis, and the visibility relation is universal.

There are three types of goals for ROVERS tasks. The first two types require obtaining soil analyses and rock analyses of certain waypoints. These are essentially *visitation goals* as in the ROUTE domain. The third type of goals are *image goals*, which are in a certain sense *disjunctive* visitation goals: To take an image of an objective, a rover equipped with a suitable camera must visit *any* waypoint from which the objective can be seen.

The domain is complicated further by the fact that not all rovers can conduct soil analyses or rock analyses, that not all rovers carry cameras, and that not all cameras are suitable for each kind of image goal (a camera must support the desired *mode* of the image). Moreover, cameras need to be calibrated against a certain objective before they can be used. A final complication is that collecting the data with the rovers is not enough to solve the task; it must also be *communicated* to the lander, which can only be done from waypoints that can see the lander location.

We remark that in the existing ROVERS benchmarks, the visibility relation between rovers and image objectives has a limited structure: For any two waypoints u and v, either u can see all objectives that v can see, or vice versa. This simplifies planning somewhat because there is no need to visit one waypoint u for calibrating a camera and another waypoint v for taking an image: Either the camera can also be calibrated from v, or the image can also be taken from u. However, we choose not to model the restriction because it is unclear whether or not it is intentional and we will see that it does not affect complexity.

A slight oddity of the IPC ROVERS domain which our definition does not duplicate is that it does not require the visibility relation for waypoints to be reflexive. In particular, in the IPC tasks the lander location cannot be seen from itself, which means that a rover cannot communicate with the lander if they are at the same location. However, this does not make a difference for approximation complexity.

We now formalize the semantics of ROVERS tasks.

Definition 5.7.2. ROVERS *Domain*

The ROVERS *domain maps* ROVERS *tasks with waypoints V, rovers R, cameras C, modes M and objectives O to state spaces as follows:*

STATES: 11-*tuples* $\langle l, V^s, V^r, D^s, D^r, D^i, R_F, C_c, V^s_G, V^r_G, I_G \rangle$, *where* $l :$
               $R \to V$ *is called the* **rover location** *function,* $V^s \subseteq V$
               and $V^r \subseteq V$ *are the sets of* **soil sample waypoints** *and*
               rock sample waypoints, $D^s \subseteq R \times V$, $D^r \subseteq R \times V$ *and*
               $D^i \subseteq R \times O \times M$ *are the sets of* **available soil data**, **available rock data** *and* **available image data**, $R_F \subseteq R$ *is the*
               set of **full** *rovers,* $C_c \subseteq C$ *is the set of* **calibrated** *cameras,*
               and $V^s_G \subseteq V$, $V^r_G \subseteq V$ *and* $I_G \subseteq O \times M$ *are the* **remaining**
               soil sample goals, **remaining rock sample goals** *and re-*
               maining image goals.
               We say that rover r **has soil data for** v *(similarly,* **has**
               rock data for v, **has image data for** o **in mode** m*) iff*
               $\langle r, v \rangle \in D^s$.

INITIAL STATE: $\langle l_0, V^s_0, V^r_0, \emptyset, \emptyset, \emptyset, \emptyset, \emptyset, V^s_\star, V^r_\star, I_\star \rangle$, *where* l_0, V^s_0, V^r_0, $V^s_\star$, $V^r_\star$
               and $I_\star$ *are as in the definition of* ROVERS *tasks.*

GOAL STATES: *Any state with no remaining soil sample, rock sample or image*
               goals.

OPERATORS: *A rover may* **move** *from waypoint v to waypoint w iff its*
               location is v and these waypoints are adjacent in its roadmap.
               This changes its location to w.
               A rover equipped for soil analysis may **sample soil** *iff it is*
               not full and its location v is a soil sample waypoint. This
               results in v no longer being a soil sample waypoint, the rover
               being full and the rover having soil data for v.
               A rover equipped for rock analysis may **sample rocks** *iff it*
               is not full and its location v is a rock sample waypoint. This
               results in v no longer being a rock sample waypoint, the rover
               being full and the rover having rock data for v.
               A rover may **empty its store** *iff it is full. This results in the*
               rover no longer being full.
               A rover r may **calibrate** *camera c iff c is installed in r and*
               the location of r can see the calibration target of c. This results
               in c being calibrated.
               A rover may **take an image** *of objective o in mode m with*
               camera c iff c is calibrated and installed in r and supports
               mode m and the rover location can see o. This results in the
               rover having image data for o in mode m and in c no longer
               being calibrated.
               A rover may **communicate** *a piece of soil, rock or image*
               data iff it has that data and if it can see the lander location.

> *This results in the corresponding goal (if any) being removed from the remaining goals.*

The SIMPLEROVERS *domain is the restriction of* ROVERS *to simple tasks.*

It is apparent from this definition that the ROVERS domain could be simplified in a number of ways without affecting semantics significantly. For example, the fact that rock or soil sampling fills the store of the rover is only important for better-than-constant-factor approximations. If we are content with constant factor approximations, we can automatically empty the store of a rover after each sampling operation. Similarly, calibration and image-taking could be combined into a single action at least for the kind of visibility relations present in the planning competition benchmark set, as discussed previously.

At its core, ROVERS planning is about visiting a set of waypoints with rovers of different capabilities (in particular, different roadmaps). We have seen similar tasks before when discussing the ROUTE family, and indeed the ROUTE results easily apply to ROVERS.

Theorem 5.7.3. PLAN-ROVERS $\in$ poly-APX $\setminus$ APX
The planning problem for ROVERS *is poly-approximable. However, the problem does not admit a constant-factor approximation unless* P = NP. *The latter result already holds for the restricted* SIMPLEROVERS *domain.*

Proof. Membership in poly-APX *is proved by the following greedy algorithm:*

1. *As long as there is a remaining soil sample goal u, find a rover equipped for soil analysis whose roadmap includes u and a waypoint v from which the lander location can be seen. (Actually, v can always chosen to be the lander location itself, but note our remarks on the visibility relation in the IPC benchmarks.) If no such rover exists, the task is unsolvable. Otherwise, move the rover to u along a shortest path, sample soil, empty the rover, move the rover to v and communicate the soil data.*
2. *Solve rock sample goals in an analogous fashion.*
3. *As long as there is a remaining image goal, find a rover with a camera supporting the required mode whose roadmap includes a waypoint u which can see the calibration target of the camera, a waypoint v which can see the image objective, and a waypoint w which can see the lander location. If no such rover exists, the task is unsolvable. Otherwise, move the rover to u along a shortest path, calibrate the camera, move the rover to v along a shortest path, take the image in the required mode, and finally move the rover to w along a shortest path and communicate the image data.*

The algorithm clearly solves the task and can easily be implemented to run in polynomial time.

To prove the hardness result, we reduce from ROUTE$_*$ *and use Theorem 4.6.1. In particular, we show that* ROUTE$_*$ $\leq_{\mathrm{OP}}$ PLAN-SIMPLEROVERS *by a reduction which preserves stretchable graph classes.*

Given a ROUTE$_*$ *task T and constant $r > 1$, we set $M = 2\left\lceil \frac{3|V_*|}{r-1} + 3|V_*| \right\rceil$, where V_* is the set of target locations of T. The corresponding* SIMPLEROVERS *task T' is obtained by stretching the roadmap graph of T by the factor M, then mapping each mobile to a rover, each mobile roadmap to an identical rover roadmap and each target location to a soil sample goal.*

The optimal measures of T and T' satisfy $m^*(T') \leq Mm^*(T) + 3|V_*|$, because any plan π for T can be transformed into a plan π' for T' by replacing each movement in π by M corresponding movements in π' and inserting three actions whenever a target (soil sample) location is first visited by some rover: sample soil, empty the store, communicate the soil data.

On the other hand, every plan π' for T' can be efficiently mapped back to a plan π for T such that $m(\pi) \leq \frac{1}{M}m(\pi')$, by mapping movements back in the obvious way (which reduces their amount by a factor of at least M) and removing all non-movement actions.

We now prove that the reduction is approximation-preserving. Given a plan π' for T' with performance ratio at most $r > 1$, the mapped-back plan π for T has the performance ratio

$$\frac{m(\pi)}{m^*(T)} \leq \frac{\frac{1}{M}m(\pi')}{\frac{1}{M}(m^*(T') - 3|V_*|)} = \frac{m(\pi')}{m^*(T') - 3|V_*|} \leq \frac{rm^*(T')}{m^*(T') - 3|V_*|}$$

$$= 1 + \frac{(r-1)m^*(T') + 3|V_*|}{m^*(T') - 3|V_*|} = 1 + (r-1)\frac{m^*(T') + \frac{3|V_*|}{r-1}}{m^*(T') - 3|V_*|}$$

$$= 1 + (r-1)\left(1 + \frac{\frac{3|V_*|}{r-1} + 3|V_*|}{m^*(T') - 3|V_*|}\right)$$

$$\overset{(*)}{\leq} 1 + (r-1)\left(1 + \frac{\frac{1}{2}M}{M - \frac{1}{2}M}\right) = 1 + (r-1) \cdot 2,$$

where inequality $(*)$ holds because we can assume that $m^*(T') \geq M$; otherwise, T is solved by the empty plan and thus trivial. Moreover, the reduction is clearly optimization-preserving because $m^*(T) \leq K$ iff $m^*(T') \leq KM + 3|V_*|$ for all $K \in \mathbb{N}_0$. □

The SIMPLEROVERS domain is hard to approximate for a reason we call *positive subgoal interaction*. There is a number of subgoals to solve (one for each soil sample goal), and for each individual goal it is easy to find an optimal (i. e., minimal length) action sequence for solving only this single goal. In other words, *subgoals are easy to solve*. Moreover, solving *all* subgoals never requires significantly more actions that the sum for all subgoal costs (at most twice as much, because it is always possible to return to the initial rover location after solving a subgoal). In other words, *subgoals do not interact negatively*. However, an optimal plan might require significantly *fewer* actions than the sum of the individual subgoal costs due to synergies: Moving a rover towards a particular sample goal may also move it closer to another sample goal, so that the combined cost of solving both goals is significantly less than the sum of the individual costs.

The hardness proof for the ROUTE domain fosters this kind of interaction by constructing, for each mobile, a roadmap which allows solving a certain set of goals at essentially the same cost as a single goal from the set. Different mobiles allow solving different sets of goals, but there is overlap between these sets. A good solution will try to minimize the number of *goal sets* that are used to solve all goals. This amounts to a general set covering problem, and general set covering problems are hard to approximate.

We emphasize this point because it emerges as a common pattern in all polynomially solvable IPC domains which are hard to approximate by a constant factor. We will now see two more examples of this phenomenon.

5.8 GRID

In the GRID domain, a robot moves along a graph with a two-dimensional grid shape (hence the name of the domain) transporting keys. It can carry one key at a time, and the goal is to move a certain set of keys to individually specified locations. So far, this is a variant of the TRANSPORT$_{11}$ domain, which we have shown to be in APX, but not in PTAS unless P = NP, although we have not proved hardness for the particular case of GRID roadmaps.

However, there is an additional twist. The GRID domain is different to the previously discussed benchmarks in that it features a *dynamic roadmap*: Locations can be initially *locked* (and hence inaccessible) and later become accessible through the invocation of *open* actions, if the mobile carries an appropriate key. Apart from these difference, GRID tasks are very much like regular TRANSPORT$_{11}$ tasks, and we define them in terms of those.

Definition 5.8.1. GRID *Tasks*
A GRID *task is defined by a 3-tuple* $\langle T, L_0, U \rangle$ *with the following components:*

- *T is a* TRANSPORT$_{11}$ *task with a grid roadmap with location set V and portable set P called the* **underlying** TRANSPORT *task. Its portables are called* **keys**, *its mobile is called the* **robot**.
- $L_0 \subseteq V$ *is the set of* **initially locked locations** *or* **doors**. *It may not contain the initial location of the robot.*
- $U \subseteq P \times L$ *is the* **unlock relation**. *We say that key* $k \in P$ **may unlock** *location* $v \in L$ *if* $\langle p, v \rangle \in L$. *For any two keys, we require that the sets of locations they may unlock are either equal or disjoint.*

A GRID *task is said to have a* **static roadmap** *if its unlock relation is empty.*

Effectively, a GRID task is a special TRANSPORT task with two additional properties: There is a set of locations that are initially inaccessible, and there is a relation that specifies which keys can be used to gain access to (some of) these locations. The constraints on the unlock relation imply that two keys are either functionally equivalent or unrelated. In the original PDDL description, this is formalized with different "shapes" for keys and locks such that each key and lock has a specific shape and keys only fit into locks with matching shapes. We now formalize the semantics of GRID planning.

Definition 5.8.2. GRID *Domain*

The GRID **domain** *maps* GRID *tasks with underlying* TRANSPORT *task T to state spaces as follows:*

STATES: *Pairs* $\langle s, L \rangle$ *where s is a state of T called the* **underlying** TRANSPORT **state** *and* $L \subseteq L_0$ *is the set of* **locked locations***.*

INITIAL STATE: $\langle s_0, L_0 \rangle$*, where* s_0 *is the initial state of T and* L_0 *is the set of initially locked locations.*

GOAL STATES: *Any state* $\langle s, L \rangle$ *where s is a goal state of the underlying* TRANSPORT *task.*

OPERATORS: **Pickup** *and* **drop** *actions have the same semantics as in T. Additionally, there exists a* **swap** *action for all keys* k_1 *and* k_2 *and locations v, which is defined as the sequential composition of "drop* k_1 *at v" and "pick up* k_2 *at v".*

Movements to a location v are only applicable if v is not locked. Apart from this restriction, they have the same semantics as in T.

The robot may **unlock** *location v iff v is locked, the current robot location is adjacent to v, and the robot carries a key which may unlock v. In the resulting state, v is no longer locked.*

In an earlier analysis, we have shown that the bounded plan existence problem for GRID is NP-complete, even when restricted to tasks without initially locked locations, which is a special case of tasks with static roadmaps [57]. From the observation that the planning problem for general GRID tasks is in NPO, it follows that the unlocking aspect does not affect the decision complexity of GRID planning. However, we will see that the same is not true when considering approximation complexity.

Theorem 5.8.3. PLAN-GRID $\in$ poly-APX

The planning problem for GRID *is poly-approximable, and it can be approximated by a constant factor when restricted to tasks with static roadmaps.*

Proof. We first prove the second result. In GRID *tasks with static roadmaps, the only difference to a* TRANSPORT$_{11}$ *tasks is the presence of* swap *actions. We say that a* GRID *plan is* **swap-free** *if it does not contain swap actions. Using a c-approximation algorithm for* TRANSPORT$_{11}$ *like the one described in Theorem 4.5.2, we can generate a plan π which is at most c times as long as an optimal swap-free plan. Moreover, if π^* is an optimal plan, then* *an optimal swap-free plan has a measure bounded by $2m(\pi^*)$ (replace each swap in π^* by a drop and a pick-up action to obtain a swap-free plan with this property). Therefore, π has a performance ratio bounded by $2c$. In other words, the* TRANSPORT$_{11}$ *algorithm is a constant-factor approximation algorithm for* GRID *tasks with static roadmaps.*

For the first result, we can reduce the general case to the static roadmap case by repeating the following operations until they can no longer be applied:

- *Find a key k that can be reached from the robot location and a locked location l that can be unlocked by k and which has an adjacent location l' that can be reached from the robot location. Stop if this is not possible.*
- *Move to the location of k, pick up k, move to l', unlock l, move to the initial location of k, drop k, move back to the initial robot location.*

Clearly, this can be done in polynomial time: Each iteration requires only polynomial time, and each iteration unlocks one of only polynomially many doors. Moreover, the resulting state after applying these actions differs from the initial state only in the fact that it has fewer locked locations, which means that it is certainly solvable if the original task was solvable. From the resulting state, it is not possible to unlock any further doors, and thus it can be solved using the techniques for GRID *tasks with static roadmaps. The overall running time is polynomial, which shows that* GRID *planning belongs to* poly-APX. □

The general GRID planning algorithm appears rather primitive, and one might hope that it is possible to improve on the performance ratios it obtains. However, the following result shows that there are limits to such improvements, placing general GRID planning in a different approximation class than the variant with static roadmaps.

Note that neither of the two earlier hardness proofs for GRID planning [57] can be easily adapted to show the following non-membership result. One of them reduces from the TRAVELLING SALESPERSON PROBLEM for grid graphs and can only be used to show non-membership in PO, as the generated GRID instances admit a polynomial-time approximation scheme. The other reduces from the 3SAT problem and maps to a subclass of GRID tasks which is 2-approximable. A new reduction is thus needed. It is based on a proof by Robert Mattmüller [89] with only minor modifications.

Theorem 5.8.4. PLAN-GRID $\notin$ APX
If $P \neq NP$, there is no constant-factor approximation algorithm for planning in the GRID *domain.*

Proof. We OP-reduce from MINIMUM SET COVER. *We are given a* MINIMUM SET COVER *instance $I = \langle S, C \rangle$ with $S = \{s_1, \ldots, s_n\}$ and $C = \{C_1, \ldots, C_k\}$ and a parameter $r > 1$. Let $w = \max\{n, k\}$, $M = (4w + 15)|S|$ and $B = 2\left\lceil \frac{M}{r-1} + M \right\rceil$. The corresponding* GRID *task T' is defined as follows:*

- *The grid has width $w + 1$ and height $B + 5$. We refer to the locations of T' as pairs $\langle x, y \rangle \in \{0, \ldots, w\} \times \{0, \ldots, B + 4\}$. The robot is initially located at $\langle 0, B \rangle$.*
- *For each subset $C_j \in C$, there is one key k_j initially located at $\langle j, 0 \rangle$. These keys are called **subset keys**.*
- *For each subset $C_j \in C$ and element $s_i \in C_j$, there is one key $k_{j,i}$ initially located at $\langle j, B + 1 \rangle$. These keys are called **subset element keys**.*
- *For each element $s_i \in S$, there is one key k_i^+ initially located at $\langle i, B + 3 \rangle$ with goal location $\langle i, B + 4 \rangle$. These keys are called **goal keys**.*
- *The initially locked locations are exactly the initial locations of the subset element keys and goal keys.*
- *Each subset key k_j may unlock location $\langle j, B + 1 \rangle$, the initial location of the corresponding subset element keys.*
- *Each subset element key $k_{i,j}$ may unlock location $\langle i, B + 3 \rangle$, the initial location of the goal key corresponding to s_i.*
- *Goal keys may not unlock any location.*

To map back a plan π' for T' to a set cover D for I, include a set C_j in D iff the corresponding subset key k_j is used to open a door in π'. To see that D constitutes a set cover, consider an arbitrary element $s_i \in S$. The initial location of the goal key k_i^+ is initially locked and must be opened in π' by some subset element key $k_{i,j}$ satisfying $s_i \in C_j$. Choose j in such a way that $k_{i,j}$ is this key. Using $k_{i,j}$ requires opening its initial location, which is only possible with k_j, and hence, by the definition of D, the set C_j is included in D. Because $s_i \in C_j$ and $s_i \in S$ was chosen arbitrarily, this shows that D is indeed a set cover.

Due to the very large distance between the initial locations of subset keys and the initial locations of subset element keys and since only one key can be carried at a time, fetching the subset keys is the bottleneck of any plan, and good plans will make use of as few subset keys as possible. Formally, for each subset $C_j \in D$, the plan π' must contain at least $2B$ movements to move from row $y \geq B$ (containing the initial robot location and subset element keys) to row 0 (containing the subset keys) and back, and hence $m(D) \leq \frac{1}{2B} m(\pi')$.

We now bound $m^*(T')$ in terms of $m^*(I)$. Let $D^* \subseteq C$ be an optimal set cover. Then the following plan π' solves T':

- *For each subset $C_j \in D^*$, do the following:*
  1. *Move from $\langle 0, B \rangle$ to $\langle j, 0 \rangle$, pick up the subset key for C_j, move to $\langle j, B \rangle$, open location $\langle j, B + 1 \rangle$, move to $\langle j, B + 1 \rangle$, drop the key.*
  2. *For each element $s_i \in C_j$ for which the goal key has not yet been moved to its goal location, pick up the element key $k_{j,i}$, move from $\langle j, B + 1 \rangle$ to $\langle i, B + 2 \rangle$, open location $\langle i, B + 3 \rangle$, move to $\langle i, B + 3 \rangle$, switch key $k_{j,i}$ with the goal key k_i^+, move to $\langle i, B + 4 \rangle$, drop the key, move back to $\langle j, B + 1 \rangle$.*
  3. *Move back from $\langle j, B + 1 \rangle$ to $\langle 0, B \rangle$.*

For each iteration of step 1., no more than $2B + w + 4$ actions are required (note that $j \leq w$), so the sum over all iterations is bounded by $(2B+w+4)|D^*|$. For each goal key in step 2., no more than $2w + 10$ steps are required (note that $i, j \leq w$), summing to $(2w + 10)|S|$ over all iterations. Finally, for each iteration of step 3., at most $w + 1$ steps are required, for a sum of $(w + 1)|D^*|$ over all iterations. We thus obtain $m^*(T') \leq m(\pi') \leq (2B + 2w + 5)|D^*| + (2w+10)|S| \leq 2B|D^*| + (4w+15)|S| = 2Bm^*(I) + M$, where the last inequality is because $|D^*| \leq |S|$.

This allows us to bound the performance ratio of the set cover D constructed from a plan π' for the $\textsc{Grid}$ task T' as follows:

$$\frac{m(D)}{m^*(I)} \leq \frac{\frac{1}{2B}m(\pi')}{\frac{1}{2B}(m^*(T') - M)} = \frac{m(\pi')}{m^*(T') - M} \leq \frac{rm^*(T')}{m^*(T') - M}$$

$$= 1 + \frac{(r-1)m^*(T') + M}{m^*(T') - M} = 1 + (r-1)\frac{m^*(T') + \frac{M}{r-1}}{m^*(T') - M}$$

$$= 1 + (r-1)\left(1 + \frac{\frac{M}{r-1} + M}{m^*(T') - M}\right)$$

$$\overset{(*)}{\leq} 1 + (r-1)\left(1 + \frac{\frac{1}{2}B}{B - \frac{1}{2}B}\right) = 1 + (r-1) \cdot 2,$$

where inequality (∗) holds because $\frac{1}{2}B \geq \frac{M}{r-1} + M$, $\frac{1}{2}B \geq M$ and we can assume that $m^*(T') \geq B$; otherwise, I is solved by the empty set cover and thus trivial. Moreover, the reduction is optimization-preserving because $m^*(I) \leq K$ iff $m^*(T') \leq 2BK + M$ for all $K \in \mathbb{N}_0$. □

Having observed that the hardness of approximately solving $\textsc{Grid}$ tasks is largely due to the necessity of opening locations, one question is just how hard $\textsc{Grid}$ planning with static roadmaps is. We know that it is not in PO unless P = NP, and we know that it is in APX. Indeed, using Arora's approximation scheme for route planning in grid graphs [6], we can devise $(2 + \varepsilon)$-approximations for all $\varepsilon > 0$, using techniques based on the approximation algorithm described by Mattmüller [89]. However, we do not know whether it is possible to improve on this result further and obtain a polynomial-time approximation scheme for $\textsc{Grid}$ with static roadmaps. At least, there appears to be no easy way to adapt Arora's technique to this problem, and we hypothesize that no such approximation scheme exists.

Besides general $\textsc{Grid}$ planning and the case of static roadmaps, another variant that is of some interest is the restriction to *uniform locks*, i.e., tasks where any given key may either unlock no door or all doors. This class of tasks is relevant because it includes all $\textsc{Grid}$ benchmarks from the first International Planning Competition. It is not hard to prove that this variant of $\textsc{Grid}$ planning still admits a constant-factor approximation algorithm [89].

5.9 DRIVERLOG

We now turn to the DRIVERLOG domain, the last proper transportation benchmark we will discuss. In a DRIVERLOG task, a fleet of trucks of unbounded capacity transport packages to their goal destinations. All trucks use the same roadmap, which can be an arbitrary connected graph. At this level of description, the domain is equivalent to TRANSPORT$_{\infty+}$. However, in addition to trucks, locations and packages, the domain also features *drivers*. Drivers are like trucks in being able to traverse locations on their own (using a different roadmap than trucks, though). However, unlike trucks, they cannot transport packages. The important property of drivers is that they can *board* trucks, and only trucks which have been boarded by a driver are able to move. We will see that this aspect of the problem leads to an increase in complexity.

Definition 5.9.1. DRIVERLOG *Task*
A DRIVERLOG *task is defined by a 7-tuple $\langle G_T, G_D, T, D, P, l_0, l_* \rangle$ with the following components:*

- $G_T = \langle V, E_T \rangle$ *is a connected graph called the **road graph**. Its vertices V are called **locations**, its edges **roads**.*
- $G_D = \langle V, E_D \rangle$ *is a connected graph called the **footpath graph**. Its edges are called **footpaths**. Note that it shares the same vertex set as the road graph.*
- *T is a finite set of **trucks**.*
- *D is a finite set of **drivers**.*
- *P is a finite set of **packages**. We require that T, D and P are disjoint.*
- *$l_0 : T \cup D \cup P \rightarrow V$ is the **initial location** function.*
- *l_* is a partial function from $T \cup D \cup P$ to V called the **goal location** function. Its domain is called the set of **goal objects**; we distinguish between **goal trucks**, **goal drivers** and **goal packages**.*

The definition contains few surprises, maybe apart from the fact that unlike the other transportation domains we study, DRIVERLOG tasks may also include goal locations for the mobiles (trucks), and for the drivers. We now define the domain.

Definition 5.9.2. DRIVERLOG *Domain*
The DRIVERLOG *domain maps* DRIVERLOG *tasks with locations V, trucks T, drivers D and packages P to state spaces as follows:*

STATES: *Functions $l : T \cup D \cup P \rightarrow V \cup T$, where only drivers and packages may be mapped to trucks. For any object $o \in T \cup D \cup P$, location $v \in V$ and truck $t \in T$ we say that o is **at** v if $l(o) = v$ and that o is **in** t if $l(o) = t$.*

INITIAL STATE: *The initial location function of the task.*

GOAL STATES: *Any state extending the goal location function of the task.*

OPERATORS: *A driver may **walk** to a location $v' \in V$ iff it is at some adjacent location in the footpath graph. This results in the driver being at v'. Walk actions have a cost of 2.*

*A driver may **board** a truck iff they are at the same location and there is no driver in the truck. This results in the driver being in the truck.*

*A driver may **leave** a truck iff he is in the truck. This results in the driver being at the same location as the truck.*

*A truck may **drive** to a location $v' \in V$ iff it is at some adjacent location in the road graph and there is some driver in it. This results in the truck being at v'.*

*A truck may **pick up** a package iff they are at the same location. This results in the package being in the truck.*

*A truck may **drop** a package iff the package is in the truck. This results in the package being at the same location as the truck.*

Our model is close to the original PDDL definition. The only difference is that the PDDL definition simulates costs of 2 for walk actions by introducing intermediate locations in the footpath graph which are never used as initial or goal locations. The definition with action costs is clearer and more concise.

We first show that DRIVERLOG planning can be solved in polynomial time.

Theorem 5.9.3. DRIVERLOG ∈ poly-APX
The planning problem for DRIVERLOG is poly-approximable,

Proof. Consider the following algorithm:

– *For each goal package which it is not at its goal location, walk to the initial location of some truck with some driver along a shortest path in the footpath graph, board the truck, drive the truck to the initial location of the package along a shortest path in the road graph, pick up the package, drive the truck to the goal location of the package along a shortest path in the road graph, drop the package, leave the truck.*

– *For each goal truck which is not at its goal location, walk to the initial location of the truck with some driver along a shortest path in the footpath graph, board the truck, drive to its goal truck location along a shortest path, leave the truck.*

– *For each goal driver who is not at his goal location, walk to that goal location along a shortest path in the footpath graph.*

The algorithm runs in polynomial time, generating a solution if the task is solvable and failing otherwise. (The latter can only happen for tasks that have no truck or no driver.)   □

We now show that generating good-quality plans in the DRIVERLOG domain is difficult.

Theorem 5.9.4. DRIVERLOG $\notin$ APX

If P $\neq$ NP, *there is no constant-factor approximation algorithm for planning in the* DRIVERLOG *domain. This is true even when restricting the problem to tasks where there is only a single driver and all goal objects are packages.*

Proof. As in the previous non-approximability proofs, we OP-reduce from MINIMUM SET COVER. *We are given a* MINIMUM SET COVER *instance* $I = \langle S, C \rangle$ *and a parameter* $r > 1$. *We first check if* I *is solvable; if not, it is mapped to some fixed unsolvable* DRIVERLOG *task. If it is, let* $M = 16|S|$ *and* $B = 2 \left\lceil \frac{M}{r-1} + M \right\rceil$ *and construct the* DRIVERLOG *task* T' *defined as follows:*

- *There is a location* v_H *called the* **hub**.
- *There are* B **startup locations** $v_0^1, \ldots, v_0^B$, *with footpaths connecting* v_0^1 *to the hub and* v_0^i *to* v_0^{i+1} *for all* $i < B$. *The only driver is originally located at* v_0^B. *The sequence of* B *footpaths connecting* v_0^B *to the hub is called the* **startup path**.
- *For each subset* $\widehat{S} \in C$, *there are* B *locations* $v_{\widehat{S}}^1, \ldots, v_{\widehat{S}}^B$. *There are footpaths connecting the hub to* $v_{\widehat{S}}^1$ *and each location* $v_{\widehat{S}}^i$ *with* $i < B$ *to the next location* $v_{\widehat{S}}^{i+1}$. *Location* $v_{\widehat{S}}^B$ *is called the* **subset location** *corresponding to* $\widehat{S}$; *the path connecting it to the central location is called the* **subset path** *for* $\widehat{S}$. *Each subset location is the initial location of a truck with no defined goal.*
- *For each subset* $\widehat{S} \in C$ *and element* $s \in \widehat{S}$, *there is a* **subset element location** $v_{\widehat{S},s}$, *connected to subset location* $v_{\widehat{S}}$ *by a road.*
- *For each element* $s \in S$, *there is an* **element start location** v_s^- *and an* **element goal location** v_s^+, *connected by a road. The element start location is connected to each subset element location* $v_{\widehat{S},s}$ *for subsets* $\widehat{S}$ *containing* s *by a footpath. Also for each element* $s \in S$, *there is a truck and a package initially located at the element start location* v_s^-. *The truck has no defined goal, and the goal location of the package is the corresponding element goal location* v_s^+.
- *Finally, there is an additional* **remote location**, *connected to all other locations through very long chains of roads and footpaths such that the distance between the remote location and the locations introduced before is at least* $4B|C|$. *(The only purpose of this location is to make the road and footpath graphs connected, as required by the definition of* DRIVERLOG *tasks.)*

To map back a plan π' *for* T' *to a set cover* D *for* I, *first check if* $m(\pi') \geq 4B|C|$. *If yes (the* **trivial case**), *map it back to the trivial set cover* C. *Otherwise (the* **non-trivial case**), *we can assume that the remote location is not visited in the plan. In this case, include a set* $\widehat{S}$ *in* D *iff the corresponding subset location* $v_{\widehat{S}}^B$ *is visited by the driver in* π'.

We now show that D is indeed a set cover and that $m(D) \leq \frac{1}{4B}m(\pi')$. In the trivial case, both statements are apparently true, so let us consider the non-trivial case.

In order to transport any of the packages in π' to their goal location, the driver must first walk along the startup path to the hub, then on to some subset location for a subset $\widehat{S} \in C$. Once the subset location is reached, the only packages that can be transported to their goal location without walking the driver back to the hub are those corresponding to the elements of $\widehat{S}$. Unless these already constitute a set cover, the driver must thus eventually return to the hub. Due to the placement of the trucks and the road layout, all trucks will be at the same locations as initially at that point. Therefore, the driver is essentially in the same situation as initially, except that some of the goal packages have been delivered. To transport further goal packages to their destinations, the driver must again walk to the subset location for a subset containing an element corresponding to such a package, from where it is possible to satisfy the goals related to the elements of that subset, and so on, visiting additional subset locations and thus adding more subsets to D until all goals are satisfied. This proves one half of our claim: D indeed constitutes a set cover.

To see that $m(D) \leq \frac{1}{4B}m(\pi')$, it is sufficient to consider the actions in π' corresponding to traversals of the startup path or subset paths. If $m(D)$ subset locations are visited in π', then the driver must traverse a subset path at least $2m(D) - 1$ times ($m(D)$ times coming from the hub and $m(D) - 1$ times returning to the hub) and traverse the startup path once. Each of these traversals requires B walk actions, each of which has a cost of 2, for a total cost of $4Bm(D)$. We thus get $m(\pi') \geq 4Bm(D)$ and hence $m(D) \leq \frac{1}{4B}m(\pi')$.

We now bound $m^*(T')$ in terms of $m^*(I)$. Let $D^* \subseteq C$ be an optimal set cover. Then the following plan π' solves T':

- *For each set $\widehat{S} \in C^*$:*
  1. *Walk to the hub and then to subset location $v_{\widehat{S}}^B$ on a shortest path.*
  2. *For each $s \in \widehat{S}$ such that the package initially located at v_s^- is not yet at its goal location, board the truck at the subset location $v_{\widehat{S}}^B$, drive to the subset element location $v_{\widehat{S},s}$, debark, walk to the element start location v_s^-, board the truck there, load the package, drive to the element goal location v_s^+, unload the package, drive back to v_s^-, debark, walk back to $v_{\widehat{S},s}$, board the truck, drive back to $v_{\widehat{S}}^B$ and debark.*

The actions in step 1. of the plan amount to a total cost of $4B|C^*| = 4Bm^*(I)$: In each iteration, B walk actions are needed to walk to the hub (either from the initial location or from some subset location), and B walk actions are needed to walk from the hub to the subset location for $\widehat{S}$. Each of these walk actions incurs a cost of 2.

The actions in step 2. of the plan amount to a total cost of 16 for each package being delivered (2 walk actions and 12 other actions), for a total cost of $16|S| = M$ for the complete plan. Thus, $m^*(T') \leq m(\pi') = 4Bm^*(I) + M$.

This allows us to bound the performance ratio of the set cover D constructed from a plan π' for the DRIVERLOG *task* T' *as follows:*

$$
\frac{m(D)}{m^*(I)} \leq \frac{\frac{1}{4B}m(\pi')}{\frac{1}{4B}(m^*(T') - M)} = \frac{m(\pi')}{m^*(T') - M} \leq \frac{rm^*(T')}{m^*(T') - M}
$$

$$
= 1 + \frac{(r-1)m^*(T') + M}{m^*(T') - M} = 1 + (r-1)\frac{m^*(T') + \frac{M}{r-1}}{m^*(T') - M}
$$

$$
= 1 + (r-1)\left(1 + \frac{\frac{M}{r-1} + M}{m^*(T') - M}\right)
$$

$$
\overset{(*)}{\leq} 1 + (r-1)\left(1 + \frac{\frac{1}{2}B}{B - \frac{1}{2}B}\right) = 1 + (r-1) \cdot 2,
$$

where inequality $(*)$ *holds because* $\frac{1}{2}B \geq \frac{M}{r-1} + M$, $\frac{1}{2}B \geq M$ *and we can assume that* $m^*(T') \geq B$; *otherwise,* I *is solved by the empty set cover and thus trivial. Moreover, the reduction is optimization-preserving because* $m^*(I) \leq K$ *iff* $m^*(T') \leq 4BK + M$ *for all* $K \in \mathbb{N}_0$. □

We remark that the hardness of DRIVERLOG planning is indeed to a major extent due to the complications introduced by drivers. Interestingly, we can prove a similar hardness result for the restriction of DRIVERLOG to the case where there are *no packages*. This only requires some minor modifications to the proof. The packages are removed. Instead, we introduce a new location $\overline{v}_s^+$ for each $s \in S$, which is adjacent to v_s^+ on the road graph, but not on the footpath graph. We call these locations the *twins* of v_s^+. There are no further roads or paths connecting twins to other locations (apart from long roads and paths to the remote location to ensure connectedness.) Moreover, in addition to the original driver, which we now call the *main driver*, we introduce two new drivers for each element s, one with v_s^+ as the initial location and its twin as a goal location (called the *goal driver* for s), and one with the twin as the initial location and v_s^+ as the goal location (called the *twin goal driver* for s).

The main ideas of the reduction apply unchanged; it is still necessary to move the main driver to enough subset locations to cover all set elements. Moreover, the length of the plan corresponding to an optimal set cover does not change significantly: the pickup and drop actions are no longer required, and instead, whenever a location v_s^+ is entered by a truck driven by the main driver, the following sequence of actions is generated: debark from the truck, board it with the goal driver, drive to the twin location, debark, board with the twin goal driver, drive back, debark, and board the truck with the main driver. To reflect these modification in the calculations, we only need to change the definition of M from $16|S|$ to $22|S|$.

5.10 AIRPORT

The last domain considered in this chapter, AIRPORT, is not a transportation domain, but it has some aspects of a route planning domain, so it is natural to discuss it here. It models ground traffic on an airport, i. e., movement of aircraft along taxiways and runways. Unlike other route planning domains, AIRPORT tasks are heavily space-constrained: Not only can any given location (called a *taxiway segment*) only be occupied by one aircraft at a time, there even exist mutual exclusion constraints between segments to the effect that at most one of them may be occupied at a given time. The purpose of these constraints is to model realistic safety conditions. Indeed, the AIRPORT domain is firmly grounded in real-world planning tasks, and some of the IPC4 benchmarks are faithful translations of realistic data from Munich Airport.

Definition 5.10.1. AIRPORT *Tasks*
*The set of **movement modes** for aircraft is defined as $M = \{$pushing, taxiing, airborne, parked$\}$.*

An AIRPORT task is defined by a 9-tuple $\langle A, S, R_T, R_P, R_B, m_0, s_0, m_\star, s_\star \rangle$ with the following components:

- *A is a finite set of **aircraft**.*
- *S is a finite set of **taxiway segments**.*
- *$R_T \subseteq S \times S$ is the **taxiway relation**.*
- *$R_P \subseteq S \times S$ is the **pushback relation**.*
- *$R_B \subseteq S \times S$ is the **blocking relation**.*
- *$m_0 : A \rightarrow \{$pushing, taxiing$\}$ is the **initial mode** function.*
- *$s_0 : A \rightarrow S$ is the **initial segment** function.*
- *$m_\star : A \rightarrow \{$airborne, parked$\}$ is the **goal mode** function.*
- *$s_\star : A \rightarrow S$ is the **goal segment** function.*

*The directed graphs $G_T = \langle S, R_T \rangle$, $G_P = \langle S, R_P \rangle$ and $G_B = \langle S, R_B \rangle$ are called the **taxiway graph**, **pushback graph** and **blocking graph** of the task.*

*The task is called **undirected** iff R_T, R_P and R_B are symmetric, **planar** iff $\langle S, R_T \cup R_P \rangle$ is a planar digraph, and **regularly constrained** iff $R_T = R_P = R_B$.*

The AIRPORT tasks of IPC4 obey two restrictions not captured by our definition. First, there are no aircraft whose initial mode is pushing and goal mode is parked. This would make little sense as the pushing mode is associated with outbound aircraft only and the parked mode is associated with inbound aircraft only. Second, the pushback graph is always a subgraph of the taxiway graph with all arcs reversed. Neither restriction has an impact on the complexity of the problem because our hardness results already hold if there are no pushing aircraft at all.

We also omitted some aspects of the PDDL definition [66] to simplify presentation. First, we do not distinguish between the *location* and *facing* of an aircraft, encoding both properties in its current *segment*. Compilations between these two representations are straightforward.

Second, the PDDL domain allows the blocking relation to depend on the aircraft (via *airplane types*); however, none of the existing benchmarks makes use of this feature. Modelling airplane types would not affect our complexity results, because we shall see that, even without airplane types, AIRPORT planning is already as hard as it is possible for a propositional PDDL planning domain.

Definition 5.10.2. AIRPORT *Domain*

The AIRPORT *domain maps* AIRPORT *tasks with aircraft A, taxiway segments S and blocking relation R_B to state spaces as follows:*

STATES: *Pairs $\langle m, s \rangle$, where $m : A \to M$ is the **current mode** function and $s : A \to S$ is the **current segment** function. We say that an aircraft **is pushing** (**is taxiing**, **is airborne**, **is parked**) iff its current mode is* pushing *(*taxiing, airborne, parked*).*

*We call a state **legal** iff it satisfies the following **blocking constraints**:*

– *If two aircraft share the same current segment, at least one must be airborne.*

– *If an aircraft with current segment u is pushing or taxiing and another aircraft with current segment v is pushing, taxiing or parked, then $\langle u, v \rangle \notin R_B$.*

INITIAL STATE: *$\langle m_0, s_0 \rangle$, where m_0 is the initial mode function and s_0 is the initial segment function of the task.*

GOAL STATES: *$\langle m_\star, s_\star \rangle$, where $m_\star$ is the goal mode function and $s_\star$ is the goal segment function of the task.*

OPERATORS: *An airplane may **move** to v if its current segment is u and either it is taxiing and $\langle u, v \rangle$ is an arc in the taxiway graph, or it is pushing and $\langle u, v \rangle$ is an arc in the pushback graph. This changes the current segment of the airplane to v.*

*An airplane may **start up** if it is pushing. This results in it taxiing.*

*An airplane may **take off** if it is taxiing. This results in it being airborne.*

*An airplane may **park** if it is taxiing. This results in it being parked.*

All these actions are only allowed if the resulting state is legal.

Our AIRPORT state spaces are defined slightly differently from the original PDDL specification, but have identical semantics. In the PDDL definition, aircraft "leave the map" when taking off, and they can only take off from specific runway segments. We ignore airborne aircraft for blocking purposes, which amounts to the same thing as having them leave the map, and while we allow take-off everywhere, it never makes sense to take off from a different segment than the goal segment of that airplane, and goal segments of outbound aircraft are always runway segments. The PDDL domain also contains a minor modelling flaw that allows aircraft to park immediately before take-off. However, this is never a useful thing to do and hence cannot affect complexity.

Theorem 5.10.3. AIRPORT *Planning Is* PSPACE-*Complete*

Plan existence and bounded plan existence for the AIRPORT *domain are* PSPACE-*complete. This is true even when only considering undirected, planar, regularly constrained tasks where all aircraft are taxiing initially and must be parked in the goal state.*

Proof. For a graph $G = \langle V, E \rangle$ *and a set of* **tokens** T, *we define a* **legal placement** *of* T *on* G *as an injective function* $\pi : T \to V$ *such that no two tokens are placed on adjacent vertices. A legal placement* π' *is a* **successor** *of another legal placement* π *iff they differ on exactly one token* $t \in T$, *for which we have* $\{\pi(t), \pi'(t)\} \in E$. *In other words, to obtain a successor of a legal placement, move a single token along an edge and verify that this results in another legal placement.*

We show PSPACE-*hardness of* AIRPORT *plan existence by polynomially reducing from the following (*PSPACE-*complete) variation of the* SLIDING TOKENS *problem [53]: Given a planar graph* G, *set of tokens* T *and legal placements* π_0, $\pi_\star$ *of* T *on* G, *is there a sequence of legal placements* $\pi_1, \ldots, \pi_M$ *such that* π_i *is a successor of* π_{i-1} *for all* $i \in \{1, \ldots, M\}$ *and* $\pi_M = \pi_\star$? *(In the original Sliding Tokens puzzle, tokens are indistinguishable and the goal has a slightly different form. Only simple adjustments to the hardness proofs are needed for the modified version; cf. Theorem 23 and Corollary 6 in the reference [53].)*

We now describe the mapping of puzzle instances to AIRPORT *tasks. Given graph* $G = \langle V, E \rangle$, *tokens* T *and placements* π_0 *and* $\pi_\star$, *we generate an* AIRPORT *task with segment set* V, *aircraft set* T, *taxiway graph* G, *pushback graph* G *and blocking graph* G, *initial segment function* π_0 *and goal segment function* $\pi_\star$. *All aircraft are taxiing in the initial state and must be parked in the goal state. Clearly, the mapping can be computed in polynomial time.*

No solution to the planning task can ever contain push, start up *or* take off *actions, so we only need to consider* move *and* park *actions. If the planning task has a solution, then the sequence of* move *actions in such a solution defines a solution to the puzzle instance. Note that if a taxiing aircraft ever moved to a segment which is adjacent to the current segment of another (taxiing or parked) aircraft, this would violate the blocking constraints. Similarly, from a solution to the puzzle we can obtain a sequence of actions that move each aircraft to its goal location without violating the blocking constraints, and from that state the task is solved by parking all aircraft. Therefore, the mapping is indeed a Karp reduction.*

Thus, plan existence for restricted AIRPORT *tasks is* PSPACE-*hard, which implies that bounded plan existence is also* PSPACE-*hard. Moreover, both problems must belong to* PSPACE *because PDDL planning in any fixed propositional domain is in* PSPACE. *This concludes the proof.*   □

The reduction implies that there exist AIRPORT tasks for which the shortest plan consists of exponentially many actions. For example, the shortest solution to the puzzle corresponding to a QBF formula with n quantifier alternations consists of $\Omega(2^n)$ many steps [53], leading to an AIRPORT task with a similarly bounded solution length.

Theorem 5.10.4. PLAN-AIRPORT $\in$ EXPO $\setminus$ NPS
The planning problem in the AIRPORT *domain can be optimally solved in exponential time, and it cannot be solved (not even sub-optimally) in sub-exponential time.*

Proof. The problem is solvable in exponential time by explicit graph search in the state space of the given task. Sub-exponential planning algorithm cannot exist because there are tasks with exponential lower bounds on shortest plan length.   □

We remark that these results for AIRPORT also extend to *parallel* planning problems. In particular, parallel solutions can only be shorter than sequential ones by a linear amount, because at most $O(n)$ many actions can be executed in parallel (one per aircraft).

As a final remark, we have also proved another polynomial reduction from the halting problem for polynomially space-constrained Turing Machines, which only generates *deterministic* AIRPORT tasks, i.e., instances where at most one action is applicable at any time. Therefore, AIRPORT planning is difficult even if no branching is involved. However, the reduction is less interesting for practical purposes because the generated tasks are neither undirected, nor planar, nor regularly constrained.

5.11 Summary

This concludes our discussion of route planning and transportation planning in the IPC benchmark suite. We showed that the planning problems for this class of domains spans almost the full range of approximation classes, including PO, APX, poly-APX, NPO and EXPO. The main results from this chapter are briefly summarized in Fig. 5.1.

Domain	Complexity
AIRPORT	EXPO \ NPS
DEPOTS	APX \ PTAS
DRIVERLOG	poly-APX \ APX
GRID	poly-APX \ APX
GRIPPER	PO
LOGISTICS	APX \ PTAS
MICONIC-10-FULLADL	NPO \ PS
MICONIC-10-SIMPLEADL	APX \ PTAS
MICONIC-10-STRIPS	APX \ PTAS
MYSTERY	NPO \ PS
MYSTERYPRIME	NPO \ PS
ROVERS	poly-APX \ APX
ZENOTRAVEL	APX \ PTAS

Fig. 5.1. Complexity results for the IPC transportation domains (if P ≠ NP)

IPC Domains: Others

In this chapter, we discuss the complexity of planning in those IPC domains which cannot be considered variants of transportation or route planning. There is no common theme to these domains, and consequently, the results presented for the different domains do not build on each other or on the results presented in the previous two chapters.

Unlike the transportation and route planning benchmarks, we will encounter comparatively few (in fact, only two) domains in which there is a difference in complexity between optimal and sub-optimal planning. Consequently, we can mostly rely on the classical methods of complexity theory throughout this chapter.

We consider nine domains, two of which (PROMELA and PIPESWORLD) are further subdivided: ASSEMBLY (Sect. 6.1), BLOCKSWORLD (Sect. 6.2), FREECELL (Sect. 6.3), MOVIE (Sect. 6.4), PIPESWORLD (Sect. 6.5), PROMELA (Sect. 6.6), PSR (Sect. 6.7), SATELLITE (Sect. 6.8), and SCHEDULE (Sect. 6.9). The chapter again ends with a brief summary of results (Sect. 6.10).

6.1 ASSEMBLY

The first domain considered in this chapter, ASSEMBLY, models the construction of *composite objects* from parts which may themselves be composite objects.

Definition 6.1.1. ASSEMBLY *Tasks*
An ASSEMBLY *task is defined by a 7-tuple* $\langle C, T, R, P_p, P_t, (\prec_c^+)_{c \in C}, (\prec_c^-)_{c \in C} \rangle$
with the following components:

- C *is a finite set of* **components**.
- T *is a finite set of* **tools**.
- $R \subseteq C \times T$ *is called the* **tool requirements** *relation. For* $\langle c, t \rangle \in R$, *we say that* c **requires** t.

- $P_p \subseteq C \times C$ is the **permanent part** relation. The associated digraph must be a directed tree, the root of which is called the **root component**. For each component c, the components c' with $\langle c, c' \rangle \in P_p$ are called the **permanent parts** of c. Each component must either have no permanent parts, in which case it is called **atomic**, or at least two permanent parts, in which case it is called **composite**.
- $P_t \subseteq C \times C$ is the **transient part** relation, which must be disjoint from the permanent part relation. The union of the two relations, called the **part relation**, must be acyclic. For each component c, the components c' with $\langle c, c' \rangle \in P_t$ are called the **transient parts** of c, and the set of components which are either permanent or transient parts of c are called **parts** of c.
- For each component c, the **assembly order** $\prec_c^+$ is a partial order on the parts of c.
- For each composite c, the **removal order** $\prec_c^-$ is a partial order on the parts of c such that $c' \prec_c^- c''$ only if c'' is a transient part of c.

Intuitively, the objective of an ASSEMBLY task is to build an object (the *root component*) from its constituent parts. This requires recursively building the constituents, down to the level of composites which only have atomic parts. Composites are assembled from their parts by adding one part at a time, with the possible orderings restricted by the *assembly order*.

In addition to being assembled, composites can be disassembled, also subject to ordering constraints. Here, we distinguish between *permanent parts* and *transient parts* of the composite. Permanent parts may only be removed if no installed part of the composite appears later in the assembly order. In other words, it is not allowed to remove a permanent part p_1 from a composite p if another part p_2 with $p_1 \prec_p^+ p_2$ is installed. Transient parts are different in that their removal is not subject to these constraints, but instead must respect different requirements: Transient part p_2 may only be removed from p if all parts p_1 with $p_1 \prec_p^- p_2$ are already installed. Another difference between transient and permanent parts is that transient parts *must* be removed for completion of the composite (justifying their name). This implies that the transient parts of a composite are again available for use as parts after the composite has been built, so that it is possible for a component to be a transient part of several composites and a permanent part of yet another.

Finally, some assemblies must have *tools* allocated to them whenever an object is installed into them or removed from them.

Definition 6.1.2. ASSEMBLY *Domain*
The ASSEMBLY *domain maps* ASSEMBLY *tasks with components C and tools T to state spaces as follows:*

STATES: *Pairs $\langle i, a \rangle$, where $i : C \to C \cup \{\bot\}$ is the **installation** function and $a : T \to C \cup \{\bot\}$ is the **tool allocation** function. We say that a component $c \in C$ is **installed in** $c' \in C$ iff $i(c) = c'$, that it is **complete** iff the components installed in*

> it are exactly its permanent parts, and that it is **available** iff
> it is complete and $i(c) = \bot$.
>
> We say that a tool $t \in T$ is **allocated to** $c \in C$ iff $a(t) = c$,
> that it is **unallocated** iff $a(t) = \bot$, and that a component
> $c \in C$ is **prepared** iff all tools it requires are allocated to it
> and $i(c) = \bot$.

INITIAL STATE: $\langle \emptyset, a_\bot \rangle$, where $a_\bot$ denotes the constant function mapping to $\bot$.

GOAL STATES: Any state where the root component is complete.

OPERATORS: A part c' of a component c may be **incorporated** into c iff c is prepared, c' is available, and all components c'' with $c'' \prec_c^+ c'$ are installed in c. In the resulting state, c' is installed in c.

A permanent part c' of a component c may be **removed** from c iff c is prepared, c' is installed in c, and no component c'' with $c' \prec_c^+ c''$ is installed in c. In the resulting state, c' is no longer installed in c.

A transient part c' of a component c may be **removed** from c iff c is prepared, c' is installed in c, and all components c'' with $c'' \prec_c^- c'$ are installed in c. In the resulting state, c' is no longer installed in c.

A tool t may be **assigned** to a component c iff it is unallocated. In the resulting state, t is allocated to c.

A tool t may be **released** from a component it is allocated to. In the resulting state, t is unallocated.

After the earlier intuitive explanation of ASSEMBLY tasks, this definition should contain few surprises. However, it significantly differs from the PDDL definition, and in a way that affects planning complexity. In particular, in the PDDL definition, once a component has become complete, it *stays complete*, even if it is disassembled into its parts.

For example, consider the (impossible) task of creating a rectangular table with four legs and an oval table with four legs out of a rectangular board, an oval board and four legs. In the IPC1 domain specification, it is possible to build the rectangular table out of the rectangular board and the four legs, remove the legs (which leaves the rectangular table in its *complete* state) and then use the legs and the oval board to build the oval table, creating two tables. This is a major modelling problem which we do not want to reproduce in our definition.

However, this means that our results are not directly applicable to the ASSEMBLY domain in the form in which it has been used as a benchmark. It turns out that this is not so problematic. First, although the domain was part of the ADL track of IPC1, neither of the two participating ADL planning systems was able to solve a single competition task, so the domain can hardly be considered an important benchmark, given the lack of empirical data available. Second, although there are now planning systems available which can solve ASSEMBLY tasks, the use of the ASSEMBLY suite as a benchmark is hampered by the fact that six of the tasks (namely tasks 7, 12–14, 19 and 27) are syntactically flawed, i. e., invalid PDDL. (Our experiments with ASSEMBLY in Part II use a modified task suite, where we tried to correct these problems in the way that seemed most reasonable to us.)

Before we discuss the complexity of ASSEMBLY planning in the "corrected" domain, let us briefly point out that the planning problem for the competition domain is at least 2-approximable: One can always solve the task by sorting the components in a way consistent with the part-of relation, then build each part in sequence by assigning all required tools to it, incorporating all parts in a suitable order, removing all parts in a suitable order, and releasing all required tools. At least all *assign* and *incorporate* actions are necessary in every solution, and there are as many actions of this type as other actions, so this plan has a performance ratio of at most 2.

We now present our main result for the ASSEMBLY domain.

Theorem 6.1.3. PLAN-ASSEMBLY $\in$ EXPO $\setminus$ NPS
The planning problem in the ASSEMBLY domain can be optimally solved in exponential time, and it cannot be solved (not even sub-optimally) in sub-exponential time.

Proof. The problem is solvable in exponential time with explicit graph search in the state space of the given task. Sub-exponential planning algorithm cannot exist because there are tasks with exponential lower bounds on shortest plan length. We prove this by presenting a family $(T_n)_{n \in \mathbb{N}_1}$ of planning tasks with task size polynomially bounded in n and $m^*(T_i) \geq c^i$ for all $i \in \mathbb{N}_1$ for some constant $c > 1$.

For all $n \in \mathbb{N}_1$, task T_n includes atomic components a_i and b_i for all $i \in \{1, \ldots, n\}$, composite components C_i for all $i \in \{1, \ldots, n\}$, and no tools. Composite C_1 has two permanent parts, a_1 and b_1. All other composites C_i have three permanent parts, a_i, b_i and C_{i-1}. Moreover, composites C_i for $i \geq 3$ have i transient parts, namely C_{i-2} and all components a_j for $j \in \{1, \ldots, i-1\}$. The assembly and removal orders are given by the transitive closure of the following orderings:

- *$a_i \prec^+_{C_i} b_i$ for all $i \in \{1, \ldots, n\}$.*
- *$b_i \prec^+_{C_i} C_{i-1}$ for all $i \in \{2, \ldots, n\}$.*
- *$C_{i-2} \prec^+_{C_i} a_i \prec^-_{C_i} C_{i-2}$ for all $i \in \{3, \ldots, n\}$.*
- *$a_j \prec^+_{C_i} b_i \prec^-_{C_i} a_j$ for all $i \in \{3, \ldots, n\}$ and all $j \in \{1, \ldots, i-1\}$.*

We now show that a shortest plan for T_n consists of at least F_n actions, where F_n is the n-th Fibonacci number.

- *Task T_1 is optimally solved by incorporating a_1 and then b_1 in C_1, leading to a plan length of $2 \geq F_1$.*

– Task T_2 is optimally solved by incorporating a_2, then a_1, then b_2 in C_2, removing a_1 from C_2, building C_1 using a plan for T_1, then incorporating C_1 in C_2, leading to a total plan length of $7 \geq F_2$.

– Task T_n for $n \geq 3$ is optimally solved by first solving T_{n-2} to complete C_{n-2}, then incorporating C_{n-2} and a_n in C_n, removing C_{n-2} from C_n, removing all components from C_j for $j \leq n - 2$ in an appropriate order to make the atomic components a_j available, incorporating all components a_j for $j < n$ in C_n, incorporating b_n in C_n, removing all components a_j for $j < n$ from C_n, solving T_{n-1} to complete C_{n-1}, and then incorporating C_{n-1} in C_n. This requires completely solving the tasks T_{n-2} and T_{n-1}, so by induction we can prove that $m^*(T_n) \geq m^*(T_{n-1}) + m^*(T_{n-2})$. Together with the lower bounds for $m^*(T_1)$ and $m^*(T_2)$, the claim follows.

Because Fibonacci numbers grow as $\Theta\left(\left(\frac{1+\sqrt{5}}{2}\right)^n\right)$, this concludes the proof.
□

Note that unlike the AIRPORT domain considered in the previous chapter, for which the planning problem also belongs to EXPO \ NPS, plan *existence* for ASSEMBLY is not at all difficult to solve. In fact, ASSEMBLY tasks are always solvable.

6.2 Blocksworld

As we observed in the introduction, the BLOCKSWORLD domain has already been extensively studied in the planning literature. In particular, Selman has shown that BLOCKSWORLD planning is 2-approximable and thus belongs to APX, but does not admit a polynomial-time approximation scheme unless P = NP [107]. Slaney and Thiébaux provide an overview of a number of approximation algorithms for BLOCKSWORLD, show how to implement them efficiently and provide empirical data for their approximation quality [108]. Owing to the extensive treatment in these two references, we do not need to discuss this classical domain further.

6.3 FreeCell

The FreeCell domain is based on the popular solitaire card game with the same name. The original card game is played with a standard deck of 52 cards, initially arranged into eight *tableau piles* of six or seven cards each. Cards can be moved between these eight tableau piles, four *free cells* and four *foundation piles* according to the following rules:

– Cards may only be picked up if they occupy a free cell or if they are the top card of a tableau pile. No more than one card can be picked up at the same time.

- Cards may only be dropped in a free cell if it does not currently hold any other card.
- Cards may only be added to a tableau pile (as its new top card) if that pile is empty, or the value of the card is one less than the value of the top card of the pile and it is of a different colour (e. g., the four of spades may only be added to tableau piles with the five of diamonds or hearts as their top card).
- Aces may be added to an empty foundation pile. Other cards may only be added to a foundation pile if their value is one higher than the value of the top card of the pile and they are of the same suit.

The objective of the game is to move all cards to foundations.

Due to the fixed deck size, the standard FREECELL game only allows for a constant number of different initial configurations. This is not very interesting from a complexity theory point of view, because problems with a finite number of instances can trivially be decided in polynomial time. Thus, it is necessary to allow varying deck sizes, either by adding new suits or by adding cards to the existing suits.

Of these, we choose the latter, because it seems the more natural choice. In fact, the planning competition tasks use the same scaling parameter, although for these, deck sizes never exceed 52 cards and in most cases, fewer cards are used. We also allow for a varying number of tableau piles and free cells, but we will see that our hardness results already hold if the number of free cells is any fixed constant.

Definition 6.3.1. FREECELL *Tasks*

*For all natural numbers $n \in \mathbb{N}_0$, $C_n = \{\Diamond, \heartsuit, \clubsuit, \spadesuit\} \times \{1, \ldots, n\}$ is called the n-**deck**. Its elements are called **cards**. For a card $\langle s, v \rangle$, s is called its **suit** and v is called its **value**. A card is called **red** if its suit is $\Diamond$ or $\heartsuit$, and **black** otherwise.*

*An n-w-**tableau** $(n, w \in \mathbb{N}_0)$ is a set of at most w non-empty sequences of cards in C_n, called **tableau piles**. Tableau piles may not contain the same card twice, and two piles of a tableau may not have common cards. The last card of a tableau pile is called its **top card**, the subsequence that is obtained by removing the top card is called the **buried part**. A card c **matches** a tableau pile iff c and the top card of the pile are of different colour and the value of c is one less than the value of the top card.*

*A FREECELL **task** is defined by a 4-tuple $\langle n, w, c, T_0 \rangle$ with the following components:*

- $n \in \mathbb{N}_0$ *is called the **suit length**.*
- $w \in \mathbb{N}_0$ *is called the **tableau width**.*
- $c \in \mathbb{N}_0$ *is called the **free cell count**.*
- *T is an n-w-tableau such that each card in C_n appears in exactly one tableau pile. It is called the **initial tableau**.*

To readers acquainted with FREECELL, the previous definition should go without much explanation. One slightly unusual aspect of the definition is that initial tableaus of FREECELL tasks are not required to contain tableau piles of (roughly) *equal size*, as is mandatory in the general game. This is not an overgeneralization, because the height of tableau piles can be equalized by adding additional cards of lowest value to the smaller piles. These will be moved to foundations immediately in any reasonable plan, resulting in the original uneven tableau. Moves of this kind are always optimal and are hence performed automatically by most FREECELL computer programs. (When computing performance ratios, one needs to be careful to properly take account of such "automatic" actions. However, for FREECELL, we shall see that deciding *plan existence* is already hard.)

We can now formally define the semantics of legal FREECELL moves by defining the domain.

Definition 6.3.2. FREECELL *Domain*
The FREECELL *domain maps* FREECELL *tasks with suit length n and tableau width w to state spaces as follows:*

STATES: *Pairs $\langle T, F \rangle$, where T is an n-w tableau simply called the **tableau** and $F \subseteq C_n$ is a set of cards called the **free cell cards**. We require that $|F|$ is bounded by the free cell count of the task and that the cards in T and the cards in F are disjoint. Cards in C_n which are neither in the tableau nor free cell cards are said to be **in foundations**.*

INITIAL STATE: *$\langle T_0, \emptyset \rangle$, where T_0 is the initial tableau of the task.*

GOAL STATES: *$\langle \emptyset, \emptyset \rangle$ (i. e., no cards are in the tableau or in free cells, so all cards are in foundations.)*

OPERATORS: *Operators are composed of **pickup subactions** and **drop subactions**.*
*There are three kinds of pickup subactions for a given card $c \in C_n$. **Pickup from free cell** is applicable iff c is a free cell card and removes the card from the set of free cell cards. **Pickup from tableau pile** is applicable iff c is the top card of a pile in the tableau and removes the card from this pile. **Pickup from tableau** is applicable iff c is the only card of a pile in the tableau and removes the pile from the tableau.*
*There are four kinds of drop subactions for a given card $c \in C_n$. Three of these are converse to the possible pickup subactions: **Drop on free cell** is applicable iff the number of free cell cards is less than the free cell count and adds the card to the set of free cell cards. **Drop on tableau pile** is applicable iff c matches some tableau pile and adds the card to this tableau pile as the new top card. **Drop on tableau** is applicable iff the number of tableau pile is less than the tableau width and introduces a new tableau pile consisting only of c.*

*The fourth drop subaction is **drop in foundations**, which is applicable iff there is no lower-value card of the same suit as c in the tableau or the set of free cell cards. This subaction adds the card to foundations.*

*The operators of the task are **movement** actions, which are defined as the compositions of pickup and drop subactions, where first an applicable pickup subaction for card c and then an applicable drop subaction for the same card c is applied to the current state.*

Note that there are two asymmetries between pickup and drop subactions: First, cards can be dropped in foundations, but not picked up from there. Second, cards that are picked up from tableau piles can only be dropped again on that tableau pile if they match it. There is no general requirement that the top card of a tableau pile matches the buried part; in particular, this is generally *not* true in the initial tableau. (And indeed, if it were, then FREE-CELL planning would be trivial, as all cards could be moved to foundations immediately.)

We will prove that FREECELL planning belongs to NPO but not to PS, unless P = NP. The inclusion in NPO holds because the domain admits short optimal plans. Non-membership in PS is due to hardness of the plan existence problem. Because both parts of the overall proof are rather involved, we present them separately, beginning with the short plan result.

Theorem 6.3.3. FREECELL *Admits Short Optimal Plans*
FREECELL *admits short optimal plans.*

Proof. We first note that there are only two kinds of subactions within FREE-CELL plans that cannot be undone immediately by applying an inverse subaction: dropping a card in foundations and picking up the top card c of a tableau pile p where c does not match the buried part of p. If m is the number of cards of the task, then any plan will contain exactly m subactions of the first and no more than m subactions of the second kind, because the number of "mismatches" in the initial tableau is bounded by m. Thus, the number of non-undoable actions is polynomial in the number of cards and thus in the size of the instance. Therefore, we only need to come up with a polynomial bound for the length of any consecutive sequence of undoable actions appearing within an optimal plan.

*To obtain this bound, we present an algorithm that, given an initial and goal state such that the goal state can be reached from the initial state by only using undoable actions, calculates a polynomial length action sequence that accomplishes this state transition. We call this subproblem without "unsafe" moves the **safe** FREECELL **planning problem**.*

*In fact, we do not bound the actual number of actions but the number of **macro moves** (or **macros**) that are part of the plan, where a macro move is a sequence of actions that moves several cards C of a tableau pile – potentially all* *of them – onto another tableau pile or to an empty position in the tableau, by making use of free cells and empty tableau positions as auxiliary storage. The component actions that form a macro only move cards from C, but individual cards in C may be moved more than once. It is possible for a macro to move only one card, so that macros generalize the regular action set.*

*Macros using only free cells as auxiliary storage (**free cell moves**) can move up to $m + 1$ cards when m free cells are available. This is accomplished by moving one card of the originating pile into each available free cell, then moving the top card of that pile to its new location, then moving the cards from the free cells to that location.*

Let $N(m, k)$ be the maximum number of cards that can be moved by a macro move if m free cells and k empty table positions are available as auxiliary storage, and let $T(m, k)$ be the maximal number of actions required for the macro. Clearly, $N(m, 0) = m + 1$ and $T(m, 0) = 2m + 1$.

Macro moves that use $k > 0$ free tableau positions to move a set of cards C can be divided into three simpler sub-macros: First, move a subset of cards $C' \subseteq C$ to an auxiliary tableau position with the first sub-macro, then move the remaining cards $C \backslash C'$ to their final destination with the second sub-macro, and finally move the cards C' to the destination with the third sub-macro.

To estimate how many cards can be moved in such a fashion, observe that the first sub-macro moves $|C'|$ cards and can use $k - 1$ or k tableau positions as auxiliary storage (depending on whether or not the overall macro moves to an existing tableau pile or moves to a new tableau pile), and the second and third sub-macro move $|C'|$ and $|C| - |C'|$ cards and can use $k - 1$ tableau positions as auxiliary storage, as the tableau pile containing the cards C' cannot be used. Thus, we must have $|C'| \leq N(m, k - 1)$ and $|C| - |C'| \leq N(m, k - 1)$ for an overall bound of $|C| \leq 2N(m, k - 1)$, so $N(m, k) \leq 2N(m, k - 1)$. On the other hand, it is easy to see that it is possible to move $2N(m, k - 1)$ in such a fashion, so we have $N(m, k) = 2N(m, k - 1)$.

To estimate how many free cell moves this takes, observe that each of the three sub-macros can use m free cells, and at least $k - 1$ auxiliary tableau positions (having potentially more tableau positions available for the first sub-macro can only reduce the amount of free cell moves required), so we have $T(m, k) \leq 3T(m, k - 1)$. Solving these two recurrences, we obtain $N(m, k) = (m + 1)2^k$ and $T(m, k) \leq (2m + 1)3^k$. For $T(m, k)$, we are only interested in an upper bound, so we set $T(m, k) = (2m + 1)3^k$.

Now assume we are applying a macro move which moves n cards. Clearly, we can assume that at most $k' = \lceil \log_2 n \rceil$ auxiliary tableau positions are used for the macro move, because this amount would be sufficient for the macro move even if no free cell is available, as $n \leq N(0, k')$. Moreover, we can assume that at most $m' = n$ free cells are used for the macro move. Thus, the total number of actions for the macro move is bounded by $\max_{0 \leq m \leq m', 0 \leq k \leq k'} T(m, k) = T(m', k') = (2n + 1)3^{\lceil \log_2 n \rceil} \leq (2n + 1)3^{\log_2 n} \cdot 3 = (6n + 3)n^{\frac{3}{2}} = O(n^{\frac{5}{2}})$. Thus, the number of actions required by a macro move is polynomially bounded in the instance size. It is thus sufficient to show *that we can solve the safe* FREECELL *planning problem with polynomially many macro moves to prove the overall result.*

A card is called **red-even** *if it is red and its value is even, or if it is black and its value is odd. Otherwise, it is called* **red-odd**. *In tableau piles, only red-even cards are added to red-even cards, and only red-odd cards are added to red-odd cards. Consequently, the same adjectives can be used when referring to tableau piles. The card on top of another card in a tableau pile is called the* **son** *of that card.*

The safe FREECELL *problem can be solved in three stages: First, compact the tableau by making all piles as big as possible, freeing up as many table positions and free cells as possible. This can clearly be done with a polynomial number of steps, because each card needs to be made the son of a given other card as the effect of a macro move at most once.*

Second, identify **bad son cards**, *i. e., cards which have a son in both the current and goal state, and the two son cards are of different suits (they must have the same value and colour). As long as there are any bad son cards, choose one of them with maximal value, apply a number of macro moves to move all cards on top of the bad son card to other locations, using up as few free cells and empty table positions as possible, then apply a macro move to move the pile of cards with the correct son on top of the bad son card and compact the tableau again. This card will now no longer be a bad son card. Clearly, this takes a polynomial number of macro moves.*

Only cards of lower value than the corrected card c are moved in this process, which means that cards of a higher value than the corrected card cannot become bad son cards as a result of these movements. We do not care if cards of a lower value are made bad son cards, because this can be changed by further iterations of the correction procedure. A card c′ of an equal value but different colour cannot be made a bad son card because if c is red-even, all cards moved are red-even and thus cannot be moved on top of c′, and if c is red-odd, all cards moved are red-odd and again c′ is unaffected.

Finally, the card c′ of an equal value and colour but different suit as c cannot be made a bad son card by this operation because if c′ has a son in both the current and goal state, then it must be the correct one: Otherwise the (newly corrected) son of c would have to be incorrect, too. Thus, after a number of iterations of this correction procedure that is limited by the total number of cards, there are no longer any bad son cards.

Once there are no more bad son cards, all that needs to be done to reach the goal state is rearranging the piles in a way that is inverse to the compacting procedure that formed the first step. The number of macro moves in this third part of the action sequence can consequently be bounded by the same polynomial as for the first part.

Putting the different parts together, we conclude that there is an overall polynomial bound on the number of actions in an optimal solution to a solvable FREECELL *task, and thus the domain admits short optimal plans.*   □

This result establishes that FREECELL planning is in NPO. We now show that this upper bound on complexity is also a lower bound.

Theorem 6.3.4. PLANEX-FREECELL *Is* NP-**Hard**
The plan existence problem for FREECELL *is* NP-*hard, even when restricted to tasks without free cells.*

Proof. We reduce from 3SAT. Let $\langle V, C \rangle$ *be a 3SAT instance, where* $V = \{v_1, \ldots, v_n\}$ *is a set of variables and* C *is a set of clauses over* V *containing exactly three literals each. We write* $l_{i,j}$ *for the j-th literal of the i-th clause. The corresponding* FREECELL *task is rather intricate and we encourage the reader to keep a look at Fig. 6.1 during the following discussion to get some intuition of how propositional formula and planning task are interrelated.*

We need some ordering for the literals over V, *so we call* v_i *the* $(2i-1)$-**th literal** *and* $\neg v_i$ *the* $2i$-**th literal** *and write* l_k *for the k-th literal. We define the* **number of occurrences** *of* l_k *as the number of pairs* $\langle i, j \rangle$ *such that* $l_{i,j} = l_k$, *and the corresponding pairs are called the* **first, second, ... occurrence** *of* l_k. *In which order the occurrences are numbered is of no importance. Additionally, we define* o_k, *the* **cumulated number of occurrences** *up to* l_k, *as the sum of the number of occurrences of* $l_{k'}$ *for all* $k' \leq k$.

Furthermore, we define the **selection value** val_S *as* $|C|+2n+2$, *the* **literal value** *of* l_k *as* $val_k = val_S + 2k + 2o_k$, *the* **clause value** val_C *as* $val_{2n} + 2$ *and the* **bottom value** val_B *as* $val_C + 6|C|$. *In the example of Figure 6.1,* $val_S = 11$, $val_1 = 15$ *(for literal* v_1*),* $val_2 = 21$ *(for* $\neg v_1$*),* $val_3 = 27$ *(for* v_2*),* $val_4 = 31$ *(for* $\neg v_2$*),* $val_5 = 37$ *(for* v_3*),* $val_6 = 41$ *(for* $\neg v_3$*),* $val_C = 43$ *and* $val_B = 61$.

The FREECELL *task has a suit length of* $val_B + 4|C| - 2$, *a tableau width of* $6|C| + 2|V| + 2$ *and a free cell count of 0. The initial tableau is arranged as follows.*

The piles of the initial tableau, depicted in Figure 6.1, fall into three groups. The first $2|V| + 1$ *piles are called the* **literal selection piles**, *depicted at the top of the figure. The next* $6|C|$ *piles are called the* **clause piles**, *organized into subgroups of six piles that each relate to a specific clause, called* **clause groups**. *The last pile, holding most of the cards, is called the* **big pile**, *at the bottom of the figure. Note that cards at the top of a pile are shown near the bottom of the picture, following the usual convention of* FREECELL *implementations on computers.*

Literal selection piles: *The first of these only contains the card* $\langle \spadesuit, val_S \rangle$. *The other piles contain three cards each. Each of them corresponds to a literal, the pile for* l_k *being defined as* $\langle \spadesuit, val_k \rangle \langle \spadesuit, val_S - k - 1 \rangle \langle \diamondsuit, val_S - k \rangle$ *if k is odd and as* $\langle \spadesuit, val_k \rangle \langle \clubsuit, val_S - k \rangle \langle \heartsuit, val_S - k + 1 \rangle$ *if k is even.*

Clause groups: *Each group is organized as follows. There are six piles corresponding to the i-th clause. We set the* bottom *value for the group as* $bottom = val_B + 4(i - 1)$ *and the* base *value for the group as* $base = val_C + 6(i - 1)$.

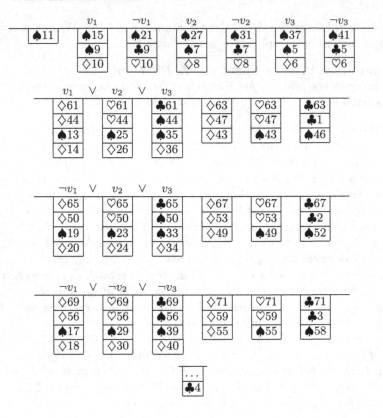

Fig. 6.1. FREECELL task corresponding to $(v_1 \lor v_2 \lor v_3) \land (\neg v_1 \lor v_2 \lor v_3) \land (\neg v_1 \lor \neg v_2 \lor \neg v_3)$

The first three piles contain four cards each. The first and second of these are of value bottom and base + 1, respectively; suit does not matter. The remaining cards depend on the literals in the clause: For $1 \leq j \leq 3$, the third and fourth cards of the j-th pile are $\langle \spadesuit, val_k - 2m \rangle$ and $\langle \diamondsuit, val_k - 2m + 1 \rangle$, where k and m are calculated such that $\langle i, j \rangle$ is the m-th occurrence of l_k.

The other three piles contain three cards each.

The fourth pile is defined as $\langle \diamondsuit, bottom + 2 \rangle \langle \diamondsuit, base + 4 \rangle \langle \diamondsuit, base \rangle$.

The fifth pile is defined as $\langle \heartsuit, bottom + 2 \rangle \langle \heartsuit, base + 4 \rangle \langle \spadesuit, base \rangle$.

The sixth pile is defined as $\langle \clubsuit, bottom + 2 \rangle \langle \clubsuit, i \rangle \langle \spadesuit, base + 3 \rangle$.

Big pile: The top card of this pile is $\langle \clubsuit, |C| + 1 \rangle$, and below this are all remaining cards, ordered such that cards of lower value are closer to the top.

We now show that this FREECELL task can be solved if and only if there is a satisfying assignment to the variables of the logical formula. First assume there is such a satisfying assignment $\alpha : V \rightarrow \{0, 1\}$. The following strategy solves the task:

For each $i \in \{1, \ldots, n\}$, move the top two cards from the literal selection piles that correspond to literals which are true under α to the first literal selection pile. This releases the bottom cards of some literal selection piles, spades cards which can then be used to move cards from the clause piles. In the example of Figure 6.1, for the assignment $\{v_1 \mapsto 1, v_2 \mapsto 1, v_3 \mapsto 0\}$ these are the 15, 27 and 41 of spades. These are called the literal choice cards.

The first three piles of each clause group relate to the literals in that clause. The top two cards of such a pile can be moved to the literal selection piles if and only if the literal choice card of the corresponding literal has been revealed.

Because we have a satisfying truth assignment, a literal is satisfied in each clause, and thus it is possible to remove the top two cards of one of the first three piles of each clause group. If the new top card is black, the top card of the fourth pile of the clause group can be moved on top of it; if it is red, the top card of the fifth pile can be moved. Thus a red card is revealed in the fourth or fifth pile, and the top card of the sixth pile can be moved on top of that.

After this has been done for all clauses, the first $|C|$ cards of clubs are available in the sixth piles of the clause groups and can be moved to foundations, allowing to move the top card of the big pile to foundations. This reveals many low-valued cards, and it is not hard to see that all cards of values up to val_S can be moved to foundations immediately. This reveals the literal choice cards of all literals that are false under the chosen assignment, allowing to move the top two cards of the clause group piles relating to unsatisfied literals to the literal selection piles as well.

After this has been done, all piles are ordered by value, with cards of lower value closer to the top, allowing to move all remaining cards to foundations, solving the task.

Now assume that the FREECELL task is solvable. It is not possible to move the bottom card of any tableau pile within the tableau before the top card of the big pile is moved, because all cards that they could be moved on top of are buried in the big pile. Before the top card of the big pile is moved, it is not possible to move the bottom card of any pile to foundations either. This implies that the first movement of the top card of the big pile cannot go to an empty tableau position.

On the other hand, it can not be moved on top of any other card as its first movement, because all possible destination cards are buried under it. Together, this implies that its first (and thus only) movement must be directly to foundations.

This in turn requires all lower-valued clubs cards to be moved to foundations first, requiring movements within the clause piles. For each clause group, the top card of the sixth pile must be moved, and it can only be moved to the second card of the fourth or fifth pile, requiring the top card of either of these piles to be moved. These in turn can only be moved on top of the second (counting from the bottom) card from any of the first three piles of that clause group. Thus, in each clause group, the top two cards of one of the first three piles must be moved somewhere else for the task to be solvable.

The only way this can be done is by uncovering the literal choice cards of corresponding literals in the way explained in the other direction of the proof. As it is not possible to uncover the literal choice card for v_i and $\neg v_i$ at the same time (for any i), this requires the existence of a satisfying assignments to the truth variables, completing the proof.   □

Putting the previous two theorems together, we get the following classification result for FREECELL.

Theorem 6.3.5. PLAN-FREECELL ∈ NPO \ PS
FREECELL *planning is in* NPO, *but not in* PS *unless* P = NP. *Hardness already holds when only considering tasks without free cells, or with an arbitrary fixed free cell count.*

Proof. Membership in NPO *and non-membership in* PS *for the case of no free cells follow from the previous two results.*

To show hardness for the case of $m > 0$ free cells, the previous proof only needs to be slightly modified to ensure that all free cells must become occupied right at the start of any plan and cannot be cleared before the top card of the big pile is moved to foundations. We refer to our earlier work on the FREECELL *domain for details [54].*   □

This result concludes our discussion of FREECELL. As our proofs show, the hardness of planning in this domain is not (or at least not exclusively) due to the difficulty in allocating free cells or empty tableau positions, but rather due to the choice of *which* card to move on top of a tableau pile when there are two possible options.

6.4 MOVIE

The objective in the MOVIE domain is to satisfy certain important prerequisites to watching a movie (such as having chips, having dip, having cheese and having the movie rewound) by actions such as getting chips, getting dip, getting cheese and rewinding the movie. Its purpose at IPC1 was to check if the performance of the competing planning systems degrades in the face of many objects in the task definitions – in a large MOVIE task, there are several dozen types of cheese, but only one of them needs to be obtained to satisfy the cheese goal. It was found that all planners could quickly solve all MOVIE tasks.

From a complexity point of view, this benchmark is not interesting. All tasks are solved by the same plan, and hence non-optimal, optimal sequential and optimal parallel planning are all constant time problems in this domain. There is little point in formally introducing the MOVIE domain, so we restrict our discussion to the following observation.

Theorem 6.4.1. PLAN-MOVIE $\in$ PO
Optimal plans in the MOVIE *domain can be generated in polynomial time.*

Proof. There is a constant bound to the number of states in a MOVIE *task, so that the planning problem can be optimally solved by explicit search in the state space.* $\square$

6.5 PIPESWORLD

The PIPESWORLD domain models the flow of oil-derivative liquids through *pipeline segments* connecting *areas*, and is inspired by applications in the oil industry [92]. Liquids are modelled as *batches* of a certain unit size. A segment must always contain a certain number of batches (i. e., it must always be full). Batches can be pushed into pipelines from either side, leading to the batch at the opposite end "falling" into the incident area. Batches have associated *product types*, and batches of certain types may never be adjacent to each other in a pipeline. Moreover, areas may have constraints on how many batches of a certain product type they can hold.

Definition 6.5.1. PIPESWORLD **Tasks**
$P = \{lco, gasoline, rat\text{-}a, oca1, oc1b\}$ *is the set of* PIPESWORLD **products**.
Two products $p, p' \in P$ *are called* **compatible** *unless* $p = rat\text{-}a$ *and* $p' \in \{oca1, oc1b\}$ *or vice versa.*

A PIPESWORLD **task** *is defined by a 10-tuple* $\langle A, B, \cdot^P, S, \cdot^-, \cdot^+, cap, C_0^S, C_0^A, C_\star^A \rangle$ *with the following components:*

- A *is a finite set of* **areas**.
- B *is a finite set of* **batches** *(of oil-derivative products).*
- $\cdot^P : B \to P$ *maps each batch* $b \in B$ *to its* **product type** *or* **type** $b^P \in P$. *We say that two batches* **can interface** *if their product types are compatible.*
- S *is a finite set of* **pipeline segments**.
- $\cdot^- : S \to A$ *maps each pipeline segment* $s \in S$ *to its* **start area** s^-.
- $\cdot^+ : S \to A$ *maps each pipeline segment* $s \in S$ *to its* **end area** s^+.
- $cap : A \times P \to \mathbb{N}_0$ *is the* **area capacity** *function.*
- *A* **segment contents function** *is a function* $C^S : S \to B^+$ *mapping each segment to a non-empty sequence of batches called the* **contents** *of that segment subject to the following restrictions:*
  - *No batch occurs in the contents of multiple segments.*
  - *No batch occurs in the contents of a segment twice.*
  - *All batches which are adjacent in the contents of some segment can interface.*
  C_0^S *is a segment contents function called the* **initial segment contents**.
- *An* **area contents function** *is a function* $C^A : A \to 2^B$ *mapping each area to a set of batches called the* **initial contents** *of that area subject to the following restrictions:*

- *The contents of different areas are disjoint.*
- *For all areas $a \in A$ and all products $p \in P$, the contents of area a include at most $cap(a, p)$ batches of type p.*
  $C_0^A : A \rightarrow 2^B$ *is an area contents function called the **initial area contents**.*
  We require that each batch either occurs in the contents of some segment in the initial segment contents, or in the contents of some area in the initial area contents, but not both.
- $C_\star^A : A \rightarrow 2^B$ *is an area contents function called the **goal area contents**.*

Note that it is possible to have multiple pipeline segments between the same pair of areas, which is why we do not model areas and pipeline segments as directed graphs. Although the terminology ("start area" vs. "end area") suggests an asymmetry, batches can be pumped into pipeline segments from either direction, so the domain is in essence undirected – we just need some formal way of distinguishing the one end of a segment from the other. We refer to the number of batches in a segment as the *length* of that segment. One important property of the PIPESWORLD domain is that segments retain their length throughout plan execution: Whenever a batch is pumped into a pipeline segment from one end, the batch at the opposite end of the segment is pushed out into the adjacent area.

Definition 6.5.2. PIPESWORLD *Domain*
The PIPESWORLD *domain maps* PIPESWORLD *tasks with areas A and segments S to state spaces as follows:*

STATES: *Pairs $\langle C^S, C^A \rangle$, where C^S is a segment contents function and C^A is an area contents function. We say that a segment is **in** a segment or area iff it is in the contents of that segment or area.*

INITIAL STATE: $\langle C_0^S, C_0^A \rangle$, *where C_0^S is the initial segment contents and C_0^A is the initial area contents of the task.*

GOAL STATES: *Any state $\langle C^S, C^A \rangle$ with $C_\star^A(a) \subseteq C^A(a)$ for all areas $a \in A$, where $C_\star^A$ is the goal contents function of the task.*

OPERATORS: *In state $\langle C^S, C^A \rangle$, if $s \in S$ is a pipeline segment with contents $b_1 \ldots b_n$ and $b \in C_A(s^-)$ is a batch that can interface with b_1, then b can be **pushed** into s. This results in a state where the new contents of segment s are $bb_1 \ldots b_{n-1}$, b is no longer in $C_A(s^-)$, and b_n is in $C_A(s^+)$.*
*Similarly, $b \in C_A(s^+)$ can be **pushed** into s if it can interface with b_n, leading to a state where the contents of s are $b_2 \ldots b_n b$, b is no longer in $C_A(s^+)$, and b_1 is in $C_A(s^-)$.*
Pushing a batch into a pipeline segment is only allowed if it results in a legal state, satisfying the constraints imposed by the area capacity function.

The PIPESWORLD domain is sometimes called PIPESWORLD-TANKAGE, to contrast it with the following PIPESWORLD-NOTANKAGE domain.

Definition 6.5.3. PIPESWORLD-NOTANKAGE *Domain*
The PIPESWORLD-NOTANKAGE *domain is the restriction of* PIPESWORLD *to those tasks where the area capacity for each area and product type is equal to the total number of batches of that product type.*

Note that with our definition of PIPESWORLD (as in the IPC4 benchmarks), the set of products and their compatibility relation is fixed. We will see that PIPESWORLD planning is already hard for this fixed compatibility relation.

Our definition of PIPESWORLD faithfully captures the PDDL specification except for one modelling flaw of the latter: In some situations, the PDDL definition allows pushing batches through a pipe even though this violates the area capacity constraints on the receiving end, making some unsolvable tasks solvable. This minor difference does not affect the applicability of our results because these already hold for the PIPESWORLD-NOTANKAGE domain, where area capacities can be ignored.

Theorem 6.5.4. PLANEX-PIPESWORLD *Is* NP-*Hard*
The plan existence problems for PIPESWORLD-TANKAGE *and* PIPESWORLD-NOTANKAGE *are* NP-*hard.*

Proof. It is sufficient to prove the result for PIPESWORLD-NOTANKAGE, *which is a special case of the other domain. We present a Karp reduction from the satisfiability problem for propositional formulae in conjunctive normal form where clauses contain at most four literals and each variable occurs in at most three clauses. This problem is known to be* NP-*hard [43, LO1]. (We could limit clauses to three literals, but then Fig. 6.3 would look less symmetric.)*

Let χ be the given formula, and let V and C be its variable and clause sets. Throughout the proof, we refer to batches of type rat-a *as white batches, batches of type* ocal *as black batches, and batches of type* gasoline *as gray batches. Observe that white batches may not interface with black batches, while gray batches may interface with anything.*

The generated PIPESWORLD-NOTANKAGE *instance is assembled from components shown in Figs. 6.2 and 6.3, where edges with differently decorated endpoints distinguish different kinds of pipeline segments. The key to these decorations is shown in Fig. 6.4. The first five kinds of segments all have length $6|V| + 1$, while the sixth has length 3. The first kind of segment is filled with $3|V|$ black batches, then a gray one to interface between black and white, and then $3|V|$ white batches. The second and third kind are completely filled with one product type, and the fourth and fifth are like the second and third except that the first batch is gray. The sixth kind is like the fifth, but only contains three batches.*

The pipe network contains one copy of the variable gadget *structure shown in Fig. 6.2 for each variable $v \in V$, and one copy of the* clause gadget *structure* *shown in Fig. 6.3 for each clause $c \in C$. The open ends to the right of the variable gadgets (dotted) are connected to the open ends to the left of the clause gadgets. In particular, if clause c contains the positive literal v, then area v' of the variable gadget is connected to any of the dangling pipeline segments of the clause gadget for c. Similarly, for negative literals in c, area $\neg v'$ is connected to a dangling pipeline segment of the clause gadget. Because every clause contains at most four literals, there are sufficiently many pipes to make these connections. Because every variable occurs in at most three clauses, at most three new pipeline segments are connected to either v' or $\neg v'$. Any pipeline segments left dangling (for clauses of size three or less) are removed.*

The areas in the variable gadgets labelled 3 are the only areas that are not initially empty, each of them containing three black batches. In each clause gadget, the goal requirement is to move the last (rightmost) black batch in the pipe connecting areas c and c_* into area c_*. We call these pipeline segments goal pipes.

This completes the description of the mapping. Clearly, the PIPESWORLD task can be generated in polynomial time. We will now show that it has a solution iff χ is satisfiable.

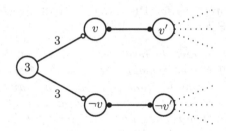

Fig. 6.2. Variable gadget

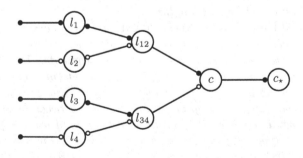

Fig. 6.3. Clause gadget

Fig. 6.4. Key to Figs. 6.2 and 6.3

First assume that χ is satisfiable, and that α is a satisfying assignment to V. For each variable $v \in V$, we push the three black batches in area 3 of the corresponding variable gadget into the pipe leading to the area denoted by a literal l satisfied by α (i. e., to area v if $\alpha(v) = 1$ and to area $\neg v$ otherwise). This pushes three batches into this area, one of which is gray. We push the gray batch, then the two white batches into the pipe leading to area l', making three black batches available there. We then push one batch into each of the pipes connecting l' to clause gadgets.

Due to the way variable gadgets are connected to clause gadgets and because α satisfies χ, this places at least one batch in one of the areas l_1, l_2, l_3 or l_4 of each clause gadget. In each clause gadget, we choose one such batch and push it into the pipe leading to l_{12} or l_{34}, placing a batch in one of these areas. This batch is then pushed into the pipe leading to c, releasing a batch there which is pushed into the goal pipe, satisfying the goal for this clause. We thus satisfy the goal in each clause gadget, which shows that the task is solvable.

Now assume that the task has a solution. Obviously, this requires that more batches are pushed into each clause gadget than pushed out of them. It is never possible to push any batch out of a clause gadget unless this batch has been previously pushed into the clause gadget through the same pipe. This is because batches moved into l_{12} from l_1 have the wrong colour to be pushed into l_2 and vice versa, and similarly there cannot be a flow between l_3 and l_4 via l_{34} or between l_{12} and l_{34} via c. (Note that there are not sufficiently many batches on the left side of the network to push a gray batch out of the pipes leading to l_1, l_3 or c.)

We can thus treat the segments connecting variable gadgets to clause gadgets as "one-way pipes", which simplifies the analysis because we can consider each variable gadget in isolation. The important property for variable gadgets is that we can either have batches in area v' or in area $\neg v'$, but never both. To see this, note that to push even a single batch into area v', we must push all three batches from area 3 into area v; otherwise we obtain only white batches in area v, which cannot be pushed into the pipe connecting to v'. Moreover, to push a batch into v', we must make use of the gray batch from the pipe be-

tween 3 *and* v' *and without pushing that gray batch back into* v *(which requires emptying area* v' *), we cannot push anything back into area* 3.

Therefore, if there is a solution, then there is one where for each variable only one of the areas v' and $\neg v'$ ever contains a batch. Define the truth assignment α so that $\alpha(v) = 1$ if area v' ever contains a batch, and $\alpha(v) = 0$ otherwise. Then a batch can only be pushed into the clause gadget area l_i if α satisfies l_i. Because at least one batch must be pushed into each clause gadget, α satisfies at least one literal in every clause, and hence χ is satisfiable. This concludes the proof.   □

Due to the NP-hardness of plan existence for PIPESWORLD, the planning problem for this domain cannot belong to PS unless P = NP. Observe that our hardness proof applies just as well to a variant of PIPESWORLD where batches of the same product type are indistinguishable, and hence the goal is expressed in term of *product types*, not batches – a more realistic model of the underlying application problem.

Unfortunately, we cannot provide an NPO membership result: the question whether or not PIPESWORLD admits short optimal plans is open.

We remark that the hardness of PIPESWORLD planning is to a large part due to the interface constraints for incompatible products, and indeed PIPESWORLD-NOTANKAGE without interface constraints (all product types are compatible) is known to admit a polynomial planning algorithm [98]. However, we can prove that bounded plan existence is still NP-hard for this PIPESWORLD variant and that it does not admit polynomial-time approximation schemes unless P = NP, placing it somewhere between APX and poly-APX. Because this domain variant is quite different from the competition domains, we abstain from presenting these detailed results.

6.6 PROMELA

PROMELA (*Process* or *Protocol Meta Language*) is the input language used by the SPIN model checker [70]. The PROMELA planning domain [33] encodes a subset of PROMELA in PDDL2.2, allowing the application of planning technology to a certain class of model-checking problems. We first introduce and discuss the general PROMELA planning domain, then the restricted subclasses PROMELA-PHILOSOPHERS and PROMELA-OPTICALTELEGRAPH, which were part of the IPC4 benchmark set.

A PROMELA task defines a distributed system consisting of a set of *processes*, modelling individual components of a distributed system, and *queues*, used for communication between processes. The goal is always to find a *deadlock* state, in which no process is able to continue its operation.

Definition 6.6.1. PROMELA *Tasks*
A PROMELA *task* is defined by an 8-tuple $\langle P, Q, \Sigma, cap, (S^p)_{p \in P}, (R^p)_{p \in P}, (W^p)_{p \in P}, (s_0^p)_{p \in P} \rangle$ *with the following components:*

- P is a finite set of **processes**.
- Q is a finite set of **queues**.
- Σ is a finite set of **messages**.
- $cap : Q \to \mathbb{N}_1$ is the **queue capacity** function.
- For each process $p \in P$:
  - S^p is a finite set of **local states**.
  - $R^p \subseteq S^p \times Q \times \Sigma \times S^p$ is a set of **read transitions**.
  - $W^p \subseteq S^p \times Q \times \Sigma \times S^p$ is a set of **write transitions**.
  - $s_0^p \in S^p$ is the **initial local state**.

We now define the PROMELA planning domain.

Definition 6.6.2. PROMELA *Domain*
The PROMELA **domain** maps PROMELA *tasks with processes P with local states $(S_p)_{p \in P}$, queues Q and messages Σ to state spaces as follows:*

STATES: *Pairs $\langle (s^p)_{p \in P}, C \rangle$, where $(s^p)_{p \in P}$ is the **local state** vector and $C : Q \to \Sigma^*$ is the **queue contents** function. For each process $p \in P$, s^p must be a local state of p called the **current local state** of the process.*

INITIAL STATE: *$\langle s_0^p, C_\epsilon \rangle$, where C_ϵ is the queue contents function which maps each queue to the empty message sequence.*

GOAL STATES: *Any state with no applicable action.*

OPERATORS: *A process $p \in P$ with read transitions R^p can **perform a read transition** $t = \langle s, q, a, s' \rangle \in R^p$ in state $\langle (s^p)_{p \in P}, C \rangle$ iff $s^p = s$ and the first element of $C(q)$ is the message a. This changes the current local state of process p to s' and removes the first element from $C(q)$.*

*A process $p \in P$ with write transitions W^p can **perform a write transition** $t = \langle s, q, a, s' \rangle \in W^p$ in state $\langle (s^p)_{p \in P}, C \rangle$ iff $s^p = s$ and $|C(q)|$ is less than the capacity of queue q. This changes the current local state of process p to s' and appends message a to $C(q)$.*

Our definition of the PROMELA domain differs from the PDDL definition in some minor ways that do not limit the applicability of our results. These are discussed towards the end of the section.

Processes can be naturally described by labelled directed graphs, where vertices correspond to process states and arcs to transitions. For a transition $t = \langle s, q, a, s' \rangle$, the graph contains an arc from s to s' with the label $q : a?$ if t is a reading transition and $q : a!$ if t is a writing transition. Figure 6.5 shows an example process from PROMELA-PHILOSOPHERS. The process corresponds to a single philosopher in the well-known *dining philosophers* problem, the queues L and R to the forks to his left and right. The intuition behind the model is that writing a message corresponds to putting a fork on the table, and reading a message corresponds to picking it up. Initial process state 1 is a set-up state in which each philosopher puts one fork on the table. After leaving this state, philosophers follow a deterministic strategy of repeatedly requesting the two forks they require in a certain order, then putting them down again.

Communicating processes are a very expressive formalism for modelling computations. This makes planning for general PROMELA tasks hard.

Theorem 6.6.3. PROMELA *Planning Is* PSPACE-*Complete.*
Plan existence and bounded plan existence for PROMELA *tasks are* PSPACE-*complete. This is true even if all queues have capacity 1 and the tasks are deterministic, i. e., at most one action is applicable in any reachable state.*

Proof. We provide a reduction that maps space-restricted Turing Machines to PROMELA *tasks such that the task has a solution iff the Turing Machine halts (starting from a blank tape).*

Let M be a Turing Machine with state set Z, including initial state $z_0 \in Z$ and accepting state $z_\star \in Z$, tape alphabet Γ, including blank symbol $\square \in \Gamma$, and transition function $\delta : (Z \setminus \{z_\star\}) \times \Gamma \to Z \times \Gamma \times \{-1, +1\}$. We assume that the machine has n tape cells, starts at the left-most one, and that attempts to move past the end of the tape in either direction is an error that terminates computation (just like reaching the accepting state).

The corresponding PROMELA *task has one* tape cell process p_i *and one* queue q_i *for each tape cell $i \in \{1, \ldots, n\}$. The set of messages is the set of Turing Machine states Z. All queues have capacity 1, all tape cell processes have state set $\Gamma \cup (\Gamma \times Z)$, and the initial state of each process is $\square$ except for process p_1 with initial state $\langle \square, z_0 \rangle$.*

For each Turing Machine transition $\delta(z, a) = \langle z', a', \Delta \rangle$ and each tape position $i \in \{1, \ldots, n\}$ where $i + \Delta \in \{1, \ldots, n\}$, tape cell process p_i has the following transitions:

- *A transition from a to $\langle a, z \rangle$ which reads z from q_i.*
- *A transition from $\langle a, z \rangle$ to a' which writes z' to $q_{i+\Delta}$.*

There is a straightforward correspondence between configurations of the Turing Machine and states of the corresponding PROMELA *task. If after k computation steps, the Turing Machine reaches state z with current tape position i and tape contents $a_1 \ldots a_n$, then after applying $2k$ actions in the* PROMELA *task, process p_i must be in state $\langle a_i, z \rangle$, each process $p_j \neq p_i$ must be in state a_j, and all queues must be empty.*

We prove this inductively. Clearly, the statement is true for $k = 0$. Assume that it is true for k. We can assume that $z \neq z_\star$ and $\delta(z, a_i) = \langle z', a', \Delta \rangle$ with $i + \Delta \in \{1, \ldots, n\}$ (otherwise the Turing Machine computation stops and there is nothing to prove). In this situation, the only process that can perform a local transition is p_i, since all other processes are in states that require reading from a queue, and all queues are empty by the induction hypothesis. Thus, a transition of p_i must be performed, and there is only one applicable transition for p_i in local state $\langle a_i, z \rangle$, which writes z' to queue $q_{i+\Delta}$ and changes the local *state of p_i to a'. After applying this action, all processes are in a local state where all outgoing transitions are read transitions, but only process $p_{i+\Delta}$ can read from a non-empty queue, so this process acts next. The only applicable transition is the one that reads message z' and changes the local state of $p_{i+\Delta}$ from $a_{i+\Delta}$ to $\langle a_{i+\Delta}, z' \rangle$. After applying this second action, all queues are empty again, and the local process states again correspond to the Turing Machine configuration as required. This concludes the inductive proof.*

The correspondence between Turing Machine configurations and trajectories in the state space of the PROMELA task implies that if the Turing Machine does not halt, then we cannot reach a deadlock in the PROMELA task. On the other hand, if the Turing Machine halts, it either does so by attempting to go past the tape boundaries or by reaching state $z_\star$. In both cases, the PROMELA task reaches a deadlock, because no local transitions are possible in the state corresponding to the last Turing Machine configuration after reaching $z_\star$ (or before going past the tape boundaries).

Thus, plan existence for PROMELA tasks is PSPACE-hard, which implies that bounded plan existence is also PSPACE-hard. Moreover, both problems must belong to PSPACE because PDDL planning in any fixed propositional domain is in PSPACE. This concludes the proof. □

This immediately gives us the following classification for the PROMELA planning problem.

Theorem 6.6.4. PLAN-PROMELA $\in$ EXPO $\setminus$ NPS
The planning problem in the PROMELA domain can be optimally solved in exponential time, and it cannot be solved (not even sub-optimally) in sub-exponential time.

Proof. The problem is solvable in exponential time with explicit graph search in the state space of the given task. Sub-exponential planning algorithm cannot exist because there are tasks with exponential lower bounds on shortest plan length. (This immediately follows from the previous proof, as there are space-restricted Turing Machines requiring an exponential number of computation steps.) □

Having established the result for the general PROMELA domain, we now turn to the PROMELA-PHILOSOPHERS and PROMELA-OPTICALTELEGRAPH domains. These domains are special cases of PROMELA where each task is characterized by a single number. In the former domain, this number defines the number of philosophers in a dining-philosophers style problem. In the latter, it defines the number of optical telegraphs in a communication protocol.

Definition 6.6.5. PROMELA-PHILOSOPHERS *Domain*
A PROMELA-PHILOSOPHERS task is given by a natural number $n \geq 2$ and denotes a PROMELA task with message set $\{fork\}$, processes p_i and queues q_i (of capacity 1) for all $i \in \{1, \ldots, n\}$. Throughout this section, process and

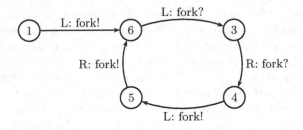

Fig. 6.5. PROMELA-PHILOSOPHERS transition graph. State numbers follow the PDDL specification

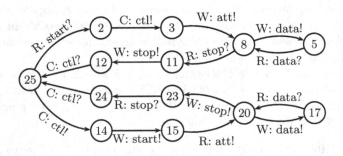

Fig. 6.6. PROMELA-OPTICALTELEGRAPH transition graph. State numbers follow the PDDL specification

queue indices of PROMELA-PHILOSOPHERS *tasks are considered modulo n. States and transitions of process p_i are given by the directed graph in Fig. 6.5, where the initial process state is state 1, L denotes the queue q_i, and R denotes the queue q_{i+1}.*

The PROMELA-PHILOSOPHERS *domain is the restriction of the* PROMELA *domain to* PROMELA-PHILOSOPHERS *tasks.*

Definition 6.6.6. PROMELA-OPTICALTELEGRAPH *Domain*

A PROMELA-OPTICALTELEGRAPH *task is given by a natural number $n \geq 2$ and denotes a* PROMELA *task with message set* $\{att, ctl, data, start, stop\}$*, processes p_i^d and p_i^u and queues q_i^c, q_i^d and q_i^u (of capacity 1) for all $i \in \{1, \ldots, n\}$. Throughout this section, process and queue indices of* PROMELA-OPTICALTELEGRAPH *tasks are considered modulo n. States and transitions of the processes are given by the directed graph in Fig. 6.6, where the initial process state is state 25. For process p_i^d, C denotes the queue q_i^c, R denotes the queue q_i^d and W denotes the queue q_i^u. For process p_i^u, C denotes the queue q_i^c, R denotes the queue q_{i+1}^u and W denotes the queue q_{i+1}^d.*

The PROMELA-OPTICALTELEGRAPH *domain is the restriction of the* PROMELA *domain to* PROMELA-OPTICALTELEGRAPH *tasks.*

Following their PDDL specification, we assume that the encoding size of PROMELA-PHILOSOPHERS and PROMELA-OPTICALTELEGRAPH tasks grows linearly in the number of processes. (Of course, more concise descriptions which only encode the scaling parameter n as a binary number are possible.)

Because of their simple scaling structure, these benchmarks are much easier to solve than general PROMELA tasks.

Theorem 6.6.7. PLAN-PROMELA-PHILOSOPHERS $\in$ PO
In the PROMELA-PHILOSOPHERS *domain, optimal plans can be generated in polynomial time.*

Proof. To reach a goal state, perform the transitions from 1 to 6 to 3 in all processes. When all processes are in state 3, they are all blocked, so this is a solution of length $2n$, if n is the number of philosophers.

We now prove optimality. Because there is only one message type and queues have size 1, queues only have two configurations, full or empty. We can verify the following invariant: Queue q_i is full iff p_i is in state 5 or 6 and p_{i-1} is in state 1, 3 or 6. Therefore p_i cannot be deadlocked in state 1 or 4 (q_i is not full if p_i is in state 1 or 4) or in state 5 (q_{i+1} is not full if p_i is in state 5). Therefore, processes can only be blocked in states 6 or 3. However, if all processes are in state 3 or 6 and p_i is in state 6, then q_i is not full and hence p_i is not blocked. Therefore, for all processes to be blocked, all of them must be in state 3. The generated plan clearly is the shortest sequence of actions achieving this. □

Theorem 6.6.8. PLAN-PROMELA-OPTICALTELEGRAPH $\in$ PO
In the PROMELA-OPTICALTELEGRAPH *domain, optimal plans can be generated in polynomial time.*

Proof. To reach a goal state, first perform the transitions from 25 to 14 to 15 in all processes p_i^{d}, then the transition from 25 to 2 in all processes p_i^{u}. Clearly, this leads to a deadlock. Optimality can be proved by similar arguments as for PROMELA-PHILOSOPHERS. □

All the results in this section easily generalize to *parallel planning*. Clearly, the general PSPACE-completeness result also applies to that setting, because PSPACE-completeness of plan existence implies PSPACE-completeness of bounded parallel plan existence for propositional PDDL domains. In the restricted domains, parallelism allows taking the transitions of each process simultaneously, so that the optimal parallel plan length for any PROMELA-PHILOSOPHERS task is 2, whereas the optimal parallel plan length for any PROMELA-OPTICALTELEGRAPH task is 3. In the latter case, observe that the process p_i^{u} can only transition to state 2 *after* p_{i-1}^{d} has transitioned to state 15.

Finally, some comments on the differences between our formalization and the actual PDDL domain. First, the PDDL definition seems to have a minor flaw in the formalization of writing to queues of capacity 2 or greater. The hardness proof does not require such queues and they do not occur in the competition domains, so this does not make a difference. Second, because of another flaw in the PDDL definition, processes can only be recognized as blocked in states with at most one outgoing transition; reaching a deadlock in which some process has two outgoing transitions in its current local state is not considered a solution, even if all those transitions are blocked. This does not affect our proofs for the competition domains, but it does mean that the PSPACE-hardness proof is not immediately applicable to the PDDL specification. However, it can be easily adjusted to work around the flaw. Finally, due to the difficulty of expressing the queue updates and dead-lock condition succinctly in PDDL, a single action in our model corresponds to a sequence of four actions in the PDDL model (assuming the PDDL formulation using ADL constructs and derived predicates – there are several alternative formulations available), and another action is needed at the end of the plan for each blocked process with an outgoing transition in the current state. Counting the number of PDDL actions, the $2n$ plan length for PROMELA-PHILOSOPHERS thus becomes $9n$ ($2n$ transitions, n processes), and the $3n$ plan length for PROMELA-OPTICALTELEGRAPH becomes $14n$ ($3n$ transitions, $2n$ processes). The optimal parallel plan lengths in the PDDL domains, following the PDDL definition of concurrency, becomes 9 for PROMELA-PHILOSOPHERS and 11 for PROMELA-OPTICALTELEGRAPH (it is not 14 since some of the "subactions" can be interleaved).

6.7 PSR

The PSR (*power supply restoration*) domain was originally introduced for planning under uncertainty [110]. The deterministic and fully observable variant described here is part of the IPC4 benchmark suite, and constitutes an extremely simplified version of the original PSR domain. The domain models a situation where parts of a power network, consisting of power sources (*circuit breakers*), switches and power lines, have turned faulty. Circuit breakers or switches can be open or closed, with open devices blocking the electrical current. The objective of a PSR task is to reconfigure the network by opening and closing devices so that as many lines as possible are fed, while avoiding to feed any faulty lines (which immediately opens all power sources feeding them).

Definition 6.7.1. PSR *Tasks*
A PSR **task** *is defined by a 6-tuple* $\langle D, C, L, F, G, O_0 \rangle$ *with the following components:*

– D *is a finite set of* **devices**.
– $C \subseteq D$ *is the set of* **circuit breakers**. *Devices which are not circuit breakers are called* **switches**.
– L *is a finite set of (power)* **lines**.

– $F \subseteq L$ is the set of **faulty lines**.
– G is a connected graph with vertex set $D \cup L$ called the **power network**. It is bipartite, with all edges connecting some device to some line. Circuit breakers must have degree 1, switches degree 1 or 2, and lines may have arbitrary degree.
– $O_0 \subseteq D \setminus C$ is the set of **initially open devices**.

Devices and lines are required to be disjoint.

The **feeder tree** of a circuit breaker c is the induced subgraph of the power network which contains c and all lines and devices reachable from c without passing through an initially open device. We require that, apart from the initially open devices, each device and each line must belong to exactly one feeder tree.

A line l is called **feedable** iff there exists a path in the power network leading from a circuit breaker to that line which does not pass through any faulty lines (including l itself).

Differently to the original PDDL definition, we do not explicitly model the *earth* device, which is always open, but rather allow to have switches with a degree of 1, which leads to the same semantics.

We now define the PSR planning domain.

Definition 6.7.2. PSR *Domain*
The PSR **domain** maps PSR tasks with devices D, circuit breakers C and lines L to state spaces as follows:

STATES: Sets $O \subseteq D$. We say that a device $d \in D$ is **open** iff $d \in O$, and that it is **closed** otherwise. We say that a line $l \in L$ is **fed** by a circuit breaker $c \in C$ iff there exists a path in the power network from c to l which does not pass through any open device (including c itself). We say that a circuit breaker is **affected** iff a faulty line is fed by it. A state is **unsafe** iff there is an affected circuit breaker, and **safe** otherwise.

INITIAL STATE: O_0, the set of initially open devices.

GOAL STATES: Safe states in which all feedable lines are fed.

OPERATORS: A device d can be **opened** iff it is closed and the state is safe. In the resulting state, d is open.
A device d can be **closed** iff it is open and the state is safe. In the resulting state, d is closed.
It is possible to **wait** iff the state is unsafe. In the resulting state, all affected circuit breakers are open.

Note that there is no requirement that a line be fed by only one circuit breaker. However, this is true for the initial state of the PDDL benchmarks due to the constraints on feeder trees, and it is also guaranteed to hold throughout plan execution by the planning algorithm we will now present. Therefore, adding a requirement that no line be fed by multiple circuit breakers would not affect our complexity results.

Somewhat surprisingly, optimal plans in PSR can be generated in polynomial time.

Theorem 6.7.3. PLAN-PSR $\in$ PO
In the PSR domain, optimal plans can be generated in polynomial time.

Proof. Solving PSR tasks requires maintaining safety and feeding all feedable lines. The safety property is monotonously increasing in the set of open devices, i. e., if $O \subseteq O'$ and state O is safe, then state O' must also be safe. (Recall that we identify states with the corresponding set of open devices.) The feeding property is monotonously decreasing in the set of open devices, i. e., if $O \subseteq O'$ and a certain line is fed in O', then it is also fed in O. Thus, the two aspects of solving a PSR task conflict in a certain way. However, as we shall see, it is possible to deal with them separately by ensuring safety first, then feeding all feedable lines.

We say that a circuit breaker is dangerous iff it is adjacent to a faulty line, and a switch is dangerous iff it is adjacent to a faulty line and to a feedable line.

If the initial state is a goal state, we return the empty plan. Otherwise, since all lines are fed initially by the circuit breaker in their feeder tree, the initial state must be unsafe, and the first action in any plan must be a wait *action, which opens all dangerous circuit breakers. We then use* open *actions to open all dangerous switches. Like the initial* wait *action, these actions must occur in any solution (although not necessarily at this point), because switches can only be opened by* open *actions (rather than by waiting, as for circuit breakers), and dangerous switches must be open in a goal state: Assume dangerous switch d were closed in a goal state. By definition, it is adjacent to a feedable line l and faulty line l'. In a goal state, l must be fed by some circuit breaker c, so l' is also fed by c, and hence c is affected and the goal state unsafe, a contradiction. It is also evident that dangerous circuit breakers must be open in a goal state to ensure safety.*

Interestingly, having all dangerous devices open is not just necessary for safety of a goal state, it is also sufficient for safety of any state. Assume that this were not the case and there were an unsafe state where all dangerous devices are open. By definition of safety, in this state there must be a path $\pi = d_1 l_1 \ldots d_n l_n$ from circuit breaker d_1 to faulty line l_n where all devices d_i are closed. We can assume that l_n is the only faulty line on the path (if l_i for $i < n$ is faulty, we consider $\pi' = d_1 l_1 \ldots d_i l_i$ instead). If $n = 1$, then the circuit breaker d_1 is dangerous and therefore not closed, a contradiction. If $n > 1$, then line l_{n-1} is feedable by the path $d_1 l_1 \ldots d_{n-1} l_{n-1}$, and hence d_n connects a feedable line to a faulty line and is dangerous and therefore not closed, a contradiction.

Therefore, we can solve the task as follows:

- *Wait, then open all dangerous switches.*
- *Compute a set of non-dangerous devices D_C of minimal cardinality such that closing D_C leads to all lines being fed. Close these devices.*

We already saw that all actions in the first step must occur in any solution. Moreover, due to the monotonicity of feeding and due to the fact that closing a device requires a single action per device (unlike opening, which can in some cases be done more efficiently with the wait *action), the generated plan is clearly optimal provided that safety is not violated by any of the closing actions. However this is ensured by the fact that having all dangerous devices open is sufficient for safety.*

Thus, we only need to show how to calculate the set D_C in polynomial time. For this purpose, we apply some transformation to the power network. First, we remove all dangerous devices along with all lines and devices that become disconnected from the circuit breakers by this operation. Clearly, since all dangerous devices are open and we are not going to close them, this is a valid operation. This results in a graph where all lines are feedable and no devices are dangerous, so we can ignore wait or open actions in the following. Second, we introduce a new (closed) main circuit breaker, *connect it to a new* main line, *and connect that line to all original circuit breakers in the network, which change status to switches (note that their degree is now 2 due to the edge from the main line). Again, this does not change the semantics of the* fed *predicate. It* does *change the semantics of affectedness for our network, but this is not a problem because our network contains no faulty lines. Third, we remove all switches with degree 1 (they are no use for solving the task) and replace all other switches with coloured (i. e., labelled) edges connecting their two neighbouring lines, using red edges for open switches and green edges for closed switches. Closing a switch thus corresponds to changing the colour of an edge to green. A line is fed iff it is reached by a path from the main circuit breaker that does not pass through any red lines, and hence all lines are fed iff the subgraph obtained by removing all red lines is connected. To achieve this with a minimal number of close actions, we can compute a spanning tree with a minimal number of red edges, or equivalently a minimal spanning tree in the weighted graph obtained by assigning weight 1 to all red edges and weight 0 to all other edges. Computing a minimal spanning tree is a polynomial time operation.* □

Some comments are appropriate at this point. First, if we use Prim's algorithm [25] for computing minimum spanning trees, it is easy to verify that the complete PSR planning algorithm amounts to the following quite simple greedy strategy:

1. If the initial state is a solution state, return the empty plan; otherwise continue.
2. Wait.
3. Open all dangerous switches.

4. Until a goal is reached, close some non-dangerous device such that closing this device leads to at least one additional line being fed.

Second, the proof critically relies on the fact that switches are connected to at most two lines, and circuit breakers only to one line. Eliminating the degree restriction for devices indeed leads to a more difficult domain variant, for which bounded plan existence is NP-complete. However, we do not prove this result here.

Finally, the problem remains easy in a parallel planning framework. In fact, according to the PDDL definition of the domain, no two PSR actions are concurrently executable, due to the conservative definition of mutexes in the presence of derived predicates [66]. Under a less strict notion of concurrency, it makes sense to allow opening several devices in parallel and closing several devices in parallel if that does not lead to any circuit breakers being affected. Using this notion, it is obvious that the optimal parallel solution length for any PSR task is 3, where the first step consists of a wait action, the second of a number of open actions, and the third of a number of close actions. This concludes our discussion of PSR.

6.8 SATELLITE

The objective of the SATELLITE domain is to plan and schedule the operation of a number of satellites gathering information about different observation targets such as stars, planets and space phenomena. Satellites are equipped with different kinds of instruments with varying capabilities.

Definition 6.8.1. SATELLITE *Tasks*
A SATELLITE *task is defined by a 10-tuple* $\langle S, I, M, D, l, C, cal, p_0, p_\star, O_\star \rangle$ *with the following components:*

- S *is a finite set of* **satellites.**
- I *is a finite set of* **instruments.**
- M *is a finite set of* **image modes** *or* **modes.**
- D *is a finite set of* **pointing directions** *or* **directions.**
- $l : I \to S$ *maps each instrument to the satellite it is* **located** *on.*
- $C \subseteq I \times M$ *is the* **instrument capabilities** *relation. If* $\langle i, m \rangle \in C$ *for an instrument* $i \in I$ *and mode* $m \in M$, *we say that* i **supports** m.
- $cal : I \to D$ *is the* **calibration target** *function.*
- $p_0 : S \to D$ *is the* **initial pointing direction** *function.*
- $p_\star : S' \to D$ *with* $S' \subseteq S$ *is the* **goal pointing direction** *function. A pair* $\langle s, d \rangle$ *with* $p_\star(s) = d$ *is called a* **pointing goal.**
- $O_\star \subseteq D \times M$ *is the set of* **image objectives.**

SATELLITE tasks resemble ROVERS tasks (Sect. 5.7) in many ways, but they are much less complicated. In particular, every satellite can immediately turn to every direction in a single action (unlike the ROVERS domain, where rovers travel between waypoints on complex maps, which differ from rover to rover), and there is no need to communicate the gathered science data.

Definition 6.8.2. SATELLITE *Domain*
The SATELLITE *domain maps* SATELLITE *tasks with satellites S, instruments I, modes M and pointing directions D to state spaces as follows:*

STATES: *4-tuples* $\langle p, I^P, I^C, O \rangle$, *where* $p : S \to D$ *is the **pointing direction** function,* $I^P \subseteq I$ *is the set of **powered-on instruments**,* $I^C \subseteq I$ *is the set of **calibrated instruments**, and* $O \subseteq D \times M$ *is the set of **remaining objectives**. We say that satellite* $s \in S$ ***points at*** *direction* $d \in D$ *iff* $p(s) = d$.

INITIAL STATE: $\langle p_0, \emptyset, \emptyset, O_\star \rangle$, *where* p_0 *is the initial pointing direction function and* $O_\star$ *is the set of image objectives of the task.*

GOAL STATES: *Any state* $\langle p, I^P, I^C, O \rangle$ *with* $p(s) = d$ *for all pointing goals* $\langle s, d \rangle$ *and* $O = \emptyset$.

OPERATORS: *A satellite* $s \in S$ *can **turn** to a direction* $d \in D$. *In the resulting state, it points at* d.
                  An instrument $i \in I$ *can be **powered on** iff no other instrument located on the same satellite is powered on. In the resulting state,* i *is added to the set of powered-on instruments.*
                  An instrument $i \in I$ *can be **powered off** iff it is currently powered on. In the resulting state,* i *is removed from the set of powered-on instruments and from the set of calibrated instruments (if present).*
                  An instrument $i \in I$ *can be **calibrated** iff it is currently powered on and the satellite it is located on points at the calibration target of* i. *In the resulting state,* i *is added to the set of calibrated instruments.*
                  An instrument $i \in I$ *can **take an image** of a pointing direction* $d \in D$ *in mode* $m \in M$ *iff it is powered on, calibrated and supports mode* m *and the satellite it is located on points at* d. *In the resulting state,* $\langle d, m \rangle$ *is removed from the set of remaining objectives (if present).*

It is easy to prove that optimal SATELLITE plans can be approximated by a constant factor.

Theorem 6.8.3. PLAN-SATELLITE $\in$ APX
PLAN-SATELLITE *Is 6-Approximable.*

Proof. A greedy planning algorithm first achieves all image objectives with a sequence of 6 actions each: After selecting an instrument supporting the required image mode (if none exists, the task has no solution), power on the instrument, point its satellite at the calibration target, calibrate the instrument, point its satellite at the image direction, take the image, and power off the instrument. After taking all images, each satellite with a pointing goal is turned to the required direction if it differs from the current direction. As a post-processing step, for each satellite that has been used for taking an image, the final powering-off action is removed.

The length of the plan is at most $6I + P$, where I is the number of image goals and P is the number of pointing goals where the goal direction is different from the initial direction. (Note that it may be the case that a satellite originally pointed at the goal direction and has been pointed away to take an image. However, in this case, the action required for pointing it back at the original direction is "paid for" in the step where extraneous power-off actions are removed.)

An optimal plan clearly contains at least $I + P$ actions, since it must contain a separate image-taking action for each image goal and a pointing action for each pointing goal where the goal direction differs from the initial direction. Because $6I + P \leq 6(I + P)$, this concludes the proof. $\qquad\Box$

This approximation algorithm is fairly simple, and it can certainly be improved to yield a lower approximation ratio. However, this ratio cannot be made arbitrarily small.

Theorem 6.8.4. PLAN-SATELLITE $\notin$ PTAS
There is no polynomial-time approximation scheme for SATELLITE *planning unless* P $=$ NP.

Proof. We OP-reduce from MINIMUM VERTEX COVER. *We are given a* MINIMUM VERTEX COVER *instance (i. e., graph)* $G = \langle V, E \rangle$ *where all vertices have degree 2 or 3 (cf. Theorem 4.4.2) and a parameter* $r > 1$ *irrelevant to the reduction.*

The corresponding SATELLITE *task* T' *is defined as follows:*

– *There is one satellite* s_v *for each vertex* $v \in V$.
– *There is one instrument* i_v *for each vertex* $v \in V$, *located on* s_v.
– *There is one mode* m_e *for each edge* $e = \{u, v\} \in E$, *supported by* i_u *and* i_v *(and by no other instrument).*
– *There are two pointing directions* d_0 *and* $d_\star$.
– *Pointing direction* d_0 *is the initial pointing direction of all satellites and the calibration target of all instruments.*
– *For each edge* $e \in E$, *there is an image objective* $\langle d_\star, m_e \rangle$. *There are no further image objectives, and no pointing goals.*

Clearly, the task can be generated in polynomial time.

To map back a plan π' for T' to a vertex cover U for G, include a vertex $v \in V$ in U iff the corresponding instrument i_v is used for taking an image in π'. This can also easily be done in polynomial time, and U is indeed a vertex cover: If it were not, then there were an edge $e = \{u, v\} \in V$ such that neither i_u nor i_v is used for taking an image in π'. But then π' cannot satisfy the image objective $\langle d_\star, m_e \rangle$, a contradiction.

The plan π' must contain at least $|E|$ image-taking actions and thus at most $m(\pi') - |E|$ other actions. For each instrument used for taking an image, at least three separate actions are needed to power on the instrument, calibrate it, and turn the satellite it is located on to $d_\star$. We thus have $m(U) \leq \frac{1}{3}(m(\pi') - |E|)$.

On the other hand, we have $m^*(T') \leq 3m^*(G) + |E|$: If U^* is an optimal vertex cover, then we can solve the SATELLITE task by powering on and calibrating all instruments i_v with $v \in U^*$ ($2m^*(G)$ actions), turning the corresponding satellites to $d_\star$ ($m^*(G)$ actions) and taking images of $d_\star$ in all modes using the calibrated instruments ($|E|$ actions). Solving the inequality for $m^*(G)$, we get $m^*(G) \geq \frac{1}{3}(m^*(T') - |E|)$.

This allows us to bound the performance ratio of the vertex cover U constructed from a plan π' for the SATELLITE task T' as follows:

$$\frac{m(U)}{m^*(G)} \leq \frac{\frac{1}{3}(m(\pi') - |E|)}{\frac{1}{3}(m^*(T') - |E|)} = \frac{m(\pi') - |E|}{m^*(T') - |E|} \leq \frac{rm^*(T') - |E|}{m^*(T') - |E|}$$

$$= 1 + \frac{(r-1)m^*(T')}{m^*(T') - |E|} = 1 + (r-1)\frac{m^*(T')}{m^*(T') - |E|}$$

$$\overset{(*)}{\leq} 1 + (r-1)\frac{m^*(T')}{m^*(T') - \frac{1}{2}m^*(T')} = 1 + (r-1) \cdot 2,$$

where inequality $(*)$ holds because $m^*(T') \geq 2|E|$: Any plan must contain at least $|E|$ image-taking actions and $3|U^*| \geq |E|$ other actions, where U^* is an optimal vertex cover. (The latter inequality holds because of the degree bounds for the vertices of G, cf. Theorem 4.4.2).

The reduction is clearly optimization-preserving because $m^*(G) \leq K$ iff $m^*(T') \leq 3K + |E|$ for all $K \in \mathbb{N}_0$. $\square$

We observe that the hardness result already holds for a fairly restricted class of SATELLITE tasks, namely those with only two pointing directions and one instrument per satellite. In fact, we could easily have used only a single pointing direction, but the existing SATELLITE benchmarks obey the restriction that no images are ever required of directions that serve as calibration targets (presumably because calibration targets are "known" objects, of which we do not require images). Our proof follows this restriction in case it is considered integral to the SATELLITE domain rather than incidental to the benchmark set.

Instead of many satellites with a single instrument each, we could also prove hardness for the case of a single satellite with many instruments. Similarly, instead of requiring many images of one direction, we could also require one image each of many directions.

6.9 SCHEDULE

For our final result, we turn our attention to the SCHEDULE domain. This domain models the processing of physical objects by a number of different machines capable of changing the appearance of objects by polishing them, painting them, punching holes into them, and similar transformations.

Definition 6.9.1. SCHEDULE *Tasks*
The set of SCHEDULE *temperatures* O_T, **surface conditions** O_{SC}, **shapes** O_S, **colours** O_C *and* **holes** O_H *and the set of* SCHEDULE **object states** O *are defined as follows:*

$$O_T = \{cold, hot\}$$
$$O_{SC} = \{rough, smooth, polished, none\}$$
$$O_S = \{cylindrical, circular, oblong\}$$
$$O_C = \{blue, yellow, red, black, none\}$$
$$O_H = \{front1, front2, front3, back1, back2, back3\}$$
$$O = O_T \times O_{SC} \times O_S \times O_C \times 2^{O_H}$$

A SCHEDULE **task** *is a finite sequence of pairs in* $O \times 2^O$. *For* SCHEDULE *tasks* $(\langle o_0^i, O_\star^i \rangle)_{i=1}^n$, *we say that* o_0^i *is the* **initial state of the** *i-th object,* *and that* $O_\star^i$ *is the* **goal specification for the** *i-th object.*

According to this definition, an object in a SCHEDULE task is characterized by its temperature, surface condition, shape, colour and set of holes, for each of which there is a finite range of possible values. A SCHEDULE task specifies a sequence of pairs $\langle o_0, O_\star \rangle$, where $o_0 \in O$ defines the initial state of a given object, and $O_\star \subseteq O$ defines its possible goal states. In the SCHEDULE domain as used at IPC2, there are further restrictions on such goal sets. For example, specifications like "This object must become smooth and oblong" are possible goals, but specifications like "This object must become red and hot or green and smooth" are not. Specifically, goals are always defined as conjunctions over assignments to individual object properties (like surface condition and shape), and there are no goals involving temperatures or holes and no goals requiring the surface condition or colour of an object to be *none*. By allowing more complex goal specifications, it seems that we make the planning problem harder, but we will shortly see that this does not affect our results. We can now define the SCHEDULE domain.

Definition 6.9.2. SCHEDULE *Domain*
The set of SCHEDULE **machinery** *is defined as* $M = \{drill\text{-}press, grinder, immersion\text{-}painter, lathe, polisher, punch, roller, spray\text{-}painter\}$.

Each machine $m \in M$ *has an associated set of* **transformations***, which are partial functions on object states. Similar to operator descriptions, we describe transformations in terms of preconditions and effects on object states:*

-   *The drill-press, polisher, punch and spray-painter may transform any object whose temperature is cold. The other machinery may transform any object.*

- *The drill-press and punch have one transformation for each hole $h \in O_H$. This transformation adds the hole h to the set of holes of the transformed object. In addition to this effect, the punch causes the surface condition of the transformed object to be rough.*
- *The immersion-painter and spray-painter have one transformation for each colour $c \in O_C \setminus \{none\}$. This transformation causes the colour of the transformed object to be c. In addition to this effect, the spray painter causes the surface condition of the transformed object to be none.*
- *The polisher has a single transformation, which causes the surface condition of the transformed object to be polished.*
- *The roll has a single transformation, which causes the surface condition and colour of the transformed object to be none, removes all holes (i. e., causes the set of holes to be the empty set), causes the temperature to be hot and the shape to be cylindrical.*
- *The lathe has a single transformation, which causes the colour of the transformed object to be none, the surface condition to be rough and the shape to be cylindrical.*
- *The grinder has a single transformation, which causes the colour of the transformed object to be none and the surface condition to be smooth.*

The SCHEDULE *planning domain maps* SCHEDULE *tasks* $(\langle o_0^i, O_\star^i \rangle)_{i=1}^n$ *to state spaces as follows:*

STATES: *3-tuples $\langle (o^i)_{i=1}^n, M_B, O_B \rangle$, where $(o^i)_{i=1}^n \in O^n$ is an n-tuple of object states whose i-th component is called the **state of the i-th object**, $M_B \subseteq M$ is called the set of **busy machines** and $O_B \subseteq \{1, \ldots, n\}$ is called the set of **busy objects**. We say that $m \in M$ is **busy** iff $m \in M_B$ and that the i-th object is **busy** iff $i \in O_B$.*

INITIAL STATE: $\langle (o_0^i)_{i=1}^n, \emptyset, \emptyset \rangle$.

GOAL STATES: *Any state where the i-th object state is in $O_\star^i$ for all $i \in \{1, \ldots, n\}$.*

OPERATORS: *For each transformation t of a machine $m \in M$ and each $i \in \{1, \ldots, n\}$, machine m may **process** the i-th object using transformation t iff t is defined for the state of the i-th object, o_i, and neither m nor the i-th object are busy. In the resulting state, the state of the i-th object is $t(o_i)$ and machine m and the i-th object are busy.*

*The **time-step** action is applicable iff at least one machine is busy. In the resulting state, the sets of busy machines and objects are empty.*

The exact definitions of the transformations are immaterial to our complexity results and are only presented for completeness of description. All that is relevant for our result is that the sets of object states, machines and transformations are of fixed size.

Plans in the SCHEDULE domain naturally fall into *episodes*, separated by *time-step* actions. During each episode, each machine can process at most one object, and hence each episode consists of at most $|M| = 8$ actions (not counting the *time-step* actions themselves). From the point of view of applications, it is more natural to optimize the number of *episodes* than the total number of actions of a plan, because it is assumed that different machines may work in parallel and we are interested in minimizing the total time spent (often called the *makespan* of the plan), rather than using a minimal number of processing step. Makespan is equivalent (plus/minus 1) to the number of *time-step* actions in a plan, so we can model makespan minimization by assigning a cost of 1 to time-step actions and a cost of 0 to all other actions. The following analysis applies to both minimization of makespan and minimization of plan length.

Theorem 6.9.3. PLAN-SCHEDULE ∈ PO
In the SCHEDULE *domain, optimal plans can be generated in polynomial time.*
    This remains true in the generalized domain where non-negative weights are assigned to all actions, as long as weights for process actions only depend on the machine, transformation and state of the object being transformed (but not on the index *of the object).*

Proof. We say that the kind *of an object in a given state of a* SCHEDULE *task is given by its current state and its goal specification.*
    Observe that if the i-th and j-th object have the same kind in state s, then there is a symmetry in the state space: In any sequence of actions that leads to a goal state from s, we can exchange all transformations of the i-th object with transformations of the j-th object to obtain another solution. Abstracting from this symmetry, we can specify actions in terms of the kind *of objects they transform, rather than their index. This leads to an alternative representation of* SCHEDULE *states: If n is the number of objects in a given* SCHEDULE *task and K is the number of object kinds, then an* abstract state *of the task is given by K numbers in the range $\{0, \ldots, n\}$ specifying how many objects of each kind are present, K numbers in the range $\{0, \ldots, |M|\}$ specifying how many objects of each kind are busy, and a subset of M specifying which machines are busy. By translating the initial state, goal specification and operators to refer to abstract states, we obtain the* abstract state space *of a* SCHEDULE *task.*
    The size of the abstract state space is clearly bounded by $(n + 1)^K \cdot (|M| + 1)^K \cdot 2^{|M|}$, which is polynomial in n and hence in the size of the SCHEDULE *task. (Recall that K and $|M|$ are constants.) Thus, an optimal abstract plan can be generated in polynomial time by explicitly constructing the abstract state space and finding a shortest path from the abstract initial state to some abstract goal state. Abstract plans can easily be translated back to concrete plans.*
    This approach generalizes to the weighted action-cost model as long as the cost of a transformation does not differ for different objects of the same kind. □

The polynomial time property of the algorithm critically relies on the fact that the number of machines and object states is constant. If the object states, machines and transformations are given as part of the individual planning tasks, then the bounded plan existence becomes NP-complete, at least in the more realistic case where only *time-step* actions are counted. This is already true if there are only three machines [43, Problem SS18]. Of course, generating bad-quality plans stays easy because objects can still be dealt with one at a time, by searching the transition graph that is defined by the set of object transformations. (We assume that these transition graphs are given explicitly, rather than in a factored representation.)

Note that the execution time of the optimal planning algorithm we presented, depending on the graph search algorithm used, grows about as quickly as n^K, where n is the input size and $K = |O \times 2^O| = 7680 \cdot 2^{7680}$, which means that this is not a practical algorithm.

As was mentioned earlier, the real competition domain does not allow for arbitrary goal descriptions, which reduces the 2^{7680} factor to 80. Still, an $O(n^{7680 \cdot 80}) = O(n^{614400})$ algorithm is not tractable in practice. Further optimizations decrease the complexity significantly, but it is not obvious what a really tractable algorithm would look like.

6.10 Summary

This concludes our discussion of the IPC benchmark suite, and thus the technical results of Part I. The main results of this chapter are briefly summarized in Fig. 6.7.

Domain	Complexity
ASSEMBLY	EXPO \ NPS
BLOCKSWORLD	APX \ PTAS
FREECELL	NPO \ PS
MOVIE	PO
PIPESWORLD	$\notin$ PS (membership in NPO open)
PROMELA	EXPO \ NPS
PROMELA-OPTICALTELEGRAPH	PO
PROMELA-PHILOSOPHERS	PO
PSR	PO
SATELLITE	APX \ PTAS
SCHEDULE	PO

Fig. 6.7. Complexity results for the non-transportation IPC domains (if P $\neq$ NP)

7

Conclusions

In the preceding chapters, we analysed the computational complexity of planning in the benchmark domains of the first four International Planning Competitions (IPC1–4) and two related families of transportation and route planning domains. In the following Sect. 7.1, we draw a number of conclusions from this analysis, and in Sect. 7.2 we close this part by pointing out what else needs to be done until we can claim to understand the problem we set out to investigate.

7.1 Ten Conclusions

We observed that a large number of IPC benchmarks encode various variants of transportation planning and route planning. Although this might sound superficial, we believe this to be the first important result of our analysis, because this prevalence of one particular kind of planning problem has – in our opinion – largely influenced the development of planning systems since the advent of the International Planning Competitions. In particular, it has strongly influenced the planning system we describe in Part II.

Conclusion 7.1.1. *Many IPC domains model transportation and route planning problems.*

Based on this observation, we introduced and analysed the TRANSPORT and ROUTE domain families. One of our intentions in looking at domain families rather than considering each competition domain in isolation was to find the *boundaries* between easy and hard planning problems. For TRANSPORT and ROUTE, one such boundary is clearly defined.

Conclusion 7.1.2. *Planning in the* TRANSPORT *and* ROUTE *domains with restricted fuel is difficult: None of these domains has a polynomial-time solution algorithm unless* P = NP.

Planning in the TRANSPORT *and* ROUTE *domains with unrestricted fuel is easy: All unrestricted-fuel domains have a polynomial-time solution algorithm.*

One possible explanation why fuel restrictions make the planning problem harder is that, by bounding the number of movements that can be performed, they effectively limit the length of the generated plans, and hence the problem of just finding *any* plan becomes akin to the problem of finding an *optimal* plan, which is unconditionally difficult.

Conclusion 7.1.3. *Optimal planning in the* TRANSPORT *and* ROUTE *domains is difficult: None of the domains admits a polynomial optimal planning algorithm unless* P = NP.

We noticed that a major source of hardness in finding optimal plans for these domains is a difficulty in *ordering*: While it is usually evident how individual portables should be transported to their destinations, the interactions between different goals make optimally solving the overall task a hard problem. Therefore, even deceivingly simple planning tasks (such as LOGISTICS tasks with a single vehicle) turn out to be NP-hard to optimize.

Due to this hardness of optimization, we considered the problem of generating *high-quality* plans. One of the outcomes of this analysis was that in none of the domains in the TRANSPORT and ROUTE families, optimal solutions can be approximated arbitrarily well within polynomial time.

Conclusion 7.1.4. *Optimal plans in the* TRANSPORT *and* ROUTE *domains cannot be approximated arbitrarily: None of the domains admits a polynomial-time approximation scheme unless* P = NP.

So is fuel the only dividing line in the TRANSPORT and ROUTE families? It turns out that it is not.

Conclusion 7.1.5. *Optimal plans in the* TRANSPORT *and* ROUTE *domains can be approximated by some constant factor if all mobiles have the same capabilities and mobiles capacities (for* TRANSPORT) *are either all 1 or all unrestricted.*

Optimal plans in the TRANSPORT *and* ROUTE *domains cannot be approximated by some constant factor if either of these restrictions is violated, unless some widely held assumptions about the class* NP *fail to be true.*

A last statement about TRANSPORT and ROUTE which we consider worth emphasizing is that two important aspects appear *not* to influence the hardness of planning.

Conclusion 7.1.6. *Whether actions have arbitrary costs or unit cost does not affect the complexity of planning in the* TRANSPORT *and* ROUTE *family.*

Conclusion 7.1.7. *Whether there is only a single mobile or multiple mobiles does not affect the complexity of planning in the* TRANSPORT *and* ROUTE *family, provided that all mobiles have the same capabilities.*

Our remaining conclusions are concerned with the IPC benchmark domains, rather than the TRANSPORT and ROUTE domain families. We have fewer general conclusions to draw for the IPC benchmarks because they specify wildly different problems, and it is difficult to summarize our findings for them in a few general statements. However, this fact in itself is already an interesting observation.

Conclusion 7.1.8. *The IPC benchmark suite is very varied, both in terms of the application problems it represents and in terms of complexity.*

We emphasize this point because, after all, we are interested in *general domain-independent planning*. It would be bad news indeed if it turned out that all IPC domains represented more or less the same problem. However, this is not the case. In particular, the IPC4 domains significantly helped increase the scope of the benchmark suite beyond transportation-style problems, while also widening the complexity spectrum of the benchmark suite by introducing the first domains with PSPACE-complete plan existence problems, namely AIRPORT and PROMELA, and the first non-trivial planning domain for which we can efficiently generate optimal plans, namely PSR. (GRIPPER and MOVIE are clearly trivial, and for SCHEDULE we cannot really claim to be able to efficiently generate optimal plans.) We believe that this is an effect of the competition organizers' focus on actively seeking for *realistic* and *structurally interesting* domains. Indeed, they mention including a PSPACE-equivalent benchmark as one of the desiderata for the IPC4 benchmark suite [37].

Despite this variety, there are a few general observations which hold across a surprisingly wide range of domains. One of them has been stated before in the context of transportation and route planning (cf. Sect. 5.7).

Conclusion 7.1.9. *When a planning domain admits polynomial-time solution algorithms but does not allow constant-factor approximations, this is usually due to positive subgoal interactions.*

Indeed, it is interesting to observe that *all* our hardness proofs for the poly-APX \ APX domains relied on the MINIMUM SET COVER problem, which exhibits positive subgoal interactions in their most purely distilled form. We believe that finding a good way of integrating heuristics for set covering problems could lead to a significant advance in plan quality of domain-independent planning systems, and might be one of the keys to developing better admissible heuristics for the general planning problem. However, we consider this a very challenging avenue of research.

Our final observation concerns domains which *do* admit constant-factor approximations.

Conclusion 7.1.10. *When a planning domain admits constant-factor approximations, then such an approximation can usually be obtained by simple greedy algorithms achieving one goal at a time.*

Indeed, from our proofs it is easy to see that for *all* IPC domains in APX except for BLOCKSWORLD, the BLOCKSWORLD-like DEPOTS domain and the PROMELA variants, polynomial-time constant-factor approximations can be obtained by using Hoffmann's *enforced hill-climbing* algorithm [68] using *the number of unsatisfied goals* as a heuristic estimator. This observation goes a large way towards explaining why planning systems based on local search such as FF or LPG often dramatically outperform more systematic search algorithm like SATPLAN.

7.2 Going Further

What to do with these results? As a chief motivation for our analysis, we quoted Pólya's belief that "First, you have to *understand* the problem." Can we claim to understand the planning benchmarks now?

Certainly, we do not understand them perfectly. To see how much more can be done, consider Slaney and Thiébaux's work on the BLOCKSWORLD domain [108], which was motivated by a similar goal as ours, namely to provide "the level of understanding required for [...] effective use as a benchmark":

> "Our results include methods for generating random problems for systematic experimentation, the best domain-specific planning algorithms against which AI planners can be compared, and observations establishing the average plan quality of near-optimal methods. We also study the distribution of hard/easy instances, and identify the structure that AI planners must be able to exploit in order to approach Blocks World successfully." [108, p. 119]

Considering these goals, at least two more steps are required until we can get a similarly good understanding of the other planning benchmarks: identifying phase transition regions to understand where the *really* hard tasks are, and having (domain-dependent) optimal solvers available to be able to realistically assess the solution quality of non-optimal planners.

Part II

Fast Downward

8

Solving Planning Tasks Hierarchically

This chapter introduces the main ideas and concepts underlying the Fast Downward planner and puts them into the context of earlier research.

In Sect. 8.1, we present a simple example to illustrate key aspects of the planner such as the exploitation of causal dependencies and the use of multi-valued state variables. Section 8.2 puts these ideas into the context of earlier research. The concluding Sect. 8.3 presents a bird's-eye view of the overall planner architecture and provides an outlook on the chapters to come.

8.1 Introduction

Consider a typical transportation planning task: The postal service must deliver a number of parcels to their respective destinations using its vehicle fleet

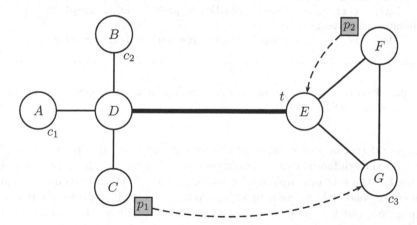

Fig. 8.1. A transportation planning task. Deliver parcel p_1 from C to G and parcel p_2 from F to E, using the cars c_1, c_2, c_3 and truck t. The cars may only use inner-city roads (thin edges), the truck may only use the highway (thick edge)

```
Variables:
    at-p1-a, at-p1-b, at-p1-c, at-p1-d, at-p1-e, at-p1-f, at-p1-g,
    at-p2-a, at-p2-b, at-p2-c, at-p2-d, at-p2-e, at-p2-f, at-p2-g,
    at-c1-a, at-c1-b, at-c1-c, at-c1-d,
    at-c2-a, at-c2-b, at-c2-c, at-c2-d,
    at-c3-e, at-c3-f, at-c3-g,
    at-t-d, at-t-e,
    in-p1-c1, in-p1-c2, in-p1-c3, in-p1-t,
    in-p2-c1, in-p2-c2, in-p2-c3, in-p2-t
Init:
    at-p1-c, at-p2-f, at-c1-a, at-c2-b, at-c3-g, at-t-e
Goal:
    at-p1-g, at-p2-e
Operator drive-c1-a-d:
    PRE: at-c1-a    ADD: at-c1-d    DEL: at-c1-a
Operator drive-c1-b-d:
    PRE: at-c1-b    ADD: at-c1-d    DEL: at-c1-b
Operator drive-c1-c-d:
    PRE: at-c1-c    ADD: at-c1-d    DEL: at-c1-c
...
Operator load-c1-p1-a:
    PRE: at-c1-a, at-p1-a    ADD: in-p1-c1    DEL: at-p1-a
Operator load-c1-p1-b:
    PRE: at-c1-b, at-p1-b    ADD: in-p1-c1    DEL: at-p1-b
Operator load-c1-p1-c:
    PRE: at-c1-c, at-p1-c    ADD: in-p1-c1    DEL: at-p1-c
...
Operator unload-c1-p1-a:
    PRE: at-c1-a, in-p1-c1    ADD: at-p1-a    DEL: in-p1-c1
Operator unload-c1-p1-b:
    PRE: at-c1-b, in-p1-c1    ADD: at-p1-b    DEL: in-p1-c1
Operator unload-c1-p1-c:
    PRE: at-c1-c, in-p1-c1    ADD: at-p1-c    DEL: in-p1-c1
...
```

Fig. 8.2. Part of a typical propositional encoding of the transportation planning task (no actual PDDL syntax)

of cars and trucks. Let us assume that a car serves all the locations of one city, and that different cities are connected via highways that are served by trucks. For the sake of simplicity, let us further assume that travelling on each segment of road or highway incurs the same cost. This is not a highly realistic assumption, but for the purposes of exposition, it will do. There can be any number of parcels, posted at arbitrary locations and with arbitrary destinations. Moreover, cities can be of varying size, there can be one or several cars within each city, and there can be one or several trucks connecting the cities.

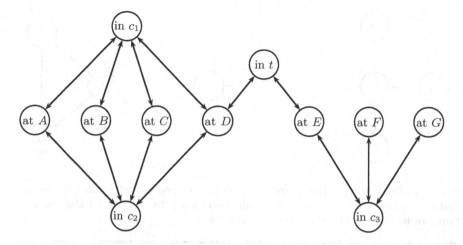

Fig. 8.3. Domain transition graph for the parcels p_1 and p_2. Indicates how a parcel can change its state. For example, the arcs between "at D" and "in t" correspond to the actions of loading/unloading the parcel at location D with the truck t

Cars will never leave a city. Figure 8.1 shows an example task of this kind with two cities, three cars and a single truck. There are two parcels to be delivered, one of which (p_1) must be moved between the two cities, while the other (p_2) can stay within its initial city.

The observant reader will have noticed by now that we are describing a TRANSPORT$_{\infty*}$ planning task. (Part of) a propositional STRIPS-like encoding is shown in Fig. 8.2.

How would human planners go about solving tasks of this kind? Very likely, they would use a hierarchical approach: For p_1, it is clear that the parcel needs to be moved between cities, which is only possible by using the truck. Since in our example each city can access the highway at only one location, we see that we must first load the parcel into some car at its initial location, then drop it off at the first city's highway access location, load it into the truck, drop it off at the other city's highway access location, load it into the only car in that city, and finally drop it off at its destination. We can commit to this "high-level" plan for delivering p_1 without worrying about "lower-level" aspects such as path planning for the cars. It is obvious to us that *any* good solution will have this structure, since the parcel can only change its location in a few clearly defined ways (Fig. 8.3). The same figure shows that the only reasonable plans for getting p_2 to its destination require loading it into the car in its initial city and dropping it off at its target location. There is no point in ever loading it into the truck or into any of the cars in the left city.

So let us assume that we have committed to the (partially ordered, as movements of the two parcels can be interleaved) "high-level plan" shown in Fig. 8.5. All we need to do to complete the plan is choose a linearization of the

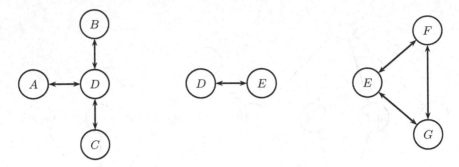

Fig. 8.4. Domain transition graphs for the cars c_1 and c_2 (left), truck t (centre), and car c_3 (right). Note how each graph corresponds to the part of the roadmap that can be traversed by the respective vehicle

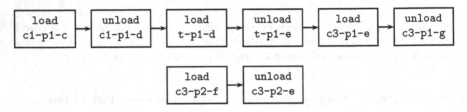

Fig. 8.5. High-level plan for the transportation planning task

high-level steps and fill in movements of the vehicle fleet between them. We have thus decomposed the planning task into a number of subtasks. The parcel scheduling task (where, and by which vehicles, a parcel should be loaded and unloaded) is separated from the path planning task for each vehicle in the fleet (how to move it from point X to Y). Both of these are graph search tasks, and the corresponding graphs are shown in Fig. 8.3 and Fig. 8.4. Graphs of this kind will be formally introduced as *domain transition graphs* in Chap. 10.

Of course these graph search tasks interact, but they only do so in limited ways: State transitions for the parcels have associated conditions regarding the vehicle fleet, which need to be considered in addition to the actual path planning in Fig. 8.3. For example, a parcel can only change state from "at location A" to "inside car c_1" if the car c_1 is at location A. However, state transitions for the vehicles have no associated conditions from other parts of the planning task, and hence moving a vehicle from one location to another is indeed as easy as finding a path in the associated domain transition graph. We say that the parcels have *causal dependencies* on the vehicles because there are operators that change the state of the parcels and have preconditions on the state of the vehicles. Indeed, these are the only causal dependencies in this task, since parcels do not depend on other parcels and vehicles do not depend on anything except themselves (Fig. 8.6). The set of causal dependencies of a planning task is visualized in its *causal graph*.

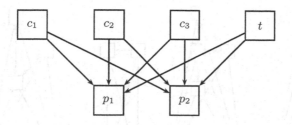

Fig. 8.6. Causal dependencies in the transportation planning task

We argue that humans often solve planning tasks in the hierarchical fashion outlined in the preceding paragraphs, and that algorithmic approaches to action planning can usefully apply similar ideas. Indeed, as we will show in the following section, we are not the first to introduce domain transition graphs and causal graphs. However, earlier work has almost exclusively focused on *acyclic* causal graphs, and for a good reason: If the causal graph of a planning task exhibits a cycle, hierarchical decomposition is not possible, because the subtasks that must be solved to achieve an operator precondition are not necessarily smaller than the original task. As far as we are aware, we were the first [58] to present a *general* planning algorithm that focuses on exploiting hierarchical information from causal graphs. However, our *causal graph heuristic* also requires acyclicity; in the general case, it considers a relaxed planning task in which some operator preconditions are ignored to break causal cycles.

Knowing that cycles in causal graphs are undesirable, we take a closer look at the transportation planning task. Let us recall our informal definition of causal graphs: The causal graph of a planning task contains a vertex for each state variable and arcs from variables that occur in preconditions to variables that occur in effects of the same operator. So far, we may have given the impression that the causal graph of the example task has the well-behaved shape shown in Fig. 8.6. Unfortunately, having a closer look at the STRIPS encoding in Fig. 8.2, we see that this is not the case: The correct causal graph, shown in Fig. 8.7, looks very messy. This discrepancy between the intuitive and actual graph is due to the fact that in our informal account of "human-style" problem solving, we made use of (non-binary) state variables like "the location of car c_1" or "the state of parcel p_1", while STRIPS-level state variables correspond to (binary) object-location propositions like "parcel p_1 is at location A". It would be much nicer if we were given a multi-valued encoding of the planning task that explicitly contains a variable for "the location of car c_1" and similar properties. Indeed, the nice looking acyclic graph in Fig. 8.6 is the causal graph of the multi-valued encoding shown in Fig. 8.8.

We have now provided enough intuition to state the design goal for the *Fast Downward* planning system, which is the topic of this second part of the volume: *To develop an algorithm that efficiently solves general proposi-*

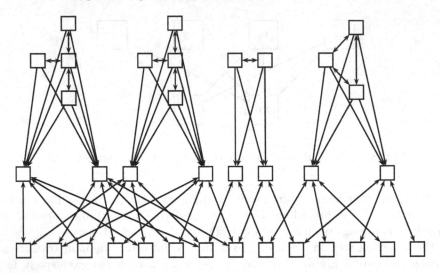

Fig. 8.7. Causal graph for the STRIPS encoding of the transportation planning task

tional planning tasks by exploiting the hierarchical structure inherent in causal graphs. We need to overcome three major obstacles in this undertaking:

– First, propositionally encoded planning tasks usually have very unstructured causal graphs. However, the intuitive dependencies often become visible in encodings with multi-valued state variables. To exploit this fact in an automated PDDL planning system, we have devised an automatic algorithm for "translating" (or reformulating) propositional tasks to multi-valued ones. The translation algorithm can be considered independently from the rest of the planner; in fact, it is now also used as part of other planning systems [111].
– Second, no matter how clever the encoding is, most planning tasks are not completely hierarchical in nature. To deal with causal cycles, we consider relaxations where some causal dependencies are ignored and use solutions to the relaxed task within a heuristic search algorithm.
– Third, even for planning tasks that can be solved hierarchically, finding such a solution is difficult (indeed, still **PSPACE**-complete). For this reason, our heuristic function only considers a fragment of a task at a time, namely subtasks induced by a single state variable and its predecessors in the causal graph. Even *this* planning problem is still **NP**-complete, so that we are content with an incomplete solution algorithm within the heuristic solver. This solution algorithm has theoretical shortcomings but never failed us in practice.

```
Variables:
    p1, p2 ∈ {at-a, at-b, at-c, at-d, at-e, at-f,at-g,
              in-c1, in-c2, in-c3, in-t}
    c1, c2 ∈ {at-a, at-b, at-c, at-d}
    c3     ∈ {at-e, at-f, at-g}
    t      ∈ {at-d, at-e}
Init:
    p1 = at-c, p2 = at-f
    c1 = at-a, c2 = at-b, c3 = at-g, t = at-e
Goal:
    p1 = at-g, p2 = at-e
Operator drive-c1-a-d:
    PRE: c1 = at-a    EFF: c1 = at-d
Operator drive-c1-b-d:
    PRE: c1 = at-b    EFF: c1 = at-d
Operator drive-c1-c-d:
    PRE: c1 = at-c    EFF: c1 = at-d
...
Operator load-c1-p1-a:
    PRE: c1 = at-a, p1 = at-a    EFF: p1 = in-c1
Operator load-c1-p1-b:
    PRE: c1 = at-b, p1 = at-b    EFF: p1 = in-c1
Operator load-c1-p1-c:
    PRE: c1 = at-c, p1 = at-c    EFF: p1 = in-c1
...
Operator unload-c1-p1-a:
    PRE: c1 = at-a, p1 = in-c1    EFF: p1 = at-a
Operator unload-c1-p1-b:
    PRE: c1 = at-b, p1 = in-c1    EFF: p1 = at-b
Operator unload-c1-p1-c:
    PRE: c1 = at-c, p1 = in-c1    EFF: p1 = at-c
...
```

Fig. 8.8. Part of an encoding of the transportation planning task with multi-valued state variables

8.2 Related Work

As a planning system based on heuristic forward search, Fast Downward is clearly related to other heuristic planners such as HSP [16] or FF [68] on the architectural level. However, in this section we focus on work that is related on the *conceptual* level, i. e., work that uses similar forms of hierarchical decomposition of causal graphs and work that uses similar forms of search in domain transition graphs.

8.2.1 Causal Graphs and Abstraction

The term *causal graph* first appears in the literature in the work by Williams and Nayak [113], but the general idea is considerably older. The approach of hierarchically decomposing planning tasks is arguably as old as the field of AI Planning itself, having first surfaced in Newell and Simon's work on the General Problem Solver [94].

Still, it took a long time for these notions to evolve to their modern form. Sacerdoti's ABSTRIPS algorithm introduced the concept of *abstraction spaces* for STRIPS-like planning tasks [106]. An abstraction space of a STRIPS task is the state space of an *abstracted task*, obtained by removing all preconditions from the operators of the original task that belong to a given set of propositions, which are *abstracted away*. (In later work by other authors, propositions which are abstracted away are also removed from the operator effects. This only makes a difference in subtle cases that require the presence of axioms; we do not distinguish between these two forms of abstraction here.) To solve a planning task, ABSTRIPS first generates a plan for an abstracted task, then refines this plan by inserting concrete plans between the abstract plan steps that "bridge the gap" between abstract states by satisfying the operator preconditions which were ignored at the abstract level. The idea is easily generalized to several levels of abstraction forming an *abstraction hierarchy*, with a very abstract level at the top where almost all preconditions are ignored, successively introducing more preconditions at every layer until the final layer of the hierarchy equals the original planning task.

One problem with this approach to planning is that in general there is no guarantee that the abstract plans bear any resemblance to reasonable concrete plans. For example, if abstraction spaces are chosen badly, it is quite possible that finding a concrete plan that satisfies the precondition of the first operator in the abstract plan is more difficult than solving the original goal at the concrete level. Such shortcomings spawned a large amount of research on the properties of abstraction hierarchies and how they can be generated automatically. Tenenberg gives one of the first formal accounts of the properties of different kinds of abstraction [109]. Among other contributions, he defines the so-called *upward solution property*, which can be informally stated as: "If there exists a concrete solution, then there also exists an abstract solution". Rather surprisingly, not all abstractions considered at the time satisfied this very basic property, without which one would be loathe to call a given state space an "abstraction" of another state space.

A limitation of the upward solution property is that it states no relationship between the concrete and abstract plan at all. For ABSTRIPS-style hierarchical planning to be successful, the abstract plan must bear some resemblance to a concrete one; otherwise there is little point in trying to refine it. Indeed, Tenenberg introduces stronger versions of the upward solution property, but more relevant to Fast Downward is Knoblock's work on the *ordered monotonicity property* [79]. An abstraction space satisfies the ordered monotonicity property if, roughly speaking, any concrete solution can be derived from some abstract solution while leaving the actions in the abstract plan intact, only inserting additional operators relevant to the concrete plan. Clearly, this is a very important property for ABSTRIPS-like hierarchical planning.

It is in Knoblock's article that causal graphs first surface (although he does not introduce a name for them). Translated to our terminology, Knoblock proves the following relationship between useful abstractions and causal graphs: *If the causal graph contains no path from a variable that is not abstracted away to a variable that is abstracted away, then the abstraction has the ordered monotonicity property.* In particular, this means that for acyclic causal graphs, it is possible to devise an abstraction hierarchy where only one new variable is introduced at each level.

Besides these theoretical contributions, Knoblock presents a planning system called ALPINE which computes an abstraction hierarchy for a planning task from its causal graph and exploits this within a hierarchical refinement planner. Although the planning method is very different, the derivation of the abstraction hierarchy is very similar to Fast Downward's method for generating hierarchical decompositions of planning tasks (Sect. 10.3).

By itself, the ordered monotonicity property is not sufficient to guarantee good performance of a hierarchical planning approach. It guarantees that every concrete solution can be obtained in a natural way from an abstract solution, but it does not guarantee that all abstract solutions can be refined to concrete ones. Such a guarantee is provided by the *downward refinement property*, introduced by Bacchus and Yang [9].

The downward refinement property can rarely be guaranteed in actual planning domains, so Bacchus and Yang develop an analytical model for the performance of hierarchical planning in situations where a given abstract plan can only be refined with a certain probability $p < 1$. Based on this analysis, they present an extension to ALPINE called HIGHPOINT, which selects an abstraction hierarchy with high refinement probability among those that satisfy the ordered monotonicity property. In practice, it is not feasible to compute the refinement probability, so HIGHPOINT approximates this value based on the notion of k-ary necessary connectivity.

8.2.2 Causal Graphs and Unary STRIPS Operators

Causal graphs are first given a name by Jonsson and Bäckström, who call them *dependency graphs* [71, 73]. They study a fragment of propositional STRIPS with negative conditions which has the interesting property that plan existence can be decided in polynomial time, but minimal solutions to a task can be exponentially long, so that no polynomial planning algorithm exists. They present an *incremental* planning algorithm with polynomial delay, i. e., a planning algorithm that decides within polynomial time whether or not a given task has a solution, and, if so, generates such a solution step by step, requiring only polynomial time between any two consecutive steps. (However, there is no guarantee that the length of the generated solution is polynomially related to the length of an optimal solution; it might be exponentially larger. Therefore, the algorithm might spend exponential time on tasks that can be solved in polynomial time.) The fragment of STRIPS covered by Jonsson and Bäckström's algorithm is called *3S* and is defined by the requirement that the causal graph of the task is acyclic and each state variable is *static, symmetrically reversible*, or *splitting*. *Static* variables are those for which it is easy to guarantee that they never change their value in any solution plan. These variables can be detected and compiled away easily. *Symmetrically reversible* variables are those where for each operator which makes them true there is a corresponding operator with identical preconditions which makes them false, and vice versa. In other words, a variable is symmetrically reversible iff its domain transition graph is undirected. Finally, a variable v is *splitting* iff its removal from the causal graph weakly disconnects its positive successors (those variables which appear in effects of operators of which v is a precondition) from its negative successors (those variables which appear in effects of operators of which $\neg v$ is a precondition).

Williams and Nayak independently prove that incremental (or, in their setting, *reactive*) planning is a polynomial problem in a STRIPS-like setting where causal graphs are acyclic and all operators are reversible [113]. If all operators are reversible (according to the definition by Williams and Nayak), all variables are symmetrically reversible (according to the definition by Jonsson and Bäckström), so this is actually a special case of the previous result. However, Williams and Nayak's work applies to a more general formalism than propositional STRIPS, so that the approaches are not directly comparable.

More recently, Domshlak and Brafman provide a detailed account of the complexity of finding plans in the propositional STRIPS (with negation) formalism with unary operators and acyclic graphs [17, 28]. (According to our formal definition of causal graphs in Sect. 10.3, operators with several effects always induce cycles in the causal graph, so *acyclic causal graph* implies *unary operators*. Some researchers define causal graphs differently, so we name both properties explicitly here.) Among other results, they prove that the restriction to unary operators and acyclic graphs does not reduce the complexity of plan existence: the problem is PSPACE-complete, just like unrestricted propositional STRIPS planning [19]. They also show that for singly connected causal graphs, shortest plans cannot be exponentially long, but the plan existence problem is still NP-complete. For an even more restricted class of causal graphs, namely polytrees of bounded indegree, they present a polynomial planning algorithm. More generally, their analysis relates the complexity of STRIPS planning in unary domains to the *number of paths* in their causal graph.

8.2.3 Multi-Valued Planning Tasks

With the exception of Williams and Nayak's paper, all the work discussed so far exclusively deals with *propositional* planning problems, where all state variables assume values from a binary domain. As we observed in the introduction, the question of propositional vs. multi-valued encodings usually has a strong impact on the connectivity of the causal graph of a task. In fact, apart from the trivial MOVIE domain, none of the common planning benchmarks exhibits an acyclic causal graph when considering its propositional representation. By contrast, the multi-valued encoding of our introductory example does have an acyclic causal graph.

Due to the dominance of the PDDL (and previously, STRIPS) formalism, non-binary state variables are not studied very often in the classical planning literature. One of the most important exceptions to this rule is the work on the SAS$^+$ planning formalism, of which the papers by Bäckström and Nebel [10] and Jonsson and Bäckström [72] are most relevant to Fast Downward. The SAS$^+$ planning formalism is basically equivalent to the *multi-valued planning tasks* we introduce in Chap. 9 apart from the fact that it does not include derived variables (axioms) or conditional effects. Bäckström and Nebel analyse the complexity of various subclasses of the SAS$^+$ formalism and discover three properties (*unariness*, *post-uniqueness* and *single-valuedness*) that together allow optimal planning in polynomial time. One of these three properties (unariness) is related to acyclicity of causal graphs, and one (post-uniqueness) implies a particularly simple shape of domain transition graphs (namely, in post-unique tasks, all domain transition graphs must be simple cycles or trees).

Bäckström and Nebel do not analyse domain transition graphs formally. Indeed, the term is only introduced in the later article by Jonsson and Bäckström, which refines the earlier results by introducing five additional restrictions for SAS$^+$ tasks, all of which are related to properties of domain transition graphs [72].

Neither of these two articles discusses the notion of causal graphs. The first earlier work we are aware of which includes *both* causal graphs and domain transition graphs as central concepts is the article by Domshlak and Dinitz on the *state-transition support* (STS) problem, which is essentially equivalent to SAS$^+$ planning with unary operators [29]. In the context of STS, domain transition graphs are called *strategy graphs* and causal graphs are called *dependence graphs*, but apart from minor details, the semantics of the two formalisms are identical. Domshlak and Dinitz provide a map of the complexity of the STS problem in terms of the shape of its causal graph, showing that the problem is NP-complete or worse for almost all non-trivial cases. One interesting result is that if the causal graph is a simple chain of n nodes and all variables are three-valued, the length of minimal plans can already grow as $\Omega(2^n)$. By contrast, *propositional* tasks with the same causal graph shape admit polynomial planning algorithms according to the result by Brafman and Domshlak [17], because such causal graphs are polytrees with a constant indegree bound (namely, a bound of 1).

Although it does not exploit the notion of causal graphs, Amir and Engelhardt's work on *factored planning* [2] is a closely related approach. In factored planning, related state variables are clustered into groups (or *factors*) for which local plans may be generated independently from one another. The approach of Amir and Engelhardt computes a global solution by combining local plans in a dynamic programming approach, traversing a tree-shaped graph of factor dependencies in bottom-up order.

The dependency graphs defined by Amir and Engelhardt are slightly different from causal graphs. In particular, they are undirected, and two factors are considered related even if they only share *precondition* variables. However, the approach has subsequently been modified by Brafman and Domshlak [18] to use causal graphs explicitly, leading to a significant improvement of worst-case performance.

To summarize and conclude our discussion of related work, we observe that the central concepts of Fast Downward and the causal graph heuristic, such as causal graphs and domain transition graphs, are firmly rooted in previous work. However, Fast Downward is the first attempt to marry hierarchical task decomposition to the use of multi-valued state variables within a general planning framework. It is also the first attempt to apply techniques similar to those of Knoblock [79] and Bacchus and Yang [9] within a heuristic search planner.

The significance of this latter point should not be underestimated: For classical approaches to hierarchical task decomposition, it is imperative that an abstraction satisfies the ordered monotonicity property, and it is important that the probability of being able to refine an abstract plan to a concrete plan is high, as the analysis by Bacchus and Yang shows. Unfortunately, non-trivial abstraction hierarchies are rarely ordered monotonic, and even more rarely guarantee high refinement probabilities. Within a heuristic approach, however, these "must-haves" turn into "nice-to-haves": If an abstraction hierarchy is not ordered monotonic or if an abstract plan considered by the heuristic evaluator is not refinable, this merely reduces the quality of the heuristic estimate, rather than causing the search to fail (in the worst case) or spend a long time trying to salvage non-refinable abstract plans (in the not much better case).

8.3 Architecture and Overview

We now describe the overall architecture of the planner and provide an overview of the remaining chapters of Part II.

Fast Downward is a classical planning system based on the ideas of heuristic forward search and hierarchical task decomposition. It can deal with the full range of propositional PDDL2.2 [38, 42], i. e., in addition to STRIPS planning, it supports arbitrary formulae in operator preconditions and goal conditions,

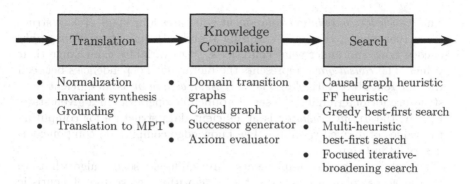

Fig. 8.9. The three phases of Fast Downward's execution

and it can deal with conditional and universally quantified effects and derived predicates (axioms).

The name of the planner derives from two sources: Of course, one of these sources is Hoffmann's very successful FF ("Fast Forward") planner [68]. Like FF, Fast Downward is a heuristic progression planner, i. e., it computes plans by heuristic search in the space of world states reachable from the initial situation. However, compared to FF, Fast Downward uses a very different heuristic evaluation function called the *causal graph heuristic*. The heuristic evaluator proceeds "downward" in so far as it tries to solve planning tasks in the hierarchical fashion outlined in the introduction. Starting from top-level goals, the algorithm recurses further and further down the causal graph until all remaining subtasks are basic graph search tasks.

Similar to FF, the planner has shown excellent performance: The original implementation of the causal graph heuristic, plugged into a standard best-first search algorithm, outperformed the previous champions in that area, FF and LPG [44], on the set of STRIPS benchmarks from the first three international planning competitions [58]. Fast Downward itself followed in the footsteps of FF and LPG by winning the propositional, non-optimizing track of IPC4.

Fast Downward solves a planning task in three phases (Fig. 8.9), to each of which we dedicate one chapter in the following:

– The *translation* component is responsible for transforming the PDDL2.2 input into a non-binary form which is more amenable to hierarchical planning approaches. It applies a number of normalizations to compile away syntactic constructs like disjunctions which are not directly supported by the causal graph heuristic and performs grounding of axioms and operators. Most importantly, it uses invariant synthesis methods to find groups of related propositions which can be encoded as a single multi-valued variable. The output of the translation component is a *multi-valued planning task*. The translation component is described in the following Chap. 9.

- The *knowledge compilation* component generates four kinds of data structures that play a central role during search: *Domain transition graphs* encode how, and under what conditions, state variables can change their values. The *causal graph* represents the hierarchical dependencies between the different state variables. The *successor generator* is an efficient data structure for determining the set of applicable operators in a given state. Finally, the *axiom evaluator* is an efficient data structure for computing the values of derived variables. The knowledge compilation component is described in Chap. 10.
- The *search* component implements three different search algorithms to do the actual planning. Two of these algorithms make use of heuristic evaluation functions: One is the well-known greedy best-first search algorithm, using the causal graph heuristic. The other is called *multi-heuristic best-first search*, a variant of greedy best-first search that tries to combine several heuristic evaluators in an orthogonal way; in the case of Fast Downward, it uses the causal graph and FF heuristics. The third search algorithm is called *focused iterative-broadening search*; it is closely related to Ginsberg and Harvey's iterative broadening [49]. It is not a heuristic search algorithm in the sense that it does not use an explicit heuristic evaluation function. Instead, it uses the information encoded in the causal graph to estimate the "usefulness" of operators towards satisfying the goals of the task. The search component is described in Chap. 11.

After describing the planning system, we evaluate its performance in Chap. 12 and provide a final discussion in Chap. 13.

With the exception of Chap. 9 discussing the translation component, this part largely follows a publication in the Journal of Artificial Intelligence Research (JAIR) [59].

9

Translation

The purpose of the translation component is to transform the input planning task, specified in the (first-order) PDDL formalism [38], into a fully instantiated multi-valued representation based on the SAS$^+$ formalism [10, 72].

PDDL and multi-valued planning tasks are introduced in Sect. 9.1, followed by an overview of the translation algorithm in Sect. 9.2. Translation is performed in four stages: *normalization* (Sect. 9.3), *invariant synthesis* (Sect. 9.4), *grounding* (Sect. 9.5), and *multi-valued planning task generation* (Sect. 9.6). The chapter ends with some notes on the performance of the translation component (Sect. 9.7).

9.1 PDDL and Multi-valued Planning Tasks

PDDL is the language in which the standard benchmarks discussed in Part I are usually expressed. In particular, the planning tasks of the international planning competitions are expressed in PDDL, so a planning system must be able to deal with this language in order to participate.

Like most current planning systems, Fast Downward is limited in scope to the *non-numerical* fragment of PDDL2.2, or what is called "level 1" of the PDDL language [42]. In other words, it does not accept PDDL tasks involving numerical state variables (introduced in PDDL level 2) or "durative actions", which allow specifying tasks that can only be solved by concurrent plans (introduced in PDDL level 3 and refined in PDDL level 4).

On the other hand, Fast Downward *can* deal with all "purely logical" aspects of the language, including arbitrary first-order formulae in action conditions and goals, universal and conditional effects, and derived predicates (axioms) introduced in PDDL2.2 [38]. To make these notions somewhat more precise, we now formally introduce the class of PDDL tasks which the planner can handle.

Definition 9.1.1. *PDDL Tasks*
*A (non-numeric, non-temporal) **PDDL task** is given by a 5-tuple $\Pi = \langle \mathcal{L}, \chi_0, \chi_\star, \mathcal{A}, \mathcal{O} \rangle$ with the following components:*

– *$\mathcal{L}$ is a finite first-order language, consisting of constant symbols (**objects**), relation symbols (**predicates**) and variable symbols. Predicates are partitioned into **fluent predicates** (affected by operators) and **derived predicates** (computed by evaluating axioms).*
– *χ_0 is a conjunction of ground atoms over objects and fluent predicates called the **initial state**.*
– *$\chi_\star$ is a closed first-order formula over $\mathcal{L}$ called the **goal formula**.*
– *$\mathcal{A}$ is a set of **schematic axioms**, which are pairs $\langle \varphi, \psi \rangle$ such that φ is an atom over $\mathcal{L}$ whose predicate symbol is a derived predicate and ψ is a formula over $\mathcal{L}$ with free(ψ) $\subseteq$ free(φ). We write the axiom $\langle \varphi, \psi \rangle$ as $\varphi \leftarrow \psi$ and call φ the **head** and ψ the **body** of the axiom.*
  We require that $\mathcal{A}$ is stratifiable, i. e., there exists a total preorder $\preceq$ on the set of derived predicates such that for each axiom where Q occurs in the head, we must have $P \preceq Q$ for all derived predicates P occurring in the body, and $P \prec Q$ for all derived predicates P occurring in a negative literal in the translation of the body to negation normal form. Intuitively, $P \prec Q$ means that the truth value of atoms over P must be computed before the truth value of atoms over Q.
– *$\mathcal{O}$ is a finite set of **schematic operators** over $\mathcal{L}$. A schematic operator $\langle \chi, e \rangle$ consists of a first-order formula χ over $\mathcal{L}$ called its **precondition** and its **effect** e. Effects are recursively defined by finite application of the following rules:*
  – *A literal l over $\mathcal{L}$, excluding derived predicates, is an effect called a **simple effect**.*
  – *If $e_1, \ldots, e_n$ are effects, then $e_1 \wedge \cdots \wedge e_n$ is an effect called a **conjunctive effect**.*
  – *If χ is a first-order formula over $\mathcal{L}$ and e is an effect, then $\chi \triangleright e$ is an effect called a **conditional effect**.*
  – *If $v_1, \ldots, v_k$ are variable symbols in $\mathcal{L}$ and e is an effect, then $\forall v_1 \ldots v_k : e$ is an effect called a **universally quantified effect** or **universal effect**.*
  Free variables of an effect are defined recursively as in first-order logic, where the set of free variables of a conditional effect is defined as free($\chi \triangleright e$) = free(χ) $\cup$ free(e).
  *The set of free variables of a schematic operator is defined as free($\langle \chi, e \rangle$) = free(χ) $\cup$ free(e). Free variables are also referred to as the **parameters** of the schematic operator.*

We assume that the reader is already familiar with PDDL semantics and point to the language definition [38, 42] for more information. Our definition allows general first-order conditions as well as (possibly nested) conditional and quantified effects and axioms. The stratifiability condition for axioms ensures that the interpretation of derived predicates is well-defined. Without this condition, there could be rules of the form "$P(x)$ is true whenever $P(x)$ is false."

Apart from syntactic differences, there are three aspects of non-numerical, non-temporal PDDL2.2 not captured by our definition:

- There are no operator names. The translator deals with operator names in such a way that the translated operators are referred to by the same name as its PDDL2.2 counterpart, so that the plans generated by the planner need not undergo any form of post-processing. This is all fairly simple, and we will not discuss this matter further.
- There is no distinction between domain constants and objects of the problem instance, or indeed between the domain and problem instance specification in general. At the level of individual problem instances at which the translator works, there is no need for such a distinction.
- There are no types. The translator compiles away types into unary predicates straight away, so a PDDL specification stating that a is an object of type vehicle is treated equivalently to an untyped specification stating that (vehicle a) is true in the initial state. Types occurring in quantified conditions or effects are translated accordingly; e.g. a precondition (exists (?v - vehicle) (empty ?v)) is translated to $\exists v : \text{vehicle}(v) \wedge \text{empty}(v)$, and an effect (forall (?v - vehicle) (empty ?v)) is translated to $\forall v : (\text{vehicle}(v) \triangleright \text{empty}(v))$.

With PDDL as a starting point, let us now introduce the kinds of planning tasks we want the translator to generate. These are based on the SAS$^+$ planning model [10, 72], extended to allow for derived predicates and conditional effects.

The definition will exhibit a number of similarities, but also a few differences between PDDL tasks and our planning model. Most notably, PDDL tasks use first-order concepts such as schematic operators whose variables can be instantiated in many different ways, while our formalism is grounded. Moreover, our formalism only allows simple conjunctions in goals, axioms and operators, and conditional effects cannot be nested. The former difference necessitates operator instantiation as part of the translation process, while the others require normalization of conditions.

Definition 9.1.2. *Multi-valued Planning Tasks (MPTs)*
A *multi-valued planning task (MPT)* is given by a 5-tuple $\Pi = \langle V, s_0, s_\star, A, O \rangle$ with the following components:

- V is a finite set of **state variables**, each with an associated finite domain $\mathcal{D}_v$. State variables are partitioned into **fluents** (affected by operators) and **derived variables** (computed by evaluating axioms). The domains of derived variables must contain the **default value** $\perp$.
  A **partial variable assignment** or **partial state** over V is a function s on some subset of V such that $s(v) \in \mathcal{D}_v$ wherever $s(v)$ is defined. A

*partial state is called an **extended state** if it is defined for all variables in
$\mathcal{V}$ and a **reduced state** or **state** if it is defined for all fluents in $\mathcal{V}$. In the
context of partial variable assignments, we write $v = d$ for the variable-
value pairing $\langle v, d \rangle$ or $v \mapsto d$.*

- *s_0 is a state over $\mathcal{V}$ called the **initial state**.*
- *$s_\star$ is a partial variable assignment over $\mathcal{V}$ called the **goal**.*
- *$\mathcal{A}$ is a finite set of (MPT) **axioms** over $\mathcal{V}$. Axioms are triples of the
form $\langle cond, v, d \rangle$, where cond is a partial variable assignment called the
condition or **body** of the axiom, v is a derived variable called the **affected
variable**, and $d \in \mathcal{D}_v$ is called the **derived value** for v. The pair $\langle v, d \rangle$
is called the **head** of the axiom and can be written as $v := d$.*

  *The axiom set $\mathcal{A}$ is partitioned into a totally ordered set of **axiom layers**
$\mathcal{A}_1 \prec \cdots \prec \mathcal{A}_k$ such that within the same layer, each affected variable
may only be associated with a single value in axiom heads and bodies. In
other words, within the same layer, axioms with the same affected variable
but different derived values are forbidden, and if a variable appears in an
axiom head, then it may not appear with a different value in a body. This
is called the **layering property**.*

- *$\mathcal{O}$ is a finite set of (MPT) **operators** over $\mathcal{V}$. An operator $\langle pre, eff \rangle$ consists
of a partial variable assignment pre over $\mathcal{V}$ called its **precondition**, and
a finite set of **effects** eff. Effects are triples $\langle cond, v, d \rangle$, where cond is a
(possibly empty) partial variable assignment called the **effect condition**,
v is a fluent called the **affected variable**, and $d \in \mathcal{D}_v$ is called the **new
value** for v.*

*For axioms and effects, we also use the notation $cond \rightarrow v := d$ in place
of $\langle cond, v, d \rangle$.*

To provide a formal semantics for MPT planning, we first need to formalize
the semantics of axioms.

Definition 9.1.3. *Extended States Defined by a State*
*Let s be a state of an MPT Π with axioms $\mathcal{A}$, layered as $\mathcal{A}_1 \prec \cdots \prec \mathcal{A}_k$.
The **extended state defined by** s, written as $\mathcal{A}(s)$, is the result s' of the
following algorithm:*

algorithm evaluate-axioms($\mathcal{A}_1, \ldots, \mathcal{A}_k, s$):

   for each variable v:

$$s'(v) := \begin{cases} s(v) & \text{if } v \text{ is a fluent variable} \\ \bot & \text{if } v \text{ is a derived variable} \end{cases}$$

   for $i \in \{1, \ldots, k\}$:

   while there exists an axiom $(cond \rightarrow v := d) \in \mathcal{A}_i$

   with $cond \subseteq s'$ **and** $s'(v) \neq d$:

   Choose such an axiom $cond \rightarrow v := d$.

   $s'(v) := d$

In other words, axioms are evaluated in a layer-by-layer fashion using fixed
point computations, which is very similar to the semantics of stratified logic

programs. It is easy to see that the layering property from Definition 9.1.2 guarantees that the algorithm terminates and produces a deterministic result. Having defined the semantics of axioms, we can now define the state space of an MPT.

Definition 9.1.4. *MPT State Transition Graph*
*The **state transition graph** of an MPT $\Pi = \langle \mathcal{V}, s_0, s_\star, \mathcal{A}, \mathcal{O} \rangle$, denoted as $\mathcal{S}(\Pi)$, is a directed graph. Its vertex set is the set of states of $\mathcal{V}$, and it contains an arc $\langle s, s' \rangle$ iff there exists some operator $\langle pre, e\!f\!f \rangle \in \mathcal{O}$ such that:*

- *$pre \subseteq \mathcal{A}(s)$,*
- *$s'(v) = d$ for all effects $cond \rightarrow v := d \in e\!f\!f$ such that $cond \subseteq \mathcal{A}(s)$, and*
- *$s'(v) = s(v)$ for all other fluents.*

Having defined state transition graphs for MPT tasks, we can introduce MPT plans, the MPT planning problem PLAN-MPT, the MPT plan existence PLANEX-MPT and the MPT bounded plan existence problem PLANLEN-MPT in a similar way to the definitions in Chap. 2. Indeed, we can consider the set of all MPTs a (very general) planning domain in the sense of Definition 2.2.2, and thus no further definitions of these minimization and decision problems are required.

The plan existence and bounded plan existence problems for MPTs are easily shown to be PSPACE-hard because they generalize the corresponding problems for propositional STRIPS, which are known to be PSPACE-complete [19]. Moreover, from Theorems 2.3.2 and 2.3.4, we know that they belong to PSPACE, so they are PSPACE-complete. Similarly, it is easy to see from Theorem 2.3.1 that PLAN-MPT belongs to EXPO\NPS because plans may be exponentially long (but not longer), and explicit search in the state transition graph can be implemented to run in exponential time.

This concludes our formal introduction of MPT planning. In the following section, we turn to the issue of generating multi-valued planning tasks from PDDL planning tasks.

9.2 Translation Overview

Translation is performed in a sequence of transformation steps. Starting from a PDDL specification, we apply some well-known logical equivalences to compile away types and simplify conditions and effects in the *normalization* step. Next, the *invariant synthesis* step computes mutual exclusion relations between atoms, which are later used for synthesizing the MPT variables. The *grounding* step performs a relaxed reachability analysis to compute the set of ground atoms, axioms and operators that are considered relevant for the planning task and computes a grounded PDDL representation. Invariant synthesis and grounding are not related to one another and could just as well be performed in the opposite order. Finally, the *MPT generation* step chooses

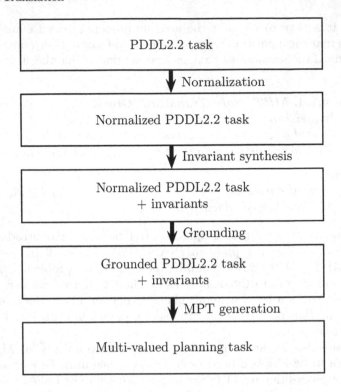

Fig. 9.1. Overview of the translation algorithm

the final set of state variables by using the information from invariants and grounding and produces the MPT output.

The translation process in outlined in Fig. 9.1. In the following sections, we will discuss the various transformation steps in sequence.

However, before we do so, we should point out that of these four steps, only three are necessary to convert a PDDL task to an MPT: the invariant synthesis step can be omitted. However, without the use of invariants, there would be a 1:1 correspondence between (relevant) ground atoms of the PDDL task and state variables of the MPT; in particular, all state variables in the generated MPT would be binary. Recalling the motivating example from the previous chapter, this would imply that the causal graph of the resulting MPT would have the undesirable form shown in Fig. 8.7, rather than the much more structured form shown in Fig. 8.6.

9.3 Normalization

The normalization step has three responsibilities: Compiling away types, simplifying conditions, and simplifying effects. The current implementation of the translator cannot handle PDDL types in their full generality: Type inheritance and the `either` construct are not supported. It would not be difficult to add these to the mix, but these seem to be unused language features, and we did not want to waste implementation effort on them. So we only deal with primitive types and the built-in standard type `object`, to which all objects belong.

9.3.1 Compiling Away Types

As indicated earlier, types are compiled away as soon as the planning task is read in. For each type occurring in the input, and for the type `object`, we introduce a new unary predicate with the same name. Typed constructs occur in PDDL2.2 specifications in a semantically meaningful way in three places:

1. Definition of domain constants and objects of the task (*typed objects*).
2. Definition of formal parameters of schematic operators (*typed operators*).
3. Definition of quantified variables in existential and universal conditions and universal effects (*typed quantifiers*).

Typed objects are translated into new atoms for the initial state. For example, the specification `someobj - sometype` leads to a new initial atom (`sometype someobj`). Moreover, for each object `someobj`, we introduce an initial atom (`object someobj`).

Typed operators are transformed by introducing new preconditions. For example, for an operator with parameter specification `:parameters (?par1 - type1 ?par2 - type2)` and precondition φ, the parameter specification is replaced by `:parameters (?par1 ?par2)` and the precondition is replaced by (`and (type1 ?par1) (type2 ?par2)` φ).

Typed quantifiers in conditions are compiled away with the usual first-order logic idioms, so that condition (`exists (?v - type)` φ) translates to (`exists (?v) (and (type ?v)` φ)) and condition (`forall (?v - type)` φ) translates to (`forall (?v) (imply (type ?v)` φ)).

Similarly, typed quantifiers in effects are compiled into conditional effects, so that the effect (`forall (?v - type)` e) becomes (`forall (?v) (when (type ?v)` e)).

After types have been eliminated, we are left with a PDDL task in the sense of Definition 9.1.1. We will thus use the more concise logical notation from that definition in the following, rather than the more lengthy PDDL syntax. For example, we will write $\varphi \vee \psi$ instead of (`or` φ ψ) and $\varphi \triangleright e$ instead of (`when` φ e).

9.3.2 Simplifying Conditions

In PDDL tasks, general first-order formulae may occur in many places: goal formula, axiom bodies, operator preconditions and conditions of conditional effects. Our aim is to replace all these with simple conjunctions of literals.

Towards this goal, we first eliminate implications with the equivalence $\varphi \to \psi \equiv \neg\varphi \lor \psi$ and translate the resulting conditions into first-order negation normal form using de Morgan's laws for first-order logic.

The next step is slightly tricky. If there are any universally quantified conditions, we rewrite the outermost universal quantification in all conditions with the equivalence $\forall x \varphi \equiv \neg \exists x \neg \varphi$. This might seem somewhat silly because this transformation destroys negation normal form, so after the rewrite, we introduce a new axiom for the subformula that violates the normal form property, $\exists x \neg \varphi$. Formally, if free$(\exists x \neg \varphi) = \{v_1, \ldots, v_k\}$, we introduce a new derived predicate **new-pred** of arity k, defined by the axiom $\text{new-pred}(v_1, \ldots, v_k) \leftarrow \psi$, where ψ is the translation of $\exists x \neg \varphi$ to negation normal form. We can then replace the original condition $\forall x \varphi$ by $\neg\text{new-pred}(v_1, \ldots, v_k)$. If several variables are universally quantified together within the same expression, we transform them together, introducing only one new derived predicate for the quantifier group. We repeat this step until there are no more universally quantified conditions. Note that only universally quantified *conditions* are translated, not universal *effects*, which also use the $\forall$ notation. Universal effects cannot simply be compiled away, so we deal with them separately in a later stage.

If after elimination of universal quantifiers the goal condition is *not* a simple conjunction, we replace it by a new axiom, since the following transformations sometimes require splitting several conditions into two, which is easy to do for axiom bodies, operator preconditions and effect conditions, but not possible in our formalism for goal conditions, of which there can be only one. So for example, if the goal is $\varphi \lor \psi$, we introduce a new parameter-less derived predicate **goal-pred** and a new axiom $\text{goal-pred} \leftarrow \varphi \lor \psi$, replacing the original goal with the atom **goal-pred**.

The next step is the elimination of disjunctions. We move disjunctions to the roots of conditions by applying the equivalences $\exists x(\varphi \lor \psi) \equiv \exists x \varphi \lor \exists x \psi$ and $\varphi \land (\psi \lor \psi') \equiv (\varphi \land \psi) \lor (\varphi \land \psi')$ and the laws of associativity and commutativity. In theory, moving disjunctions over conjunctions can lead to an exponential increase in formula size, which we could avoid by introducing new axioms for component formulae. In practice, the conditions encountered in actual planning domains are not problematic in this regard, and we decided that the potential savings in the size of the representation were not worth the overhead of maintaining the state of another derived variable during search.

After disjunctions have been moved to the root of all formulae, we can eliminate them by splitting the surrounding structures: If the disjunction $\varphi \lor \psi$ is part of an axiom body, we generate two axioms with identical head, one with body φ and one with body ψ. If the disjunction is part of an operator precondition, we replace the operator by two copies of the original, one with precondition φ and one with precondition ψ. Finally, if the disjunction is part of an effect condition, we replace the conditional effect $(\varphi \lor \psi) \rhd e$ by $(\varphi \rhd e) \land (\psi \rhd e)$.

Next, we move existential quantifiers out of conjunctions by applying the equivalence $(\exists x \varphi) \land \psi \equiv \exists x(\varphi \land \psi)$. The equivalence only holds when $x \notin$

free(ψ), so to avoid trouble here and later, we first rename all variables bound by quantifiers to some unique name.

Having moved existential quantifiers to the root of conditions, we eliminate them as follows: For axioms, we simply drop them, following the PROLOG convention that all free variables in the body that are not part of the head are implicitly existentially quantified. For operator preconditions, we also drop them, adding the existentially quantified variables to the parameter list of the schematic operator. For effect conditions, we replace $(\exists x\varphi) \rhd e$ by $\forall x : (\varphi \rhd e)$.

9.3.3 Simplifying Effects

After the somewhat laborious simplification of conditions, effect simplification is conceptually very simple. First, universal and conditional effects are moved into conjunctive effects by the equivalences $\forall x : (e \wedge e') \equiv (\forall x : e) \wedge (\forall x : e')$ and $\varphi \rhd (e \wedge e') \equiv (\varphi \rhd e) \wedge (\varphi \rhd e')$. Second, conditional effects are moved into universal effects by the equivalence $\varphi \rhd (\forall x : e) \equiv \forall x : (\varphi \rhd e)$. Finally, nested effects of the same type are flattened, i. e., conjunctive effects containing conjunctive effects are collapsed into a single conjunctive effects with more conjuncts, universal effects containing universal effects are collapsed into a single universal effect quantifying over more variables, and nested conditional effects of the type $\varphi \rhd (\psi \rhd e)$ are transformed to $(\varphi \wedge \psi) \rhd e$. Note that this latter modification preserves the previously generated normal form for effect conditions.

After these transformations, the possible nesting of effects is thus restricted to the simple chain *conjunctive effect $\succ$ universal effect $\succ$ conditional effect $\succ$ simple effect*. However, not all effect types must necessarily be present, e. g. conditional effects *need not* occur within universal effects, etc. To enforce a regular effect structure, we replace simple effects e not surrounded by conditional effects by $\top \rhd e$ ($\top$ is seen as the empty conjunction, so this condition is in normal form), conditional effects e not surrounded by universal effects by $\forall : e$ (quantifying over zero variables), and universal effects e not surrounded by conjunctive effects by a conjunctive effect containing the singleton e.

As a result, after normalization each operator has a list (conjunction) of effects, each a simple effect with an associated set of universal quantifiers and an associated condition, both of which can be trivial. Thus it is not necessary to store normalized operator effects in a tree structure; a flat vector is sufficient.

9.3.4 Normalization Result

This concludes the normalization step. In Fig. 9.1, we referred to the output of the normalization phase as a *normalized PDDL2.2 task*. Let us formalize this notion here for the benefit of further discussion:

Definition 9.3.1. *Normalized PDDL Tasks*
A *normalized PDDL task* is a PDDL task that satisfies the following structural restrictions:

- *The goal formula is a conjunction of literals.*
- *All axiom bodies are conjunctions of literals (except for the possible implicit
  existential quantification of free variables not occurring in the axiom head).*
- *All operator preconditions are conjunctions of literals.*
- *All effect conditions are conjunctions of literals.*
- *All operator effects are conjunctions of universally quantified conditional
  simple effects.*

In the following, we will refer to the individual simple effects of an operator
in a normalized PDDL task as being arranged in an *effect list*. For the simple
effect e occurring within the universal conditional effect $\forall vars : \varphi \triangleright e$, we will
refer to *vars* as the set of *bound variables* of e and to φ as the *condition* of e.
If e is a positive literal, we will call it an *add effect*, otherwise a *delete effect*.

9.4 Invariant Synthesis

An invariant is a property of a world state in a planning task which is satisfied
by all world states that are reachable from the initial state. Many invariants
are uninteresting; for example, the property "At least five state variables are
true" is an invariant in most propositional STRIPS planning tasks, but does
not seem to entail a useful piece of information for a planner. Other invariants
would be useful to know but are too difficult to verify. For example, "This
state is not a goal state" is an invariant iff the planning task is not solvable,
so confirming the invariance of that state property is PSPACE-hard.

Nevertheless, invariants are a useful tool for many planning systems, which
is why they have been studied by many researchers in a variety of contexts [41, 45, 100, 101], often involving SAT-based planning. For the purposes
of translating propositional planning tasks to a multi-valued formalism, *mutual exclusion* (*mutex*) invariants are especially interesting. A mutex invariant
states that certain propositions can never be true at the same time. This affects translation because a set of propositions which are pairwise mutually
exclusive can be easily encoded as a *single* state variable whose value specifies
which of the propositions is true (if any is true at all), rather than as a number
of state variables encoding the truth value for each proposition individually.

Invariance is usually proven inductively: First, one shows that a hypothe-
sized property is true in the initial state. Then, one shows that if the property
is true in some state, it must also be true in all successor states. Together, this
implies that the property is true in all reachable states, and thus an invariant.

As mentioned before, the automatic discovery of invariants is a hard problem in general, but for many relevant types of state properties, sufficient conditions exist that can be checked quickly. Still, synthesizing invariants is costly,

and for this reason, we are interested in algorithms working directly with the first-order PDDL description of a planning task, not on a grounded representation. Indeed, our algorithm goes beyond this requirement by not relying on the information in the *task* file of the PDDL input at all, solely exploiting information present in the *domain* file. This is a valuable feature, but it rules out the possibility of proving mutex conditions, because a mutex cannot be established without checking the initial state. Instead, we use a slight generalization of mutexes.

Definition 9.4.1. *Monotonicity Invariant Candidates*
*A **monotonicity invariant candidate** for a PDDL task Π is given by a pair $\mathcal{P} = \langle V, \Phi \rangle$, where V is a set of first-order variables called the **parameters** of the candidate, and Φ is a set of atoms. Variables occurring freely in Φ which are not parameters are called **counted variables** of the candidate.*

For $V = \{v_1, \ldots, v_m\}$ and $\Phi = \{\varphi_1, \ldots, \varphi_k\}$, we write $\mathcal{P}$ symbolically as $\forall v_1 \ldots v_m \; \varphi_1 + \cdots + \varphi_k \downarrow$. In the special case $V = \emptyset$, we write $\forall \cdot \; \varphi_1 + \cdots + \varphi_k \downarrow$.

In the following, we will mostly refer to monotonicity invariant candidates as *invariant candidates* or simply *candidates*; we do not consider other kinds of invariant candidates.

The preceding definition defines the syntax for invariant candidates; we now have to provide the semantics. Since this is somewhat involved, we provide an example from a transportation domain first. Consider the invariant candidate $\langle \{p\}, \{\mathtt{at}(p, l), \mathtt{in}(p, v)\}\rangle$, where p, l and v are variable symbols. We write this as $\forall p \, \mathtt{at}(p, l) + \mathtt{in}(p, v) \downarrow$ and read it as "For all packages p, the number of locations l such that $\mathtt{at}(p, l)$ is true plus the number of vehicles v such that $\mathtt{in}(p, v)$ is true, is non-increasing." In our terminology, p is the parameter of the candidate, while l and v are the counted variables. This invariant candidate is an actual invariant – it *does* hold in all reachable states – and it is one of the invariants found by our algorithm in a LOGISTICS-like domain. Let us now formalize what it means for a candidate to be an invariant.

Definition 9.4.2. *Monotonicity Invariants*
Let Π be a PDDL task and let $\mathcal{P} = \langle V, \Phi \rangle$ be a monotonicity invariant candidate for Π.

*An **instance** of $\mathcal{P}$ is a function α mapping the variables in V to objects of the planning task Π.*

*The set of **covered facts** of an instance α is the set of all ground atoms of the planning task Π which unify with some $\varphi \in \Phi$ under α, i.e., the set of all ground atoms φ_0 of Π for which there exists a variable map $\beta \supseteq \alpha$ such that $\beta(\varphi) = \varphi_0$ for some $\varphi \in \Phi$.*

*The **weight** of an instance α in a state s is the number of covered facts of α which are true in s.*

*The monotonicity invariant candidate $\mathcal{P}$ is called a **monotonicity invariant** iff for all instances α of $\mathcal{P}$, all states s reachable from the initial state of Π and all successor states s' of s, the weight of α in s' is no greater than the weight of α in s.*

The definition is probably best understood by considering the previously discussed example invariant. Similar to our convention for invariant candidates, we usually refer to monotonicity invariants simply as *invariants*.

As hinted before, monotonicity invariants are useful for grouping a number of related propositions into a single multi-valued variable: If we have found an invariant for a planning task *and* a given instance of that invariant has weight 1 in the initial state, then the facts covered by that instance are pairwise mutually exclusive. This is how the synthesized invariants are utilized during the later stages of translation.

So how do we generate invariants? Since the number of feasible candidates is too high for a guess-and-check algorithm, we follow a *guess, check and repair* approach: Starting from a set of a few simple initial elements, we try to prove that the candidates are indeed invariants. Whenever this is the case, we keep the invariant and do not consider it further. However, when the proof fails, we try to detect *why* this is the case and refine the candidate to generate more candidates that do not fail for the same reason (although they might fail for other reasons). From a high-level perspective, this is basically a search problem, and indeed we solve it using standard breadth-first search with a closed list. What remains to be said is how the search space of the algorithm is defined:

- *Initial states:* What are the initial candidates?
- *Termination test:* How do we prove that a candidate is an invariant?
- *Successor set:* How do we refine a candidate for which this proof fails?

We will deal with these questions in the following.

9.4.1 Initial Candidates

Before starting the actual invariant synthesis algorithm, we check which predicates are affected by operators at all: Some predicates, including but not limited to those representing types, are *constant* in the sense that atoms over these predicates have the same truth values in all states. Such predicates are no longer needed after grounding, so we need not consider them for invariant candidates. Of course, a constant predicate trivially satisfies a monotonicity invariant, but these are not very useful.

Therefore, we limit the set of interesting predicates to all *modifiable fluent predicates*, i.e., predicates which occur within operator effects (as part of a simple effect, not merely as part of an effect condition). Note that this also excludes derived predicates. In theory, there is no reason why there should be no monotonicity invariants involving derived predicates, but in practice we have not seen examples of this, and detecting them would require a more global view of the task definition and hence more effort than we would like to spend. We will come back to the issue of derived predicates when discussing our method for proving invariance.

The set of initial invariant candidates consists of all those candidates (up to isomorphism, i. e., renaming of variables) which contain at most one counted variable and exactly one atom, over a modifiable fluent predicate, whose parameters are distinct variables. In our experience, invariants with several counted variables per atom are exceedingly rare; in fact, we have not seen an example in practice.

To illustrate the initialization of invariant candidates, we show the three candidates generated for the binary at predicate in the LOGISTICS domain:

$$\forall x \, \mathsf{at}(x, l) \downarrow \tag{9.1}$$

$$\forall l \, \mathsf{at}(x, l) \downarrow \tag{9.2}$$

$$\forall x, l \, \mathsf{at}(x, l) \downarrow \tag{9.3}$$

Similar candidates are introduced for the in predicate. Intuitively, the first candidate states that no object can be at more locations in the successor state than in the current state, the second candidate states that no location can be occupied by more objects in the successor state than in the current state, and the third candidate states that a given object cannot occupy a given location in the successor state if this is not the case in the current state.

Candidates (9.2) and (9.3) are obviously not invariants. Candidate (9.1) is not an invariant either because an object which is currently inside a vehicle can be at some location in the successor state while being at no location in the current state. However, we will see that we can refine (9.1) into an invariant.

9.4.2 Proving Invariance

When is an invariant candidate an invariant? We stated that invariants are usually proved by establishing their truth in the initial state and using inductive arguments for the effects of operator application. For monotonicity invariants, only the inductive step is necessary; there is nothing special to prove about the initial state. So in order to prove that a given invariant candidate is an invariant, we must show that no operator can increase the weight of any of its instances. An operator increases the weight of some instance of an invariant candidate iff the number of covered facts that it makes true is greater than the number of covered facts that it makes false. If an operator does not increase the weight of any instance, then we say that it is *balanced* with regard to the invariant.

Ultimately, we are interested in instances of monotonicity invariants that give rise to mutexes, so that only instances of weight 1 are relevant for us. For this reason, we use the following condition which is slightly stronger than balance.

Definition 9.4.3. *Threatened Invariant Candidates*
An invariant candidate $\mathcal{P}$ is **threatened** *by a schematic operator iff one of the following two conditions holds:*

- *The operator has an add effect that can increase the weight of an instance of $\mathcal{P}$ in some state, but no delete effect that is guaranteed to decrease the weight of the same instance in the same state. In this case, we say that the operator is **unbalanced** with regard to $\mathcal{P}$.*
- *When ignoring delete effects, the operator can increase the weight of some instance of $\mathcal{P}$ in some state by at least 2. In this case, we say that the operator is **too heavy** for $\mathcal{P}$.*

Clearly, not being threatened by any schematic operator is a sufficient condition for being a monotonicity invariant. The definition gives rise to the algorithm shown in Fig. 9.2. Most of the actual work is in unifying operator parameters and quantified variables of universal conditions; the algorithm simplifies significantly in STRIPS domains. We do not want to discuss the algorithm in all detail, instead focusing on two points that require some explanation, namely the satisfiability and entailment tests that occur towards the end of algorithms *check-operator-too-heavy* and *check-operator-unbalanced*.

For the heaviness test, two add effects can only lead to an operator being too heavy if the operator is actually applicable (o'.precondition is true), both add effects apply (e.condition and e'.condition are true) and the add effects actually add propositions that were not true previously (e.atom and e'.atom are false). For the imbalance test, an add effect is unbalanced by default. However, it becomes balanced if whenever the operator is actually applicable (o'.precondition is true), the add effect triggers (e.condition is true) and actually adds something (e.atom is false), then something is deleted at the same time, which means that the delete effect triggers (e'.condition is true) and deletes something that was previously true (e'.atom is true).

Coming back to the earlier LOGISTICS example, all three initial candidates are threatened by the same operator **unload-truck**, whose add effect $\mathbf{at}(x, l)$ is not balanced. Thus, as indicated before, none of (9.1)–(9.3) is an invariant.

There are a few subtleties about the algorithm which we want to point out briefly:

- We duplicate universal effects at the beginning of *check-operator-too-heavy* so that we can detect if two different instantiations of the same universal effect can simultaneously increase the weight of some instance of the invariant candidate.
- Where Fig. 9.2 contains statements like "Let o' be a copy of o where variables are renamed so that...", the question arises whether such a renaming is uniquely determined, and what to do if it is not. Indeed, renamings are unique (and easy to compute) as long as all atoms of the candidate refer to *different* predicates, which is usually the case. However, the algorithm generalizes to invariant candidates with several occurrences of the same predicate, like $\forall x \, \mathbf{at}(x, y) + \mathbf{at}(y, x) \downarrow$. This requires that all possible (non-isomorphic) renamings must be considered for o' in algorithm *check-operator-unbalanced*. In our experience, invariants of this type are not very useful, although Fast Downward implements them correctly.

---

algorithm prove-invariant(V, Φ):
        for each schematic operator o with add effect for some predicate in Φ:
                call check-operator-too-heavy(o, V, Φ).
                call check-operator-unbalanced(o, V, Φ).
        accept candidate as an invariant.

{ In the following, the *variables* of an operator include both its operator
parameters and quantified variables of its effects. We assume that all variable
names are unique, and that whenever a variable is renamed, the change is
immediately reflected in program variables referring to effects of the operator.
For example, if $e = \mathbf{at}(p, l)$ is an effect of operator o and p is renamed to v
within o, then e becomes $\mathbf{at}(v, l)$. }

algorithm check-operator-too-heavy(o, V, Φ):
        Let o' be a copy of o.
        Duplicate all (non-trivially) quantified effects of o'.
        Assign unique names to all quantified variables in effects of o and o'.
        for each pair (e, e') of add effects of o' that affect a predicate in Φ:
            if the variables of operator o' can be renamed so that
                (e.atom $\neq$ e'.atom **and**
                covers(V, Φ, e.atom) **and** covers(V, Φ, e'.atom) **and**
                o'.precondition$\wedge e$.condition$\wedge e'$.condition$\wedge \neg e$.atom$\wedge \neg e'$.atom
                is satisfiable):
                reject candidate. { The operator is too heavy. }

algorithm check-operator-unbalanced(o, V, Φ):
        for each add effect e of o that affects a predicate in Φ:
            Let o' be a copy of o where the variables are renamed so that
            covers(V, Φ, e.atom) is true. Do not rename two variables
            to the same variable except when forced.
            for each delete effect e' of o' that affects a predicate in Φ:
                if the quantified variables of e' can be renamed in o'
                so that (covers(V, Φ, e'.atom) **and**
                      o'.precondition $\wedge$ e.condition $\wedge$ $\neg e$.atom $\models$
                      o'.precondition $\wedge$ e'.condition $\wedge$ e'.atom):
                continue with next add effect e.
                { This add effect is balanced. }
            reject candidate. { The operator is unbalanced. }

function covers(V, Φ, ψ):
        for each $\varphi \in \Phi$:
            if the counted variables in φ (those not in V)
            can be renamed so that $\varphi = \psi$:
                return true.
        return false.

Fig. 9.2. Algorithm for proving that an invariant candidate $\langle V, \Phi \rangle$ is an invariant

- We have noted before that we do not consider invariants involving derived predicates. This is because axioms correspond to operators that have a single add effect, but no delete effect. Invariant candidates including derived predicates can thus never be balanced, except if the axiom body is unsatisfiable, which is not a very interesting case. Since we do not consider derived predicates within invariants, we can ignore axioms completely during invariant synthesis.
- Instead of using full-blown satisfiability and entailment tests, more limited tests are possible if they only err in the "conservative" direction. In practice, Fast Downward only employs simple structural entailment tests. However, this is due to scarcity of development time, not to conserve runtime, and we intend to extend the test to more complete logical reasoning.

One final subtlety concerns the semantics of PDDL operators with conflicting effects. Note that our balance test does not special-case the possibility that e.atom equals e'.atom, i.e., that the same atom is added and deleted. For the PDDL semantics that we adhere to, an operator which would add and delete the same atom would be invalid and thus inapplicable, not threatening any invariant candidates. We call this the *consistent effect semantics*. However, under another commonly accepted semantics, the add effect would "win" in such a case. We call this the *add-after-delete semantics*. Using the *add-after-delete* semantics, we would need to add the condition e.atom $\neq e'$.atom before the entailment test in *check-operator-unbalanced*. We believe that there is no commonly agreed "correct" semantics for PDDL with regard to this issue: For some standard benchmarks, such as MYSTERYPRIME, only the consistent effect semantics are reasonable, while for others, such as ROVERS, only the add-after-delete semantics make sense. Without going into further details, we note that it is not too difficult to adjust the algorithm to use the add-after-delete semantics.

9.4.3 Refining Failed Candidates

As indicated in the overview of the invariant synthesis algorithm, we do not give up immediately if we cannot prove a given candidate to be an invariant. Instead, we try to *refine* it by adding atoms that can restore balance. In algorithmic terms, whenever we reject an invariant candidate $\langle V, \Phi \rangle$, we try to generate a set of new candidates of the form $\langle V, \Phi \cup \{\varphi'\} \rangle$.

Whether or not this is promising depends on the reason why the candidate was rejected. If it was rejected because an operator is too heavy, then no possible refinement that adds an atom to the candidate can change this fact, and we give up on the candidate completely. If, however, it was rejected because of unbalanced operators, there is hope that we can deal with the flaw by adding an atom that can match some delete effect of the threatening operator, balancing the unbalanced add effect.

The refinement algorithm is shown in Fig. 9.3. The actual implementation in Fast Downward does not generate all possible refining atoms φ' naïvely,

---

algorithm refine-candidate(V, Φ):

    Select some schematic operator o such that
      check-operator-unbalanced(o, V, Φ) fails.
    for each atom φ' over variables from V and at most one other variable
        for which covers(V, Φ, φ') is not true:
        $\Phi' := \Phi \cup \{\varphi'\}$
        Simplify Φ' by removing atoms from Φ that are covered by φ'.
        (These cannot contribute to the weight of an instance of $\langle V, \Phi' \rangle$.)
        Simplify Φ' by removing unused parameters.
        if check-operator-too-heavy(o, V, Φ') does not fail:
            Add $\langle V, \Phi' \rangle$ to the set of invariant candidates.

Fig. 9.3. Algorithm for refining an unbalanced invariant candidate $\langle V, \Phi \rangle$

but rather uses information from the set of delete effects of the threatening operator o and the failed call to *check-operator-unbalanced* to only create candidates for φ' for which there is a chance that the new balance check will succeed. Since this is conceptually straight-forward, we do not go into more detail about this technique.

Instead, let us return to the LOGISTICS example. Recall that candidate (9.1), $\forall x\, \mathtt{at}(x, l) \downarrow$, is threatened by the operator unload-truck, whose add effect $\mathtt{at}(x, l)$ is unbalanced. The operator has only one delete effect, namely $\neg\mathtt{in}(x, t)$. Indeed, $\mathtt{in}(x, t)$ is a suitable refinement atom for φ' without further variable renaming, since the unload-truck operator is balanced with regard to the refined candidate $\forall x\, \mathtt{at}(x, l) + \mathtt{in}(x, t) \downarrow$. So we add this candidate to the set of currently considered candidates. At a later stage, it will be considered by *prove-invariant*, which will show that it is indeed an invariant.

By contrast, the other two candidates cannot be suitably refined. In order to refine (9.3), $\forall x, l\, \mathtt{at}(x, l) \downarrow$ to balance the drive-truck operator, we would need to add the atom $\mathtt{at}(x, l')$, which is the only delete effect of that operator. However, this atom covers the original atom $\mathtt{at}(x, l)$ (note that the converse is not true, because only l' is a counted variable), leading to the candidate $\forall x, l\, \mathtt{at}(x, l')$ where parameter l is unnecessary, so that it simplifies to $\forall x\, \mathtt{at}(x, l')$. This candidate is isomorphic to (9.2) and hence not considered again.

Considering candidate (9.2) and the drive-truck operator, the only possible refinement is $\forall \cdot\, \mathtt{at}(x, l') \downarrow$ ("The total number of at propositions is non-increasing"), which turns out to be violated by the unload-truck operator, but can be further refined to $\forall \cdot\, \mathtt{at}(x, l') + \mathtt{in}(x, l') \downarrow$ ("The total number of at and in propositions is non-increasing"). This latter candidate is actually an invariant. However, its only instance clearly has a weight greater than 1 in the initial state of any non-trivial LOGISTICS task and thus turns out not to provide any mutex information.

LOGISTICS $\forall x$ at(x,l) + in(x,t) ↓

BLOCKSWORLD $\forall \cdot$ handempty() + holding(b) ↓
              $\forall b$ holding(b) + clear(b) + on(b',b) ↓
              $\forall b$ holding(b) + ontable(b) + on(b,b') ↓

GRID $\forall \cdot$ armempty() + holding(k) ↓
                  $\forall \cdot$ at-robot(l) ↓
                  $\forall \cdot$ open(d) + locked(d) ↓
                  $\forall \cdot$ locked(d) ↓
                  $\forall d$ open(d) + locked(d) ↓
                  $\forall d$ locked(d) ↓
                  $\forall k$ holding(k) + at(k,l) ↓

Fig. 9.4. Invariants found in some standard benchmark domains

9.4.4 Examples

This concludes our description of the invariant synthesis algorithm. To give
an impression of the kind of invariants it generates, Fig. 9.4 shows some of the
results obtained on IPC domains. The invariants found in the GRID domain
are most interesting, as they include some monotonicity information that is
not a mutex: The third GRID invariant states that the total number of open
and locked doors never increases, the fourth invariant states that the number
of locked doors never increases, and the sixth invariant states that a door
which is not locked can never become locked.

9.4.5 Related Work

Before moving on to the next translation step, we should point out that the
algorithm described in this section is not the only approach to invariant synthesis proposed in the literature. Therefore, we now provide a brief comparison
to four other approaches, sorted in decreasing order of relatedness:

– Edelkamp and Helmert's algorithm [34] proposed for the MIPS planner
   [35, 36],
– Gerevini and Schubert's DISCOPLAN [45, 46],
– Rintanen's invariant synthesis algorithm [101], and
– Fox and Long's TIM [27, 41].

We point out that apart from the first algorithm in the list, all of these
were developed independently from ours, although all but the last one follow
very similar ideas.

Edelkamp and Helmert's algorithm is the most closely related approach.
In fact, Fast Downward's algorithm can be considered as the extension of
the MIPS algorithm to non-STRIPS domains. Compared to the original algorithm, Fast Downward's invariant synthesis incorporates some cosmetic and

performance improvements, but the main difference is the coverage of universal and conditional effects. On STRIPS domains, both algorithms generate the same set of invariants.

DISCOPLAN uses a very similar guess, check and repair approach. However, the method for refining invariant candidates appears to be quite different, although this is somewhat difficult to assess because the algorithm is not completely described in the literature and source code of an implementation is not available. One major difference is that Fast Downward's algorithm immediately refines an invariant as soon as an operator is discovered which threatens it. DISCOPLAN, on the other hand, first collects all threats to an invariant for *all* operators, and only then generates refinements, which attempt to address all these threats at the same time. On the one hand, collecting threats across operators allows making more informed choices in invariant refinement. On the other hand, it appears that this approach incurs a performance penalty. For example, while Fast Downward's invariant synthesis algorithm always terminates in very short time for all IPC benchmark tasks, DISCOPLAN fails on 46 of the 50 IPC4 AIRPORT tasks by running out of time. Another drawback of DISCOPLAN is that, although it is not limited to STRIPS, it can only deal with a subset of ADL features, which is not sufficient for the IPC benchmarks. Finally, also for STRIPS domains, there are some invariants important for an efficient MPT encoding which our algorithm discovers but DISCOPLAN misses. For example, in the DRIVERLOG domain, our approach can prove that a given driver can only be at one place or inside one truck at the same time, which allows encoding driver location in a single variable. An encoding based on the invariants found by DISCOPLAN would need to introduce a separate state variable for each driver-location and driver-truck pair. On the positive side, DISCOPLAN can generate many classes of invariants beside mutexes; however, these are not relevant to PDDL-to-MPT translation.

Rintanen's algorithm follows the same guess-check-repair structure as our algorithm and DISCOPLAN. One main difference (and advantage) of Rintanen's algorithm is that its "check" step uses the information from *all* current invariant candidates, rather than just the one currently being considered, to strengthen the induction hypothesis. An interesting difference is that, unlike our algorithm, it always proceeds from stronger invariant candidates to weaker ones. Note that for inductive proofs, both strengthening and weakening an invariant candidate can be a promising refinement strategy. In particular, weaker statements are not necessarily easier to prove than stronger ones because the induction hypothesis is also weaker. A problem of Rintanen's algorithm is that it is limited to STRIPS and that it is not sufficiently efficient for many of the IPC benchmarks. For this reason, we have not made a detailed comparison regarding the kinds of invariants it can or cannot find; from our limited experience, we believe the approaches to be comparable in this respect, at least for the mutexes we are interested in. Like DISCOPLAN, Rintanen's approach can find more general classes of invariants.

Finally, Fox and Long's TIM (for *type inference module*) is (or can be interpreted as) an invariant synthesis algorithm which follows a conceptually very different approach to the other algorithms described here, focusing on the notion of *property spaces* which are generated from the type structure of the task, which is in turn based on a *type inference* technique which gives the system its name. TIM was originally [41] limited to STRIPS and thus not directly usable for us. It has since been extended to handle ADL constructs [27] in parallel to the development of our invariant synthesis algorithm.

9.5 Grounding

Having computed monotonicity invariants, the next translation step is to obtain a grounded representation of the normalized PDDL task.

Definition 9.5.1. *Grounded PDDL Tasks*
*A **grounded PDDL task** is a PDDL task such that all literals occurring in the goal formula, axioms and operators are ground literals.*

Before performing the actual axiom and operator instantiation that yields the grounded representation, we try to determine which ground atoms of the PDDL task can actually become true. In a typical planning task, most ground atoms can never be true, either because they are not type-correct (for example, at(vehicle1, vehicle2)), or for more subtle reasons (for example, at(vehicle1, loc1) where there is no path from the initial location of vehicle1 to loc1). Instantiating operators or axioms in such a way that their preconditions or bodies are necessarily false in every reachable state would be wasteful.

Determining whether or not a given atom can ever be true is as difficult as planning itself, but an over-approximation of the set of reachable atoms can be computed efficiently based on the idea of *relaxed planning tasks* in the sense of HSP and FF [16,68]. Instead of computing the set of reachable atoms of a PDDL task Π itself, we thus compute the reachable atoms of a relaxed planning task $\mathcal{R}(\Pi)$, which differs from Π as follows:

– Negative literals in axiom bodies, operator preconditions, effect conditions and goal condition are assumed to be always true.
– Delete effects of operators are ignored.

It is easy to see that the set of reachable atoms of $\mathcal{R}(\Pi)$ is a superset of the set of reachable atoms of Π, so any ground atoms not reachable in $\mathcal{R}(\Pi)$ need not be represented in the grounded version of Π.

The nice property of relaxed planning tasks is that computing their reachable atoms is conceptually simple. Nevertheless, this step is the most time-critical part of the whole translation component, because the set of reachable atoms can be huge in some of the benchmark domains, especially those with a comparatively simple logical structure like LOGISTICS or SATELLITE. Therefore, it is important to compute reachable atoms efficiently. This is what the *Horn exploration* algorithm is designed for.

9.5.1 Overview of Horn Exploration

The idea of Horn Exploration is to encode the atom reachability problem for relaxed planning tasks as a set of logical facts and rules, i. e., as a logic program. This allows us to efficiently compute the set of reachable atoms by computing the *canonical model* of that logic program, which is the set of ground atoms implied by the program. The algorithm consists of three steps: Generating the logic program, translating it into a normal form, and computing its canonical model. Before going into detail for each of these steps, let us formally define what we mean by a logic program:

Definition 9.5.2. *Positive Logic Programs*
Let $\mathcal{L}$ be a first-order language.

*A **positive Horn clause** over $\mathcal{L}$ is a formula of the form $\varphi_1 \wedge \cdots \wedge \varphi_k \to \psi$ ($k \geq 0$), where φ_i and ψ are (usually not ground) atoms over $\mathcal{L}$. It can be written as $\psi \leftarrow \varphi_1, \ldots, \varphi_k$. Using this notation, ψ is called the **head** and $\varphi_1, \ldots, \varphi_k$ is called the **body** of the clause. Positive Horn clauses are usually assumed to be universally quantified. For a given positive Horn clause χ with $free(\chi) = \{v_1, \ldots, v_k\}$, we define $\chi_\forall = (\forall v_1 \ldots v_k : \chi)$. Similarly, for a set of positive Horn clauses $\mathcal{R}$, we define $\mathcal{R}_\forall = \{ \chi_\forall \mid \chi \in \mathcal{R} \}$.*

*A **positive logic program** over $\mathcal{L}$ is a pair $\langle \mathcal{F}, \mathcal{R} \rangle$, where $\mathcal{F}$ is a set of ground atoms over $\mathcal{L}$ called the set of **facts** and $\mathcal{R}$ is a set of positive Horn clauses over $\mathcal{L}$ called **rules**.*

*The **canonical model** of a positive logic program $\langle \mathcal{F}, \mathcal{R} \rangle$ is the set of all ground atoms φ with $\mathcal{F} \cup \mathcal{R}_\forall \models \varphi$.*

Next, we show how to translate the reachability problem into a positive logic program. Afterwards, we demonstrate how to translate this logic program into a particularly simple form and how to compute the canonical model of the simplified logic program efficiently.

9.5.2 Generating the Logic Program

Due to the fact that the PDDL task has been normalized, generating the logic program is conceptually easy. A ground atom is reachable in the relaxed task iff it is true in the initial state or there exists some axiom or operator of the relaxed task that can make it true. Therefore, the set of facts of the logic program is formed by the atoms in the initial state of the planning task, and the set of rules is derived from the axiom and operator definitions. Additionally, we introduce a rule for the goal of the planning task to detect solvability of the relaxed task; if it is unsolvable, the original task is unsolvable too, which we can report immediately and stop planner execution.

Recall from Sect. 9.3 that at this stage, all conditions occurring in the PDDL task are conjunctions of literals. For such conjunctions φ, we denote the conjunction of all *positive* literals of φ by φ^+. In the context of logic programs, we follow the PROLOG convention of using uppercase letters for first-order variables and lower-case letters for constants. The exploration rules for a normalized PDDL tasks are generated as follows:

– *Axioms:* For schematic axioms $a = \varphi \leftarrow \psi$ with $\psi^+ = \psi_1^+ \wedge \cdots \wedge \psi_m^+$ and $\mathrm{free}(\varphi) \cup \mathrm{free}(\psi) = \{X_1, \ldots, X_k\}$, we generate the *axiom applicability rule*

$$a\text{-applicable}(X_1, \ldots, X_k) \leftarrow \psi_1^+, \ldots, \psi_m^+.$$
and the *axiom effect rule*
$$\varphi \leftarrow a\text{-applicable}(X_1, \ldots, X_k).$$
– *Operators:* For schematic operators o with parameters $\{X_1, \ldots, X_k\}$ and precondition φ with $\varphi^+ = \varphi_1^+ \wedge \cdots \wedge \varphi_m^+$, we generate the *operator applicability rule*
$$o\text{-applicable}(X_1, \ldots, X_k) \leftarrow \varphi_1^+, \ldots, \varphi_m^+.$$
and for each add effect e of o adding the atom ψ with quantified variables $\{Y_1, \ldots, Y_l\}$ and effect condition φ with $\varphi^+ = \varphi_1^+ \wedge \cdots \wedge \varphi_m^+$, we generate the *effect trigger rule*
$$e\text{-triggered}(X_1, \ldots, X_k, Y_1, \ldots, Y_l)$$
$$\leftarrow o\text{-applicable}(X_1, \ldots, X_k), \varphi_1^+, \ldots, \varphi_m^+.$$
and *effect rule*
$$\psi \leftarrow e\text{-triggered}(X_1, \ldots, X_k, Y_1, \ldots, Y_l).$$
– *Goal rule:* For the goal φ with $\varphi^+ = \varphi_1^+ \wedge \cdots \wedge \varphi_m^+$, we generate the *goal rule*
$$\text{goal-reachable}() \leftarrow \varphi_1^+, \ldots, \varphi_m^+.$$

The correctness of these rules should be evident. The reader might wonder why we sometimes introduce new predicates that do not seem necessary. For example, axiom applicability rule and axiom effect rule could be combined into a single rule without introducing the auxiliary predicate a-applicable. The purpose of these predicates is to track which axioms and operators must be instantiated when grounding the PDDL task. For example, in the LOGISTICS domain, we will not generate a ground operator (fly-airplane plane1 loc1 loc3) if loc3 is not an airport location, since in this case the canonical model of the logic program does not include the atom fly-airplane-applicable(plane1, loc1, loc3). The operator applicability predicates serve the additional purpose of "factoring out" common subexpressions. Without them, all operator preconditions would need to be repeated in each effect trigger rule (or effect rule, if effect trigger rules were similarly eliminated).

9.5.3 Translating the Logic Program to Normal Form

After the logic program has been generated, it is translated into the following normal form:

Definition 9.5.3. *Normal Form for Positive Logic Programs*
*An atom in first-order logic is called **variable-unique** if it does not contain two occurrences of the same variable. (For example, an atoms like $P(X, Y, X)$ is not variable-unique because variable X occurs twice. Repetitions of constants are allowed.)*

*A rule of a positive logic program is called **variable-unique** if the head and all atoms of the body are variable-unique.*

*A rule of a positive logic program is called a **projection** **rule** if it is variable-unique and it is of the form $\varphi \leftarrow \varphi_1$ with $free(\varphi) \subseteq free(\varphi_1)$. In other words, projection rules are unary rules where all variables in the head occur in the body.*

*A rule of a positive logic program is called a **join** **rule** if it is variable-unique and it is of the form $\varphi \leftarrow \varphi_1, \varphi_2$ with $free(\varphi_1) \cup free(\varphi_2) = free(\varphi) \cup (free(\varphi_1) \cap free(\varphi_2))$. In other words, join rules are binary rules where all variables occurring in the head occur in the body, and all variables occurring in the body but not in the head occur in* both *atoms of the body.*

*A positive logic program is **in normal form** if all rules are either projection rules or join rules.*

The names of the rule types in Definition 9.5.3 are reminiscent of the related database-theoretic operations from relational algebra: Projection rules correspond to the projection operator π and join rules correspond to the natural join operator $\bowtie$ (or strictly speaking, a combination of natural join and projection). We will now describe how to convert the positive logic program from the previous section into normal form.

First, we eliminate duplicate variable occurrences as follows: If any rule contains atoms with duplicate occurrences of the same variable X, we change one occurrence of X in any such atom into a new variable X' and add the atom `equals`(X, X') to the body of the rule. We repeat until no further such transformations are possible. If we needed to introduce any `equals` atoms, we add the fact `equals`(o, o) to the initial state for each object o of the planning task.

Second, for any variable X that occurs in the head but not in the body of a rule, we add the atom `object`(X) to the rule body. (Remember from Sect. 9.3.1 that `object`(o) is true for any object o of the planning task.)

Third, all rules with an empty body are converted into facts. Their heads must be ground atoms because all variables occurring in the head must occur in the (in this case, empty) body after the previous transformation.

After these transformations, all remaining unary rules are projection rules; we still need to normalize rules with two or more atoms in the body. As a first step towards this goal, we determine if the body of such a rule contains any

---

algorithm greedy-join(*rule*):
    while |*rule*.body| > 2:
        Choose φ, $\varphi' \in$ *rule*.body such that $\varphi \neq \varphi'$ and
          join-cost(*rule*, φ, φ') is minimal.
        $X_1, \ldots, X_k$:= join-vars(*rule*, φ, φ')
        Generate a new predicate symbol p with arity k.
        Generate a new join rule $p(X_1, \ldots, X_k) \leftarrow \varphi, \varphi'$.
        rule.body := *rule*.body $\setminus \{\varphi, \varphi'\} \cup \{p(X_1, \ldots, X_k)\}$

function join-vars(*rule*, φ, φ'):
    { Compute the relevant variables for the predicate generated
      by joining φ and φ'. }
    return free($\{\varphi, \varphi'\}$) $\cap$ free($\{$*rule*.head$\} \cup$ *rule*.body $\setminus \{\varphi, \varphi'\}$).

function join-cost(*rule*, φ, φ'):
    new-arity := |join-vars(*rule*, φ, φ')|
    max-old-arity := max(|free(φ)|, |free(φ')|)
    min-old-arity := min(|free(φ)|, |free(φ')|)
    return (new-arity $-$ max-old-arity, new-arity $-$ min-old-arity, new-arity).
    { Cost estimates are triples which are compared lexicographically.
    We prefer joins where "new arity" $-$ "max old arity" (the increase
    in arity) is small and consider the other criteria only in case of ties. }

Fig. 9.5. The greedy join algorithm for decomposing a rule into join rules

variables that occur in no other atom of the rule, neither in the body nor in the head. If this is the case, such variables are projected away as follows: We are given the rule $\varphi \leftarrow \varphi_1, \ldots, \varphi_k$, where free($\varphi_i$) = $\{X_1, \ldots, X_k\}$ contains variables not present in any of the other atoms, say $\{X_{j+1}, \ldots, X_k\}$. Then we introduce a new predicate p and replace the original rule by the two rules $\varphi \leftarrow \varphi_1, \ldots, \varphi_{i-1}, p(X_1, \ldots, X_j), \varphi_{i+1}, \ldots, \varphi_k$ and $p(X_1, \ldots, X_j) \leftarrow \varphi_i$.

After this transformation, all binary rules are valid join rules. In the last normalization step, we split rules with $m > 2$ atoms in the body into $m - 1$ join rules by applying the *greedy join algorithm*, illustrated in Fig. 9.5. The algorithm iteratively picks two atoms from the rule body and joins them, introducing a new predicate for the result of the join and replacing the two atoms in the rule body by an instance of that new predicate. This process is repeated until the body of the rule no longer contains more than two atoms.

The order in which atoms are joined is critical for the speed of evaluating the join rules. To see this, consider the rule $p(X) \leftarrow q(X), s(X, Y), t(Y)$. One possible decomposition into join rules yields the rules $u(X) \leftarrow s(X, Y), t(Y)$ and $p(X) \leftarrow q(X), u(X)$. Another possible decomposition yields the rules $v(X, Y) \leftarrow q(X), t(Y)$ and $p(X) \leftarrow s(X, Y), v(X, Y)$. We can expect that the canonical model of the first decomposition contains relatively few instances of the intermediate predicate u, maybe about as many as it contains instances of t. On the other hand, the canonical model of the second decomposition contains as many instances of the intermediate predicate v as the product of the number of instances of q and t, which can be much higher.

Since the performance of our algorithm for computing the canonical model of a logic program is closely related to the model size, we prefer to generate smaller intermediate results. The greedy join algorithm tries to achieve this goal by preferring to join atoms that contain many common variables and lead to intermediate predicates of low arity.

9.5.4 Computing the Canonical Model

Having translated the logic program into normal form, we are ready to compute the canonical model. We use a queue-based approach, distinguishing between reachable atoms that have already been processed, which means that the consequences of their being reachable have already been evaluated (*closed atoms*), and reachable atoms that still need to be processed (*open atoms*). Open atoms are those that are currently stored in the queue, while closed atoms are those that were enqueued once, but no longer are.

Our algorithm, shown in Fig. 9.6, stores open atoms in the *queue* variable, while both open and closed atoms are stored in the result variable *canonical-model*. Additionally, it uses the following data structures:

- *Rule matcher*: A rule matcher is an indexing structure that supports efficient unification queries on the bodies of logic programs. When given a ground atom a, the rule matcher determines all projection rules $\varphi \leftarrow \varphi_1$ and join rules $\varphi \leftarrow \varphi_1, \varphi_2$ such that φ_1 or φ_2 unifies with a, i.e., such that it is possible to substitute objects for variables in φ_1 or φ_2 in such a way that a is obtained. The rule matcher reports the matched rules and whether φ_1 or φ_2 was matched (if both unify with a, two matches are generated).

  Note that matching ground atoms to the rules they can trigger is simple if the rules do not contain constants in the body. Unfortunately, some of the IPC4 benchmarks contain a huge number of operator schemas involving constants (most importantly, the STRIPS formulation of the AIRPORT domain), and an efficient indexing structure is important for those. Rule matchers are implemented as decision-tree like data structures very similar to *successor generators*, which are discussed in Chap. 10. Because of that similarity and because they are not central to the instantiation algorithm, we do not discuss rule matchers further.

- *Join rule indices:* Each join rule $r = \varphi \leftarrow \varphi_1, \varphi_2$ maintains two hash tables $r.\text{index}_1$ and $r.\text{index}_2$ that map instantiations of the common variables of φ_1 and φ_2 to instantiations of the variables of φ_1 and φ_2, respectively.

  At any time (except during updates) and for any assignment *key* to the common variables of φ_1 and φ_2, $r.\text{index}_1[\text{key}]$ contains those variable mappings $\alpha \supseteq \text{key}$ for the variables of φ_1 for which $\alpha(\varphi_1)$ belongs to the closed

```
algorithm calculate-canonical-model(F, R):
        for each join rule r ∈ R:
                r.index₁ := make-empty-hashtable()
                r.index₂ := make-empty-hashtable()
        rule-matcher := build-rule-matcher(R)
        queue := make-queue(F)
        canonical-model := F
        { In the following, enqueuing a fact means adding it to queue and
          canonical-model if it is not yet an element of canonical-model. }
        while queue is not empty:
                current-fact := queue.pop()
                for each match m ∈ rule-matcher.match(current-fact):
                        if m refers to φ₁ in a projection rule r = φ ← φ₁:
                                Let α be the variable assignment
                                   for which α(φ₁) = current-fact.
                                Enqueue α(φ).
                        else if m refers to φ₁ in a join rule r = φ ← φ₁, φ₂:
                                Let α be the variable assignment
                                   for which α(φ₁) = current-fact.
                                key := α restricted to free(φ₁) ∩ free(φ₂)
                                Add α to r.index₁[key].
                                Enqueue (α ∪ β)(φ) for each β ∈ r.index₂[key].
                        else if m refers to φ₂ in a join rule r = φ ← φ₁, φ₂:
                                { Handled analogously to the previous case. }
```

Fig. 9.6. Computing the canonical model of a positive logic program $\langle \mathcal{F}, \mathcal{R} \rangle$ in normal form

set. Similarly, $r.\text{index}_2[key]$ contains those variable mappings $\beta \supseteq key$ for the variables of φ_2 for which $\beta(\varphi_2)$ belongs to the closed set.

This information can be exploited for quickly determining all possible instantiations of φ_2 that match a given instantiation of φ_1, or vice versa, as is done in the algorithm. Note that the variable assignment $\alpha \cup \beta$ considered in the algorithm is indeed a function, since α and β agree on all variables for which they are both defined.

To motivate the soundness of *compute-canonical-model*, we state an important invariant which holds before and after each iteration of the **while** loop: *All non-closed atoms which can be derived in one step from the closed atoms using the rules of the logic program* $\langle \mathcal{F}, \mathcal{R} \rangle$ *are open atoms.* This implies that upon termination of the algorithm, when there are no more open atoms and hence *canonical-model* holds exactly the set of closed atoms, the model is closed under application of $\mathcal{R}$. Because it also contains all facts from $\mathcal{F}$ and only contains facts that can be derived from $\mathcal{F}$, it thus contains exactly the canonical model of $\langle \mathcal{F}, \mathcal{R} \rangle$.

The invariant is obviously true initially, since there are no closed atoms at the beginning of the algorithm. With our descriptions of the data structures of *compute-canonical-model*, the reader should have no trouble verifying that it remains true after each iteration of the **while** loop.

This concludes our discussion of the Horn exploration algorithm. One final word on performance: If we assume that the arity of predicates in the logic program is bounded by a constant, then all basic operations of *calculate-canonical-model* can be performed in constant time. The runtime of the algorithm then typically scales roughly linearly in the combined size of its input and output (the computed canonical model). However, runtime can be worse if there are many situations where the algorithm tries to enqueue an atom that is already part of the canonical model.

9.5.5 Axiom and Operator Instantiation

With the help of the canonical model, instantiating axioms and operators is very straight-forward. To compute the grounded representation, we scan through the set of ground atoms in the canonical model in the order in which they were generated, creating axiom and operator instances as follows:

- When encountering atoms of the form $a\text{-applicable}(x_1, \ldots, x_k)$ where a is a schematic axiom, we generate a ground instance of a with the parameters substituted with $x_1, \ldots, x_k$.
- When encountering atoms of the form $o\text{-applicable}(x_1, \ldots, x_k)$ where o is a schematic operator, we generate a ground instance of o without effects. Like in the case of axioms, the parameters of the operator are substituted with $x_1, \ldots, x_k$, and the precondition is instantiated accordingly.
- When encountering atoms of the form $e\text{-triggered}(x_1, \ldots, x_k, y_1, \ldots, y_l)$ where e is an effect of some operator o, we look up the set of already generated ground operators to find the operator $o(x_1, \ldots, x_k)$. This operator must have been generated previously because an $e\text{-triggered}$ atom can only be derived after the corresponding $o\text{-applicable}$ atom. Having found the ground operator, we attach to it the effect obtained by instantiating the variables in e with $y_1, \ldots, y_l$.

After a single pass through the canonical model, we have thus generated a grounded PDDL task which is equivalent to the normalized PDDL task we started with.

9.6 Multi-valued Planning Task Generation

Together with the invariants synthesized earlier, the grounded PDDL task generated in the previous stage provides all the information we need for transforming the STRIPS task into a multi-valued planning representation, which constitutes the final translation step.

```
algorithm compute-mutex-groups(invariants, P_f, s_0):
    for each invariant I ∈ invariants:
        for each instance α of I:
            if weight(α, s_0) = 1:
                Create a mutex group containing all atoms in P_f
                covered by α.
```

Fig. 9.7. Computing mutex groups from the set of monotonicity invariants *invariants*, the set of reachable atoms P_f and the initial state s_0

Recall from Definition 9.1.2 that a multi-valued planning task (MPT) is given by a 5-tuple $\Pi = \langle V, s_0, s_\star, A, O \rangle$ of variables V, each with an associated finite domain, initial state s_0 and goal $s_\star$, axioms A and operators O. We start by defining suitable variables and variable domains; everything else then more or less falls into place.

9.6.1 Variable Selection

Each variable of the generated MPT corresponds to one or more (reachable) ground atoms of the STRIPS task. We start by enumerating the set P of all such atoms, partitioned into atoms P_f which are instances of *modifiable fluent predicates* or *derived predicates* and atoms P_c which are instances of *constant predicates* (cf. Sect. 9.4.1).

We want to represent as many ground atoms by a single state variable as possible. To achieve this, we first determine the set of *mutex groups* induced by the computed invariants. Mutex groups are computed in a straightforward manner by instantiating the monotonicity invariants in all possible ways, checking for each if it has weight 1 in the initial state, and if so, which atoms from P_f it covers. The algorithm is shown in Fig. 9.7; the actual implementation in Fast Downward uses an indexing structure for efficiently determining the set of reachable atoms covered by a given invariant instance.

Normally, not every mutex group will correspond to an MPT state variable, since the same atom can be part of several mutex groups, but of course only needs to be encoded once. As an example of this phenomenon, consider Fig. 9.8, which shows the mutex groups of a BLOCKSWORLD task with four blocks. If, for example, we decide to encode mutex groups (1)–(4) with four multi-valued state variables, then we only need to encode one atom from each of the other groups, since all instance of the on and holding predicates are already represented. Therefore, the translator would first generate four state variables with domains consisting of seven values each, namely holding(x), clear(x), on(a, x), on(b, x), on(c, x), on(d, x) and the seventh option "none of the other six is true". (Of these seven values, two – block x being on top of itself and none of the six atoms being true – are actually impossible.) After-

(1) $\{\text{holding}(a), \text{clear}(a), \text{on}(a, a), \text{on}(b, a), \text{on}(c, a), \text{on}(d, a)\}$
(2) $\{\text{holding}(b), \text{clear}(b), \text{on}(a, b), \text{on}(b, b), \text{on}(c, b), \text{on}(d, b)\}$
(3) $\{\text{holding}(c), \text{clear}(c), \text{on}(a, c), \text{on}(b, c), \text{on}(c, c), \text{on}(d, c)\}$
(4) $\{\text{holding}(d), \text{clear}(d), \text{on}(a, d), \text{on}(b, d), \text{on}(c, d), \text{on}(d, d)\}$
(5) $\{\text{holding}(a), \text{ontable}(a), \text{on}(a, a), \text{on}(a, b), \text{on}(a, c), \text{on}(a, d)\}$
(6) $\{\text{holding}(b), \text{ontable}(b), \text{on}(b, a), \text{on}(b, b), \text{on}(b, c), \text{on}(b, d)\}$
(7) $\{\text{holding}(c), \text{ontable}(c), \text{on}(c, a), \text{on}(c, b), \text{on}(c, c), \text{on}(c, d)\}$
(8) $\{\text{holding}(d), \text{ontable}(d), \text{on}(d, a), \text{on}(d, b), \text{on}(d, c), \text{on}(d, d)\}$
(9) $\{\text{holding}(a), \text{holding}(b), \text{holding}(c), \text{holding}(d), \text{handempty}()\}$

Fig. 9.8. Mutex groups for a BLOCKSWORLD task with four blocks. Some atoms, e. g. $\text{on}(a, a)$, are reachable in the relaxed task although they are never true in the "real" task

wards, it would encode the truth values of the remaining atoms $\text{ontable}(x)$ and $\text{armempty}()$ with binary state variables.

In this case, there was at least one atom in each mutex group that was unique to this particular group, so that the resulting encoding is not much better than an encoding which simply takes all mutex groups and introduces a state variable for each. However, in other cases, one group can be completely covered by others; examples of this can be found in the AIRPORT domain. In this case we would like to cover the set of reachable atoms with as few state variables as possible.

Unfortunately, as we have seen in Part I, set cover problems of this kind are NP-complete [43, problem SP5] and indeed not even c-approximable [7], so we limit our covering efforts to the greedy algorithm shown in Fig. 9.9, which is the best approximation algorithm known for this problem, achieving an $O(\log n)$-approximation [7]. Iteratively, we pick a mutex group P of maximal cardinality and introduce a new MPT state variable with domain $P \cup \{\bot\}$, where $\bot$ stands for "none of the elements of P is true". We then remove all covered elements from all other mutex groups, removing groups that no longer contain more than one element. This process is repeated until all mutex groups have been removed. At this stage, the remaining uncovered atoms p are represented by binary variables with domain $\{p, \bot\}$.

After execution of the algorithm, for each reachable atom $p \in \mathcal{P}_f$ there is exactly one MPT variable whose domain includes p. The translation will ensure that this variable, which we denote as $var(p)$ in the following, assumes the value p in an MPT state iff p is true in the corresponding state of the PDDL task. With this information, we can now go about converting the rest of the PDDL task to the MPT representation.

9.6.2 Converting the Initial State

We start by converting the initial state, which is the easiest step. For each atom $p \in \mathcal{P}_f$ that is in the initial state, we set the initial value of $var(p)$ to

---

algorithm choose-variables($\mathcal{P}_f$, *mutex-groups*):
    uncovered := $\mathcal{P}_f$
    while *mutex-groups* $\neq \emptyset$:
        Pick a mutex group P of maximal cardinality.
        Create an MPT variable v with domain $\mathcal{D}_v = P \cup \{\bot\}$.
        uncovered := *uncovered* $\setminus P$
        mutex-groups := $\{\ P' \setminus P \mid P' \in \textit{mutex-groups}\ \}$
        mutex-groups := $\{\ P' \mid P' \in \textit{mutex-groups} \wedge |P'| \geq 2\ \}$
    Create an MPT variable v with domain $\{p, \bot\}$ for all remaining
    elements of *uncovered*.

Fig. 9.9. Greedy algorithm for computing the MPT variables and variable domains

p. MPT variables for which there is no initial state atom p with $var(p) = p$ are initialized to $\bot$. Note that different initial state atoms p, p' must satisfy $var(p) \neq var(p')$, because p and p' could only be represented by the same MPT variable if they were mutually exclusive, which implies their not being in the initial state together. Therefore, the converted initial state is indeed well-defined.

9.6.3 Converting Operator Effects

Translating the state changes incurred by operator effects requires some care. For add effects setting an atom p to true, conversion is easy: Such an effect is always translated to an MPT effect setting $var(p)$ to p, because we know p to be true after operator application if the effect fires.

However, for delete effects setting an atom p to false, the correct translation is not as clear. We cannot simply set $var(p)$ to $\bot$ ("none of the variables represented by $var(p)$ is true") unconditionally, because this is not always correct: It could be the case that another effect of the same operator triggers simultaneously and adds another atom represented by the same variable, or that p *was not true* when the operator was applied, but some other atom represented by $var(p)$ was.

Therefore, the correct translation is to set $var(p)$ to $\bot$ only if we know that p was previously true and that no effect adding an atom represented by $var(p)$ triggers simultaneously, and not to do anything if this is not the case. If the other effects of the operator that add atoms represented by $var(p)$ have effect conditions $\chi_1, \ldots, \chi_k$, then this is achieved by adding $p \wedge \neg\chi_1 \wedge \cdots \wedge \neg\chi_k$ to the effect condition of the delete effect.

If some of the formulas χ_i are proper conjunctions (i. e., neither constant true nor singleton literals), this results in an effect condition which is not a conjunction of literals. In this case, we introduce a new derived variable v_i that evaluates to true whenever $\neg\chi_i$ is true, and use v_i in the effect condition instead.

All things considered, this conversion of delete effects looks very complicated, and indeed in most cases easier translations are possible. For this purpose, we detect two common special cases, with which we deal differently:

- If we see that whenever the delete effect triggers, some add effect affecting the same variable must trigger as well, because it has the same or a more general effect condition, then we do not need to represent the delete effect in the MPT at all. The add effect will take care of the value change of its affected variable.
- On the other hand, if we see that no add effect affecting the same variable can trigger at the same time, because no such effect exists or each of their effect conditions is inconsistent with the condition of the delete effect, then we can convert the delete effect to an effect setting $var(p)$ to $\bot$. If p is not already part of the operator precondition or effect condition, we must add it to the effect condition to make sure that $var(p)$ is only cleared if it was previously set to p.

In most cases, translating delete effects is straight-forward because the two simpler cases are by far more common than the general case. In particular, for operators without conditional effects, one of the special cases always applies.

9.6.4 Converting Conditions

The third major translation step is the conversion of grounded conditions of the PDDL task, which occur in the goal, in operator preconditions and effect conditions and in axiom bodies.

To translate a grounded condition, we first check if it contains any atoms not in $\mathcal{P}_f$. These have constant truth values, so that the condition can be simplified accordingly. If this leads to a constant false condition, we react accordingly (for the goal, we report that the task is unsolvable; for axiom bodies, operator preconditions or effect conditions, we remove the axiom, operator or effect).

Having considered trivially false conditions, we translate each positive literal p in the condition to the pairing $var(p) = p$. Translating negative literals $\neg p$ is slightly more tricky. Recall the BLOCKSWORLD example discussed earlier, where we generated the MPT state variable v with $\mathcal{D}_v = \{\texttt{holding(a)},$ $\texttt{clear(a)}, \texttt{on(a,a)}, \texttt{on(b,a)}, \texttt{on(c,a)}, \texttt{on(d,a)}, \bot\}$, and consider a condition including the atom $\neg\texttt{on(c,a)}$. If the condition also contains some positive literal concerning variable v, for example the atom $\texttt{clear(a)}$, then we do not need to encode $\neg\texttt{on(c,a)}$ at all, because it is implied by the other literal. However, otherwise there is no simple way to represent $\neg\texttt{on(c,a)}$ as an MPT condition. We would need to write something like $v \neq \texttt{on(c,a)}$, but conditions of this form are not supported by the representation.

Therefore, in situations like this, similar to what we did when translating difficult effect conditions that arise for complicated delete effects, we introduce a new derived variable $\textit{not-}p$ with domain $\{\top, \bot\}$ and generate an axiom

$(v = d) \rightarrow (\texttt{not-}p := \top)$ for each value $d \in \mathcal{D}_v \setminus \{p\}$. The pairing $\texttt{not-}p = \top$ can then serve as a translation of the literal $\neg p$.

If we wanted to avoid introducing new axioms, we could further normalize the PDDL task so that no negative literals occur in conditions. There are well-known translation methods to achieve such a normal form, but for our purposes, our method has the advantage that no new non-derived state variables are introduced, keeping the memory requirements of search states small.

9.6.5 Computing Axiom Layers

As a final translation step, we must compute the axiom layers for the MPT representation so that the semantics match with those of stratified logic programs (cf. Definitions 9.1.1 and 9.1.2).

This is done as follows: Whenever the body of an axiom a includes the condition $v = \bot$ for some derived variable v, then all axioms a' with affected variable v must be evaluated before a, i. e. we introduce an ordering constraint $a' \prec a$. If the axiom definitions of the original PDDL task corresponded to a stratifiable logic program, then the graph containing all such ordering constraints will be acyclic. Thus, we can use a topological sort algorithm to assign the individual axioms to axiom layers: The first axiom layer contains all axioms without predecessors in the graph, the second axiom layer contains all axioms whose predecessors belong to the first layer, and so on, until layers are assigned to all axioms.

9.6.6 Generating the Output

Having partitioned the axioms into layers, we have finished translating the PDDL task. Before generating output, the translator applies a few post-processing techniques to simplify the generated task where possible.

Most importantly, if there are two axioms with the same head, $a = (cond \rightarrow v := d)$ and $a' = (cond' \rightarrow v := d)$ with $cond \subset cond'$, then a is triggered whenever a' is triggered, so a' is unnecessary. In such a case, which occurs frequently in domains where axioms encode transitive closures, we say that a *dominates* a' and only keep a. Similarly, we do not keep several copies of the same axiom which only differ in the order in which the conditions are listed.

Once post-processing is completed, the generated MPT is written to disk in a simple text format suitable for easy parsing by the other components of the planner.

9.7 Performance Notes

Before moving on to the other components of Fast Downward, let us briefly discuss the performance of the translation component and compare it to the related MIPS system.

9.7.1 Relative Performance Compared to MIPS Translator

As discussed in detail in Sect. 8.2, there are a number of other approaches to planning that use multi-valued planning tasks or similar formalisms as their base input. However, there exists only one earlier approach exploiting multi-valued planning tasks within a *PDDL planner* and thus requiring the sort of translation described in this chapter, namely the MIPS planning system [35]. In many ways, the Fast Downward translator can be considered a further development of the MIPS translator, which is described in an article by Edelkamp and Helmert [34].

The main difference between the MIPS translator as described in that article and the Fast Downward translator described in this chapter is that the latter is more general. The original MIPS algorithm cannot deal with ADL-style conditions or effects, with derived predicates, or with schematic operators involving constants. Moreover, its runtime typically scales exponentially in the number of schematic operators. This is not very problematic for constant-free STRIPS domains which typically exhibit a small number of operators, constant across the tasks of the domain. However, some of the IPC4 domains contain partially pre-instantiated operators leading to very large domain specifications. For example, 15 of the 50 AIRPORT tasks (STRIPS formulation) of IPC4 contain more than 1300 schematic operators each. Dealing better with such high numbers of schematic operators was one of the key motivations for our new developments in invariant synthesis (Sect. 9.4) and grounding (Sect. 9.5). We point out that Edelkamp has independently extended the translation algorithm of MIPS since IPC2. However, there is no published work on these efforts, so we do not provide a comparison.

To compare the relative performance of the original MIPS translator and Fast Downward's translator, we applied both to those 566 IPC1–4 benchmark tasks that the MIPS translator can handle, i.e., pure STRIPS tasks without domain constants. This is a somewhat unfair problem suite for the Fast Downward translator because its key performance improvements are in efficiently dealing with complex domain descriptions during grounding – a complication which does not arise for these benchmarks. We should also point out that the MIPS translator is implemented in C++, whereas Fast Downward's translator is implemented in the higher-level Python language. It is reasonable to expect that a C++ reimplementation of the translator could lead to a speedup of at least one order of magnitude on large tasks.

The overall result of our comparison is that in general, the MIPS translator is clearly the faster of the two systems. To discount the impact of very

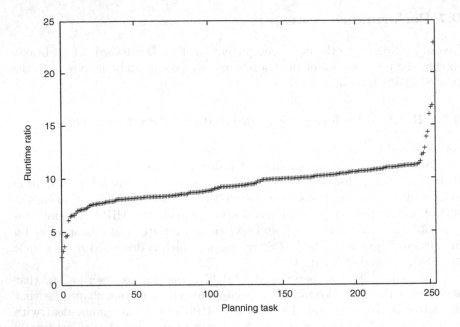

Fig. 9.10. Runtime comparison between the MIPS translator (faster) and Fast Downward translator (slower). Each data point corresponds to one planning task. Data points are sorted by runtime ratio

easy tasks, we further limited the benchmark set to those tasks for which either translator required at least one second of runtime, obtaining a total of 254 data points. For all of these, the MIPS translator was the faster of the two algorithms (or implementations). Figure 9.10 shows the ratio between the runtimes of the two translators on these 254 benchmark tasks. In 243 out of 254 cases, this ratio is between 6 and 13, with five outliers below and six outliers above this region. Note that on the horizontal axis, task are sorted by the observed runtime ratio, not by task size or any other scaling measure, so the upward slope of the curve cannot be interpreted as any kind of asymptotical scaling behaviour. This mode of display was chosen because there was no visible correlation between the observed speedup and task size (or some other apparent measure of task complexity). In other words, the algorithms appear to scale equally well, which is no surprise given that our translation algorithm is very similar to the MIPS translator on this fragment of PDDL.

Regarding the kinds of translations they generate, the two translators are interchangeable. For this reason, as a speed optimization, Fast Downward can be configured to use the C++-based MIPS translator on those domains for which it is applicable. (The performance results in the following section are all with respect to the Python-based Fast Downward translator, in order to allow comparisons across domains.)

measure	mean ± dev.	25%	median	75%	max.
translation time	12.35 ± 82.52 s	0.39 s	1.19 s	4.28 s	1960.0 s
size of output	672 ± 3253 KB	23 KB	93 KB	326 KB	81748 KB
state variables	565 ± 3176	25	55	240	61842
operators/axioms	7046 ± 38308	289	1009	3469	989250

Fig. 9.11. Some statistics on the performance and output characteristics of the translation component. The first column shows the aspect of the task being measured, the second column shows the mean and standard deviation for the respective measure, and the remaining columns show the 25%, 50% (median), 75% and 100% (maximum) percentiles. Statistics are based on the 1442 propositional benchmarks from the fully automated tracks of IPC1–4

9.7.2 Absolute Performance

In addition to comparisons to other similar techniques, it is of course also of importance how fast the translator computes its result in *absolute* terms. On a state-of-the-art computer, it is sufficiently efficient to generate MPT encodings for all 1442 IPC1–4 benchmark tasks. Moreover, compared to the time required in the search component, translation time is essentially negligible in the vast majority of cases. (Some exceptions to this exist in "structurally simple" domains like SATELLITE and LOGISTICS.) A short summary of "average" performance for the translator (for different notions of average) is provided in Fig. 9.11. All experiments were conducted on a machine with a 3.066 GHz Intel Xeon CPU, setting a memory limit of 2 GB.

To get an impression of the size and translation cost of a "typical" task, the mean values, which are heavily influenced by some very large PSR tasks, are misleading. The comparatively high standard deviations show that the distributions are highly irregular, so the percentile information is probably more meaningful than the mean values. To get an impression of what very large planning tasks look like, Fig. 9.12 provides information about the "largest" five input tasks according to each of the four measures *translation time, encoding size of translated task, no. of state variables in translated tasks* and *no. of operators and axioms in translated tasks.* As an extreme example, the largest PSR instance from IPC4, task PSR-LARGE #50, could only just be translated within the memory bound, consuming more than 1.9 GB of RAM before completing translation after 32:40 minutes. However, with 61842 relevant state variables, almost all of them derived variables, this task is far from being solvable with current domain-independent planning technology anyway. For comparison, at IPC4, Fast Downward could only solve the PSR-LARGE instances up to #31. The largest solved PSR instance comprises 5807 relevant MPT state variables and needed 39 seconds for translation. Of the other competitors, the best system could solve 11 tasks of the PSR-LARGE benchmark set, of which the largest comprises 527 state variables. MPT translation took 4 seconds for this instance.

translation time	task	time
	PSR-LARGE #50	1 960.00 s
	SATELLITE #33	1 355.50 s
	PSR-LARGE #48	1 145.80 s
	PSR-LARGE #46	920.45 s
	SATELLITE #32	634.80 s
size of output	task	size
	SATELLITE #33	81 748 KB
	SATELLITE #32	52 032 KB
	SATELLITE #36	34 758 KB
	SATELLITE #31	29 997 KB
	SATELLITE #35	27 672 KB
state variables	task	amount
	PSR-LARGE #50	61 842
	PSR-LARGE #48	48 210
	PSR-LARGE #46	40 357
	PSR-LARGE #49	36 790
	PSR-LARGE #44	32 174
operators/axioms	task	amount
	SATELLITE #33	989 250
	SATELLITE #32	638 665
	SATELLITE #36	428 109
	SATELLITE #31	368 990
	SATELLITE #35	342 193

Fig. 9.12. Statistics for the five largest planning tasks in each of the four categories

Knowledge Compilation

The purpose of the knowledge compilation component, described in this chapter, is to build a number of data structures that capture the central concepts of Fast Downward's *causal graph heuristic* and that facilitate efficient state expansion in its various search algorithms.

The following Sect. 10.1 provides a short overview of this planner phase, followed by the in-depth treatment of *domain transition graphs* (Sect. 10.2), *causal graphs* (Sect. 10.3), and finally *successor generators* and *axiom evaluators* (Sect. 10.4).

10.1 Overview

The knowledge compilation component sets the stage for the search algorithms by compiling the critical information about the planning task into a number of data structures for efficient access. In other contexts, computations of this kind are often called *preprocessing*. However, "preprocessing" is such a nondescript word that it can mean basically anything. For this reason, we prefer a term that puts a stronger emphasis on the role of this module: To rephrase the critical information about the planning task in such a way that it is directly useful to the search algorithms. Of the three building blocks of Fast Downward (translation, knowledge compilation, search), it is the least time-critical part, always requiring less time than translation and being dominated by search for all but the most trivial tasks.

Knowledge compilation comprises three items. First and foremost, we compute the *domain transition graph* of each state variable. The domain transition graph for a state variable encodes under what circumstances that variable can change its value, i.e., from which values in the domain there are transitions to which other values, which operators or axioms are responsible for the transition, and which conditions on other state variables are associated with the transition. Domain transition graphs are described in Chap. 11. They are a central concept for the computation of the causal graph heuristic, described in Sect. 11.2.

Second, we compute the *causal graph* of the planning task. Where domain transition graphs encode dependencies between values for a given state variable, the causal graph encodes dependencies between different state variables. For example, if a given location in a planning task can be unlocked by means of a key that can be carried by the agent, then the variable representing the lock state of the location is dependent on the variable that represents whether or not the key is being carried. This dependency is encoded as an arc in the causal graph. Like domain transition graphs, causal graphs are a central concept for the computation of the causal graph heuristic, giving it its name. The causal graph heuristic requires causal graphs to be acyclic, a requirement which is rarely satisfied in practice. For this reason, the knowledge compilation component also generates an acyclic subgraph of the real causal graph when cycles occur. This amounts to a relaxation of the planning task where some operator preconditions are ignored. In addition to their usefulness for the causal graph heuristic, causal graphs are also a key concept of the *focused iterative-broadening search* algorithm introduced in the following chapter. We discuss causal graphs in Sect. 10.3.

Third, we compute two data structures that are useful for any forward-searching algorithm for MPTs, called *successor generators* and *axiom evaluators*. Successor generators compute the set of applicable operators in a given world state, and axiom evaluators compute the values of derived variables for a given reduced state. Both are designed to do their job as quickly as possible, which is especially important for the focused iterative-broadening search algorithm, which does not compute heuristic estimates and thus requires the basic operations for expanding a search node to be implemented efficiently. These data structures are discussed in Sect. 10.4.

10.2 Domain Transition Graphs

The domain transition graph of a state variable is a representation of the ways in which the variable can change its value, and of the conditions that must be satisfied for such value changes to be allowed. Domain transition graphs were introduced by Jonsson and Bäckström in the context of SAS$^+$ planning [72]. Our formalization of domain transition graphs generalizes the original definition to planning tasks involving axioms and conditional effects.

Definition 10.2.1. *Domain Transition Graphs*
Let $\Pi = \langle \mathcal{V}, s_0, s_\star, \mathcal{A}, \mathcal{O} \rangle$ be a multi-valued planning task, and let $v \in \mathcal{V}$ be a state variable of Π.

*The **domain transition graph** of v, in symbols $DTG(v)$, is a labelled directed graph with vertex set $\mathcal{D}_v$. If v is a fluent, $DTG(v)$ contains the following arcs:*

– *For each effect cond → v := d' of an operator o with precondition pre such that pre ∪ cond contains some condition v = d, an arc from d to d' labelled with pre ∪ cond \ {v = d}.*
– *For each effect cond → v := d' of an operator o with precondition pre such that pre ∪ cond does not contain the condition v = d for any d ∈ 𝒟ᵥ, an arc from each d ∈ 𝒟ᵥ \ {d'} to d' labelled with pre ∪ cond.*

If v is a derived variable, DTG(v) contains the following arcs:

– *For each axiom cond → v := d' ∈ 𝒜 such that cond contains some condition v = d, an arc from d to d' labelled with cond \ {v = d}.*
– *For each axiom cond → v := d' ∈ 𝒜 such that cond does not contain the condition v = d for any d ∈ 𝒟ᵥ, an arc from each d ∈ 𝒟ᵥ \ {d'} to d' labelled with cond.*

Arcs of domain transition graphs are called **transitions**. Their labels are referred to as the **conditions** of the transition.

Domain transition graphs can be weighted, in which case each transition has an associated non-negative integer weight. Unless stated otherwise, we assume that all transitions derived from operators have weight 1 and all transitions derived from axioms have weight 0.

The definition is somewhat lengthy, but its informal content is easy to grasp: The domain transition graph for v contains a transition from d to d' if there exists some operator or axiom that can change the value of v from d to d'. Such a transition is labelled with the conditions on *other* state variables that must be true if the transition shall be applied. Multiple transitions between the same values using different conditions are allowed and occur frequently.

We have already seen domain transition graphs in the introduction in Chap. 8 (Figs. 8.3 and 8.4), although they were only introduced informally and did not show the arc labels usually associated with transitions. Figure 10.1 shows some examples from a simple task in the GRID domain, featuring a 3×2 grid with a single initially locked location in the centre of the upper row, unlockable by a single key. In the MPT encoding of the task, there are three state variables: variable r with $\mathcal{D}_r = \{\, \langle x, y \rangle \mid x \in \{1, 2, 3\},\ y \in \{1, 2\} \,\}$ encodes the location of the robot, variable k with $\mathcal{D}_k = \mathcal{D}_r \cup \{carried\}$ encodes the state of the key, and variable d with $\mathcal{D}_d = \{closed, open\}$ encodes the state of the initially locked grid location.

If all operators of an MPT are unary (i. e., only have a single effect) and we leave aside axioms for a moment, then there is a strong correspondence between the state space of an MPT and its domain transition graphs. Since vertices in domain transition graphs correspond to values of state variables, a given state is represented by selecting one vertex in each domain transition graph, called the *active vertex* of this state variable. Applying an operator means changing the active vertex of some state variable by performing a transition in the corresponding domain transition graph. Whether or not such a transition is allowed depends on its condition, which is checked against the

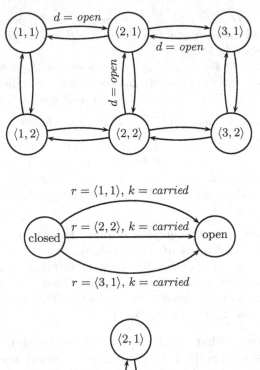

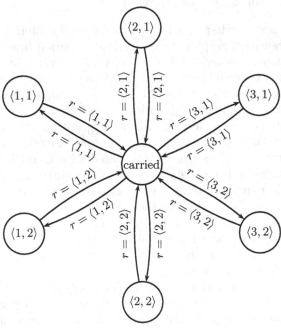

Fig. 10.1. Domain transition graphs of a GRID task. Top: $DTG(r)$ (robot); middle: $DTG(d)$ (door), bottom: $DTG(k)$ (key)

active vertices of the other domain transition graphs. We will use the GRID example to motivate the importance of this correspondence. Consider an initial state where the robot is at location $\langle 1, 1 \rangle$, the key is at location $\langle 3, 2 \rangle$, and the door is locked. We represent this by placing pebbles on the appropriate vertices of the three domain transition graphs. We want to move the pebble in the domain transition graph of the key to location $\langle 2, 1 \rangle$. This can be done by moving the robot pebble to vertex $\langle 1, 2 \rangle$, then $\langle 2, 2 \rangle$, then $\langle 3, 2 \rangle$, moving the key pebble to the vertex *carried*, moving the robot pebble back to vertex $\langle 2, 2 \rangle$, moving the door pebble to *open*, moving the robot pebble to vertex $\langle 2, 1 \rangle$ and finally moving the key pebble to vertex $\langle 2, 1 \rangle$.

The example shows how plan execution can be viewed as simultaneous traversal of domain transition graphs [29]. This is an important notion for Fast Downward because the causal graph heuristic computes its heuristic estimates by solving subtasks of the planning task by looking for paths in domain transition graphs in basically the way we have described.

As mentioned before, this view of MPT planning is only completely accurate for unary tasks without axioms, for which the domain transition graphs are indeed a complete representation of the state space. For non-unary operators, we would need to "link" certain transitions in different domain transition graphs which belong to the same operator. These could then only be executed together. For axioms, we would need to mark certain transitions as "mandatory", requiring that they be taken whenever possible. (This is only intended as a rough analogy and leaves out details like layered axioms.)

In earlier work [58], we have successfully applied this view of planning to STRIPS tasks. Extending the notion to plans with conditional effects provides no challenges because domain transition graphs always consider planning operators one effect at a time, in which case the effect condition can simply be seen as part of the operator precondition. However, axioms provide a challenge that is easily overlooked. If we want to change the value of a fluent from d to d', the domain transition graph contains all the important information; just find a path from d to d' and try to find out how the associated conditions can be achieved. Consider the same problem for a derived state variable. Let us assume that unlocking the location in the GRID example leads to a drought, causing the robot to freeze if it enters a horizontally adjacent location. We could encode this with a new derived variable f (for *freezing*) with domain $\mathcal{D}_f = \{\top, \bot\}$, defined by the axioms $d = open, r = \langle 1, 1 \rangle \rightarrow f := \top$ and $d = open, r = \langle 3, 1 \rangle \rightarrow f := \top$. The domain transition graph $DTG(f)$ is depicted in Fig. 10.2 (left).

The problem with that domain transition graph is that it does not tell us how we can change the state of variable f from $\top$ to $\bot$. In general, in MPTs derived from STRIPS tasks where derived predicates occur negatively in any condition, the domain transition graph does not contain sufficient information for changing the value of a derived variable from "true" to "false". Derived variables never assume the value $\bot$ due to a *derivation* of this value; because of negation as failure semantics, they only assume the value *by default* if *no*

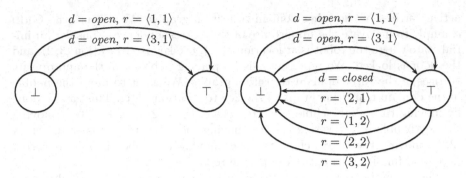

Fig. 10.2. Domain transition graphs for the *freezing* variable in the GRID task, normal (left) and extended (right). Note that only the extended graph shows how to change state from "freezing" ($\top$) to "not freezing" ($\bot$)

other value can be derived. If we want to reason about ways of setting the value of a derived variable to $\bot$, we will need to make this information explicit.

In logical notation, whether or not a derived variable assumes a given value by triggering an axiom at a given layer is determined by a formula in disjunctive normal form, with one disjunct for each axiom setting the value. For example, our axioms $d = open, r = \langle 1, 1 \rangle \rightarrow f := \top$ and $d = open, r = \langle 3, 1 \rangle \rightarrow f := \top$ correspond to the DNF formula $(d = open \land r = \langle 1, 1 \rangle) \lor (d = open \land r = \langle 3, 1 \rangle)$. If we want to know when these rules do *not* trigger, we must negate this formula, leading to the CNF formula $(d \neq open \lor r \neq \langle 1, 1 \rangle) \land (d \neq open \lor r \neq \langle 3, 1 \rangle)$. To be able to encode this information in the domain transition graph, we need to replace the inequalities with equalities and translate the formula back to DNF. Since such transformations can increase the formula size dramatically, we apply simplifications along the way, removing duplicated and dominated disjuncts. The result in this case is the DNF formula $d = closed \lor r = \langle 2, 1 \rangle \lor r = \langle 1, 2 \rangle \lor r = \langle 2, 2 \rangle \lor r = \langle 3, 2 \rangle$.

A domain transition graph for a derived variable which has been enriched to contain the possible ways of causing the variable to assume the value $\bot$ is called an *extended domain transition graph*, shown for the example of the freezing variable in Fig. 10.2 (right). Since computing the extended domain transition graph can be costly and is not always necessary, the knowledge compilation component scans the conditions of the planning task (axioms, operator preconditions and effect conditions, goal) for occurrences of pairings of the type $v = \bot$ for derived variables v. Extended domain transition graphs are only computed for those derived variables for which they are required.

Note that negative occurrences of derived variables can cascade: If u, v and w are derived variables with domain $\{\top, \bot\}$ and the condition $v = \bot$ is present in some operator precondition, and moreover v is defined by the axiom $u = \top, w = \top \rightarrow v := \top$, then v assumes the value $\bot$ whenever u or

w do, so we would require extended domain transition graphs for u and w as well.

On the other hand, multiple layers of negation as failure can cancel each other out: If derived variable v only occurs in conditions of the form $v = \bot$ but never in positive form and is defined by the axiom $u = \bot, w = \bot \rightarrow v := \top$, then we do not necessarily require extended domain transition graphs for u and w.

In general, whether or not we need extended domain transition graphs for a derived variable is determined by the following rules:

- If v is a derived variable for which the condition $v = d$ for $d \neq \bot$ appears in an operator precondition, effect condition or in the goal, then v *is used positively*.
- If v is a derived variable for which the condition $v = \bot$ appears in an operator precondition, effect condition or in the goal, then v *is used negatively*.
- If v is a derived variable for which the condition $v = d$ for $d \neq \bot$ appears in the body of an axiom whose head is used positively (negatively), then v is used positively (negatively).
- If v is a derived variable for which the condition $v = \bot$ appears in the body of an axiom whose head is used positively (negatively), then v is used negatively (positively).

The knowledge compilation component computes extended domain transition graphs for all derived variables which are used negatively and (standard) domain transition graphs for all other state variables. Normal domain transition graphs are computed by going through the set of axioms and the set of operator effects following Definition 10.2.1, which is reasonably straightforward; the computation of extended domain transition graphs has been outlined above. Therefore, the algorithmic aspects of this topic should not require further discussion.

10.3 Causal Graphs

Causal graphs have been introduced informally in the introduction. Here is our formal definition.

Definition 10.3.1. *Causal Graphs*
*Let Π be a multi-valued planning task with variable set $\mathcal{V}$. The **causal graph** of Π, in symbols $CG(\Pi)$, is the directed graph with vertex set $\mathcal{V}$ containing an arc $\langle v, v' \rangle$ iff $v \neq v'$ and one of the following conditions is true:*

- *The domain transition graph of v' has a transition with some condition on v.*
- *The set of affected variables in the effect list of some operator includes both v and v'.*

In the first case, we say that an arc is induced by a **transition condition.**
In the second case, we say that it is induced by **co-occurring effects.**

Of course, the set of arcs induced by transition conditions and the set of arcs induced by co-occurring effects are not mutually exclusive. The same causal graph arc can be generated for both reasons.

Informally, the causal graph contains an arc from a source variable to a target variable if changes in the value of the target variable can depend on the value of the source variable. Such arcs are included also if this dependency is of the form of an *effect* on the source variable. This agrees with the definition of *dependency graphs* by Jonsson and Bäckström [73], although these authors distinguish between the two different ways in which an arc in the graph can be introduced by using labelled arcs.

Whether or not co-occurring effects should induce arcs in the causal graph depends on the intended semantics: If such arcs are not included, the set of causal graph ancestors $anc(v)$ of a variable v are precisely those variables which are relevant if our goal is to change the value of v. Plans for this goal can be computed without considering any variables outside $anc(v)$, by eliminating all variables outside $anc(v)$ from the planning task and simplifying axioms and operators accordingly. We call this the *achievability definition* of causal graphs, because causal graphs encode what variables are important for achieving a given assignment to a state variable.

However, with the achievability definition, a planner that only considers $anc(v)$ while generating an action sequence that achieves a given valuation for v may modify variables outside of $anc(v)$, i.e., the generated plans have side effects which could destroy previously achieved goals or otherwise have a negative impact on overall planning. Therefore, we prefer our definition, which we call the *separability definition* of causal graphs.

10.3.1 Acyclic Causal Graphs

Following the separability definition of causal graphs, solving a subtask over variables $anc(v)$ is always possible without changing any values outside of $anc(v)$. This leads us to the following observation.

Observation 10.3.2. *Acyclic Causal Graphs and Strongly Connected Domain Transition Graphs*
Let Π be an MPT such that $CG(\Pi)$ is acyclic, all domain transition graphs are strongly connected, there are no derived variables, and no trivially false conditions occur in operators or goals. Then Π has a solution.

By *trivially false* conditions, we mean conditions of the kind $\{v = d, v = d'\}$ for $d \neq d'$. Note the similarity of Observation 10.3.2 to the results of Williams and Nayak [113] on planning in domains with unary operators, acyclic causal graphs and reversible transitions. Under the separability definition of causal graphs, acyclic causal graphs imply unariness of operators because operators with several effects introduce causal cycles. Moreover, strong connectedness of domain transition graphs is closely related to Williams' and Nayak's reversibility property, although it is a weaker requirement.

The truth of the observation can easily be seen inductively: If the planning task has only one state variable and the domain transition graph is strongly connected, then any state (of the one variable) can be transformed into any other state by applying graph search techniques. If the planning task has several state variables and the causal graph is acyclic, we pick a sink of the causal graph, i.e., a variable v without outgoing arcs, and check if a goal is defined for this variable. If not, we remove the variable from the task, thus reducing the problem instance to one with fewer state variables, solved recursively. If yes, we search for a path from $s_0(v)$ to $s_\star(v)$ in the domain transition graph of v, which is guaranteed to exist because the graph is strongly connected. This yields a "high-level plan" for setting v to $s_\star(v)$ which can be fleshed out by recursively inserting the plans for setting the variables of the predecessors of v in the causal graph to the values required for the transitions that form the high-level plan. Once the desired value of v has been set, v can be eliminated from the planning task and the remaining task can be solved recursively.

The algorithm is shown in Fig. 10.3. Although it is backtrack-free, it can require exponential time to execute because the generated plans can be exponentially long. This is unavoidable; even for MPTs that satisfy the conditions of Observation 10.3.2, shortest plans can be exponentially long. An example of such planning tasks is given in the proof of Theorem 4.4 in the article by Bäckström and Nebel [10].

This method for solving multi-valued planning tasks is essentially *planning by refinement*: We begin by constructing a very abstract skeleton plan, which is merely a path in some domain transition graph, then lower the level of abstraction by adding operators to satisfy the preconditions required for the transitions taken by the path. Strong connectedness of domain transition graphs guarantees that every abstract plan can actually be refined to a concrete plan. This is precisely Bacchus and Yang's [9] *downward refinement property* (cf. Sect. 8.2.1).

10.3.2 Generating and Pruning Causal Graphs

The usefulness of causal graphs for planning by refinement is not limited to the acyclic case. Consider a subset $\mathcal{V}'$ of the task variables which contains all its causal graph descendants. In general, if we restrict the task to $\mathcal{V}'$ by removing all occurrences of other variables from the initial state, goal, operators and axioms, we obtain an abstraction of the original task which satisfies Knoblock's [79] ordered monotonicity property (Sect. 8.2.1).

Unfortunately, one major problem with this approach is that the requirement to include all causal graph descendants is quite limiting. It is not uncommon for the causal graph of a planning task to be strongly connected, in which case this technique will not allow us to abstract away any variables

```
algorithm solve-easy-MPT(𝒱, s₀, s⋆, 𝒪):
    if s⋆ = ∅:
            { The goal is empty: the empty plan is a solution. }
            return ⟨⟩.
    else:
            Let v ∈ 𝒱 be a variable not occurring in preconditions
                or effect conditions in 𝒪.
            { Such a variable exists if the causal graph is acyclic. }
            𝒱' := 𝒱 \ {v}.
            𝒪' := { o ∈ 𝒪 | o does not affect v }.
            plan := ⟨⟩
            if s⋆(v) is defined:
                    Let t₁, ..., tₖ be a path of transitions in DTG(v)
                        from s₀(v) to s⋆(v).
                    { t₁, ..., tₖ is a "high-level plan" that reaches the goal
                        for v, but ignores preconditions on other variables. }
                    for each t ∈ {t₁,...,tₖ}:
                            { Recursively find a plan that achieves
                                the conditions of t. }
                            Let cond and o be the condition
                                and operator associated with t.
                            Let s₀' be the state reached after executing plan,
                                restricted to 𝒱'.
                            Extend plan by solve-easy-MPT(𝒱', s₀', cond, 𝒪').
                            Extend plan by o.
            { After dealing with v, recursively plan for goals on
                the remaining variables. }
            Let s₀' be the state reached after executing plan, restricted to 𝒱'.
            s⋆' := s⋆ restricted to 𝒱'.
            Extend plan by solve-easy-MPT(𝒱', s₀', s⋆', 𝒪').
            return plan
```

Fig. 10.3. Planning algorithm for MPTs with acyclic causal graph and strongly connected domain transition graphs

at all. However, in a heuristic approach, we are free to simplify the planning task. In particular, by ignoring some operator preconditions for the purposes of heuristic evaluation, we can make an arbitrary causal graph acyclic. Clearly, the more aspects of the real task we ignore, the worse we can expect our heuristic to approximate the actual goal distance. Considering this, our aim is to ignore as little information as possible. We will now explain how this is done.

The knowledge compilation component begins its causal graph processing by generating the "full" causal graph (Definition 10.3.1). One consequence of the separability definition of causal graphs is that all state variables which are not ancestors of variables mentioned in the goal are completely irrelevant. Therefore, having computed the graph, we then compute the causal graph ancestors of all variables in the goal. Any state variables which are not found to be goal ancestors are eliminated from the planning task and causal graph. Associated operators and axioms are removed. (This simplification is closely related to Knoblock's criterion for the *problem-specific* ordered monotonicity property [79].) Afterwards, we compute a *pruned causal graph*, an acyclic subgraph of the causal graph with the same vertex set. We try do this in such a fashion that "important" causal dependencies are retained whenever possible. More specifically, we apply the following algorithm.

First, we compute the strongly connected components of the causal graph. Cycles only occur within strongly connected components, so each component can be dealt with separately. Second, for each connected component, we compute a total order $\prec$ on the vertices, retaining only those arcs $\langle v, v' \rangle$ for which $v \prec v'$. If $v \prec v'$, we say that v' has a *higher level* than v. The total order is computed in the following way:

1. We assign a weight to each arc in the causal graph. The weight of an arc is n if it is induced by n axioms or operators. The lower the cumulated weight of the incoming arcs of a vertex, the fewer conditions are ignored by assigning a low level to this vertex.
2. We then pick a vertex v with minimal cumulated weight of incoming arcs and select it for the lowest level, i. e., we set $v \prec v'$ for all other vertices v' in the strongly connected component.
3. Since v has been dealt with, we remove the vertex and its incident arcs from consideration for the rest of the ordering algorithm.
4. The remaining task is solved by iteratively applying the same technique to order the other vertices until only a single vertex remains.

The reader will notice that the pruning choices within a strongly connected component are performed by a greedy algorithm. We could also try to find sets of arcs of minimal total weight such that eliminating these arcs results in an acyclic graph. However, this is an NP-equivalent problem, even in the case of unweighted graphs [43, problem GT8].

After generating the pruned causal graph, we also prune the domain transition graphs by removing from the transition labels of $DTG(v)$ all conditions on variables v' with $v \prec v'$. These are the conditions that are ignored by the heuristic computation. Finally, we simplify the domain transition graphs by removing *dominated transitions*: If t and t' are transitions between the same two values of a variable, and the condition of t is a proper subset of the condition of t', then transition t is easier to apply than t', so that we remove t'. Similarly, if there are several transitions with identical conditions, we only keep one of them.

10.3.3 Causal Graph Examples

To give some impression of the types of causal graphs typically found in the standard benchmarks and the effects of pruning, we show some examples of

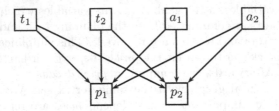

Fig. 10.4. Causal graph of a LOGISTICS task. State variables t_i and a_i encode the locations of trucks and airplanes, state variables p_i the locations of packages

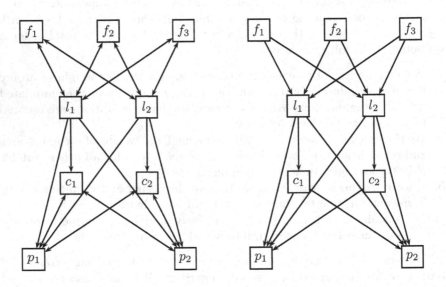

Fig. 10.5. Causal graph of a MYSTERY task (left) and of a relaxed version of the task (right). State variables f_i encode the fuel at a location, state variables l_i and c_i encode the locations and remaining capacities of trucks, and state variables p_i encode the locations of packages

increasing graph complexity. Figure 10.4 depicts an example task from the LOGISTICS domain, featuring two trucks, two airplanes and two packages. As can be seen, the graph is acyclic, so it requires no pruning for the causal graph heuristic. Since LOGISTICS tasks also feature strongly connected domain transition graphs, they can even be solved by *solve-easy-MPT*.

The next figure, Fig. 10.5, shows an example from the MYSTERY domain with three locations, two trucks and two packages. The causal graph contains a number of cycles, but these are mostly local. By pruning arcs from vertices l_i to f_j, we ignore the fact that we must move trucks to certain locations if we want to use up fuel at that location. As using up fuel is not a very useful thing

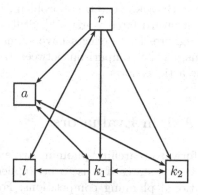 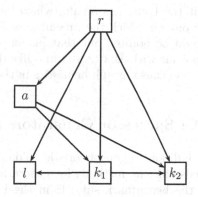

Fig. 10.6. Causal graph of a GRID task (left) and of a relaxed version of the task (right). State variable r encodes the location of the robot, a encodes the status of the robot arm (empty or carrying a key), l encodes the status of the locked location (locked or open), and k_1 and k_2 encode the locations of the two keys

to do, this is not a big loss in information. By pruning arcs from vertices p_i to c_j, we ignore the fact that vehicles can only increase or decrease their current capacity by unloading or loading packages. Compared to heuristics based on ignoring delete effects, this is not a great loss in information, since ignoring delete effects in the MYSTERY domain almost amounts to ignoring capacity and fuel constraints altogether. By pruning just these arcs, we can eliminate all cycles in the causal graph, so the MYSTERY domain can be considered fairly well-behaved.

A worse case is shown in Fig. 10.6, which shows an example from the GRID domain with an arbitrary number of locations, of which a single one is locked. There are two keys, one of which can unlock the locked location. Eliminating cycles here requires a few minor relaxations regarding the status of the robot arm (empty or non-empty), but also one major simplification, namely the elimination of the arc from l to r representing the fact that the robot can only enter the locked location if it has been unlocked.

As a (nearly) worst-case example, consider a task in the BLOCKSWORLD domain (no figure). A typical MPT encoding uses one state variable h for encoding whether or not the hand is empty and two state variables per block in the task: For the i-th block, t_i encodes whether or not the block is lying on the table, and b_i encodes which block is lying on top of it, or if it is clear or being held by the arm. In the causal graph of such a task, variable h has ingoing arcs from and outgoing arcs to all other state variables, and all state variables b_i are connected to each other in both directions. Only the state variables t_i have a slightly simpler connection structure, being only connected to h and to b_i for the same value of i. Any relaxation of the task that eliminates cycles from the causal graph loses a large amount of information, and it is not surprising that the DEPOTS domain, which includes a BLOCKSWORLD subproblem, is the one for which the precursor of Fast Downward fared worst [58]. Still, it should be pointed out that planners that ignore delete effects have similar problems with BLOCKSWORLD-like domains, as the comparison between the FF and causal graph heuristics in the same article shows.

10.4 Successor Generators and Axiom Evaluators

In addition to good heuristic guidance, a forward searching planning system needs efficient methods for generating successor states if it is to be applied to the benchmark suite from the international planning competitions. For some domains, our causal graph heuristic or other popular methods like the FF heuristic provide excellent goal estimates, yet still planning can be too time-consuming because of very long plans and vast branching factors.

The variant of best-first search implemented in Fast Downward does not compute the heuristic estimate for each state that is generated. Essentially, heuristic estimates are only computed for closed nodes, while computation is deferred for nodes on the search frontier. (Our best-first search variant is discussed in detail in the following chapter.) For domains with strong heuristic guidance and large branching factors, the number of nodes on the frontier can by far dominate the number of nodes in the closed set. As a case in point, consider the problem instance SATELLITE #29. For solving this task, the default configuration of Fast Downward only computes heuristic estimates for 67 597 world states while adding 107 233 381 states to the frontier. Clearly, determining the set of applicable operators quickly is of critical importance in such a scenario.

In some SATELLITE tasks, there are almost 1 000 000 ground operators, so we should try to avoid individually checking each operator for applicability. Similarly, in the biggest PSR tasks, more than 100 000 axioms must be evaluated in each state to compute the values of the derived variables, so this computation must be made efficient. For these purposes, Fast Downward uses two data structures called *successor generators* and *axiom evaluators*.

10.4.1 Successor Generators

Successor generators are recursive data structures very similar to decision trees. The internal nodes have associated conditions, which can be likened to the decisions in a decision tree, and the leaves have associated operator lists which can be likened to a set of classified samples in a decision tree leaf. They are formally defined as follows.

Definition 10.4.1. *Successor Generators*
*A **successor generator** for an MPT $\Pi = \langle \mathcal{V}, s_0, s_\star, \mathcal{A}, \mathcal{O} \rangle$ is a tree consisting of **selector nodes** and **generator nodes**.*

*A selector node is an internal node of the tree. It has an associated variable $v \in \mathcal{V}$ called the **selection variable**. Moreover, it has $|\mathcal{D}_v| + 1$ children accessed via labelled edges, one edge labelled $v = d$ for each value $d \in \mathcal{D}_v$, and one edge labelled $\top$. The latter edge is called the **don't care edge** of the selector.*

*A generator node is a leaf node of the tree. It has an associated set of operators from $\mathcal{O}$ called the set of **generated operators**.*

Each operator $o \in \mathcal{O}$ must occur in exactly one generator node, and the set of edge labels leading from the root to this node (excluding don't care edges) must equal the precondition of o.

Given a successor generator for an MPT Π and a state s of Π, we can compute the set of applicable operators in s by traversing the successor generator as follows, starting from the root:

- At a selector node with selection variable v, follow the edge $v = s(v)$ and the don't care edge.
- At a generator node, report the generated operators as applicable.

To build a successor generator for Π, we apply a top-down algorithm which considers the task variables in an arbitrary order $v_1 \prec v_2 \prec \cdots \prec v_n$. At the root node, we choose v_1 as selection variable and classify the set of operators according to their preconditions with respect to v_1. Operators with a precondition $v_1 = d$ will be represented in the child of the root accessed by the edge with the corresponding label, while operators without preconditions on v_1 will be represented in the child of the root accessed by the don't care edge. In the children of the root, we choose v_2 as selection variable, in the grandchildren v_3, and so on. There is one exception to this rule to avoid creating unnecessary selection nodes: If no operator in a certain branch of the tree has a condition on v_i, then v_i is not considered as a selection variable in this branch. The construction of a branch ends when all variables have been considered, at which stage a generator node is created for the operators associated with that branch.

10.4.2 Axiom Evaluators

Axiom evaluators are a simple data structure used for efficiently implementing the well-known *marking algorithm* for propositional Horn logic [30], extended and modified for the layered logic programs that correspond to the axioms of an MPT.

They consist of two parts. First, an indexing data structure maps a given variable/value pairing and a given axiom layer to the set of axioms in this layer in whose body the pairing appears. Second, a set of counters, one for each axiom, counts the number of conditions in the body of the axiom which have not yet been derived.

---

algorithm evaluate-axiom-layer(s, $\mathcal{A}_i$):
    for each axiom $a \in \mathcal{A}_i$:
        a.counter := $|a.\text{cond}|$
    for each variable v:
        for each axiom $a \in \mathcal{A}_i$ with a condition $v = s(v)$ in the body:
            a.counter := a.counter $- 1$
    while there is a not yet considered axiom $a \in \mathcal{A}_i$ with a.counter = 0:
        Let a be such an axiom.
        $\langle v, d \rangle$:= a.head
        if $s(v) \neq d$:
            $s(v) := d$
            for each axiom $a \in \mathcal{A}_i$ with condition $v = d$ in the body:
                a.counter := a.counter $- 1$

Fig. 10.7. Computing the values of the derived variables in a given planning state

Within Fast Downward, axioms are evaluated in two steps. First, all derived variables are set to their default value $\perp$. Second, algorithm *evaluate-axiom-layer* is executed for each axiom layer in sequence to determine the final values of the derived variables. This algorithm performs the necessary fixed point computation to determine which axioms are triggered in this layer (Fig. 10.7). By maintaining a queue or stack of triggered axioms, into which an axiom is inserted as soon as its counter reaches 0, the algorithm can easily be implemented to run in linear time in the size of its input.

The actual implementation of *evaluate-axiom-layer* in Fast Downward is slightly more efficient than indicated by the pseudo-code, since it avoids looking at each variable for each layer in the initialization step of the algorithm. However, this is a minor technical detail, so we turn to the remaining piece of Fast Downward's architecture, the search component.

11

Search

After the preparational steps of translation and knowledge compilation, the search component described in this chapter performs the actual work in finding a plan.

The following Sect. 11.1 presents a high-level overview of the search component. After this, we introduce the two heuristics used by Fast Downward, namely the causal graph heuristic (Sect. 11.2) and FF heuristic (Sect. 11.3). The rest of the chapter is dedicated to the search algorithms used, emphasizing how (and why) they differ from more traditional techniques. We discuss our modification of greedy best-first search in Sect. 11.4, multi-heuristic best-first search in Sect. 11.5 and focused iterative-broadening search in Sect. 11.6.

11.1 Overview

Unlike the translation and knowledge compilation components, for which there is only a single mode of execution, the search component of Fast Downward can perform its work in various alternative ways. There are three basic search algorithms to choose from:

1. *Greedy best-first search:* This is the standard textbook algorithm [105], modified with a technique called *deferred heuristic evaluation* to mitigate the negative influence of wide branching. We have also extended the algorithm to deal with *preferred operators*, similar to FF's helpful actions [68]. We discuss greedy best-first search in Sect. 11.4. Fast Downward uses this algorithm together with the causal graph heuristic, discussed in Sect. 11.2.
2. *Multi-heuristic best-first search:* This is a variation of greedy best-first search which evaluates search states using multiple heuristic estimators, maintaining separate open lists for each. Like our variant of greedy best-first search, it supports the use of *preferred operators*. Multi-heuristic best-first search is discussed in Sect. 11.5. Fast Downward uses this algorithm together with the causal graph and FF heuristics, discussed in Sects. 11.2 and 11.3.

3. *Focused iterative-broadening search:* This is a simple search algorithm that does not use heuristic estimators, and instead reduces the vast set of search possibilities by focusing on a limited operator set derived from the causal graph. It is an experimental algorithm; in the future, we hope to further develop the basic idea of this algorithm into a more robust method. Focused iterative-broadening search is discussed in Sect. 11.6.

For the two heuristic search algorithms, a second choice must be made regarding the use of *preferred operators*. There are five options supported by the planner:

1. Do not use preferred operators.
2. Use the *helpful transitions* of the causal graph heuristic as preferred operators.
3. Use the *helpful actions* of the FF heuristic as preferred operators.
4. Use helpful transitions as preferred operators, falling back to helpful actions if there are no helpful transitions in the current search state.
5. Use both helpful transitions and helpful actions as preferred operators.

Each of these five options can be combined with any of the two heuristic search algorithms, so that there is a total of eleven possible settings for the search component, ten using one of the heuristic algorithms and one using focused iterative-broadening search.

In addition to these basic settings, the search component can be configured to execute several alternative configurations in parallel by making use of an internal scheduler. Both configurations of Fast Downward that participated in IPC4 made use of this feature by running one configuration of the heuristic search algorithms in parallel with focused iterative-broadening search. As its heuristic search algorithm, the configuration *Fast Downward* employed greedy best-first search with helpful transitions, falling back to helpful actions when necessary (option 4.). The configuration *Fast Diagonally Downward* employed multi-heuristic best-first search using helpful transitions and helpful actions as preferred operators (option 5.). To avoid confusion between the complete Fast Downward planning system and the particular configuration called "Fast Downward", we will refer to the IPC4 planner configurations as *FD* and *FDD* in the following. The name of the planning system as a whole is never abbreviated.

11.2 The Causal Graph Heuristic

The *causal graph heuristic* is the centrepiece of Fast Downward's heuristic search engine. It estimates the cost of reaching the goal from a given search state by solving a number of subtasks of the planning task which are derived by looking at small "windows" of the (pruned) causal graph. For some additional intuitions about the design of the heuristic and a discussion of theoretical aspects, we refer to the article in which we first introduced the heuristic [58].

11.2.1 Conceptual View of the Causal Graph Heuristic

For each state variable v and each pair of values $d, d' \in \mathcal{D}_v$, the causal graph heuristic computes a heuristic estimate $cost_v(d, d')$ for the cost of changing the value of v from d to d', assuming that all other state variables carry the same values as in the current state. (This is a simplification. Cost estimates are not computed for state variables v or values d for which they are never required. We ignore this fact when discussing the heuristic on the conceptual level.) The heuristic estimate of a given state s is the sum over the costs $cost_v(s(v), s_\star(v))$ for all variables v for which a goal condition $s_\star(v)$ is defined.

Conceptually, cost estimates are computed one variable after the other, traversing the (pruned) causal graph in a bottom-up fashion. By bottom-up, we mean that we start with the variables that have no predecessors in the causal graphs; we call this order of computation "bottom-up" because we consider variables that can change their state of their own accord *low-level*, while variables whose state transitions require the help of other variables have more complex transition semantics and are thus considered *high-level*. Note that in our figures depicting causal graphs, *high-level* variables are typically displayed near the *bottom*.

For variables without predecessors in the causal graph, $cost_v(d, d')$ simply equals the cost of a shortest path from d to d' in the (pruned) domain transition graph $DTG(v)$. For other variables, cost estimates are also computed by graph search in the domain transition graph. However, the conditions of transitions must be taken into account during path planning, so that in addition to counting the number of transitions required to reach the destination value, we also consider the costs for achieving the value changes of the other variables necessary to set up the transition conditions.

The important point here is that in computing the values $cost_v(d, d')$, we completely consider all interactions of the state variable v with its predecessors in the causal graph. If changing the value from d to d' requires several steps and each of these steps has an associated condition on a variable v', then we realize that v' must assume the values required by those conditions *in sequence*. For example, if v represents a package in a transportation task that must be moved from A to B by means of a vehicle located at C, then we recognize that the vehicle must first move from C to A and then from A to B in order to drop the package at B. This is very different to the way HSP- or FF-based heuristics work on such examples. However, we only consider interactions with the *immediate* predecessors of v in the causal graph. Interactions that occur via several graph layers are not captured by the heuristic estimator.

In essence, we compute $cost_v(d, d')$ by solving a particular subtask of the MPT, induced by the variable v and its predecessors in the pruned causal graph. For this subtask, we assume that v is initially set to d, we want v to assume the value d', and all other state variables carry the same value as in the current state. We call this planning task the *local subtask for v, d and d'*, or the *local subtask for v and d* if we leave the target value d' open.

---

algorithm compute-costs-bottom-up(Π, s):
        for each variable v in a bottom-up traversal of the pruned causal graph:
                Let $\mathcal{V}'$ be the set of immediate predecessors of v
                in the pruned causal graph.
                for each pair of values $\langle d, d' \rangle \in \mathcal{D}_v \times \mathcal{D}_v$:
                        Generate the following planning task $\Pi_{v,d,d'}$:
                        — *Variables:* $\mathcal{V}' \cup \{v\}$.
                        — *Initial state:* $v = d$ and $v' = s(v')$ for all $v' \in \mathcal{V}'$.
                        — *Goal:* $v = d'$.
                        — *Axioms and operators:*
                          1. Those corresponding to transitions
                             in the pruned DTG of v.
                        2. For all variables $v' \in \mathcal{V}'$ and values $e, e' \in \mathcal{D}_{v'}$,
                             an operator with precondition $v' = e$,
                             effect $v' = e'$, and cost $cost_v'(e, e')$.
                            { All variables $v' \in \mathcal{V}'$ have been considered
                             previously, so their cost values are known. }
                    Set $cost_v(d, d')$ to the cost of a plan π that solves $\Pi_{v,d,d'}$.

Fig. 11.1. The *compute-costs-bottom-up* algorithm, a high-level description of the causal graph heuristic

For a formalization of these intuitive notions of how the *cost* estimates are generated, consider the pseudo-code in Fig. 11.1. It does not reflect the way the heuristic values are actually computed within Fast Downward; the algorithm in the figure would be far too expensive to evaluate for each search state. However, it computes the same cost values as Fast Downward does, provided that the algorithm generating the plans π in the last line of the algorithm is the same one as the one used for the "real" cost estimator.

11.2.2 Computation of the Causal Graph Heuristic

The actual computation of the causal graph heuristic traverses the causal graph in a top-down direction starting from the goal variables, rather than bottom-up starting from variables without causal predecessors. In fact, this top-down traversal of the causal graph is the reason for Fast Downward's name.

Computing cost estimates in a top-down traversal implies that while the algorithm is computing plans for local subtasks of a given variable, it typically does not yet know the costs for changing the state of its causal predecessors. The algorithm *compute-costs* addresses this by evaluating the cost values of dependent variables through recursive invocations of itself.

For a given variable-value pairing $v = d$, we always compute the costs $cost_v(d, d')$ for all values of $d' \in \mathcal{D}_v$ at the same time, similar to the way Dijkstra's algorithm computes the shortest path not from a single source to a single destination vertex, but from a single source to all possible destination vertices. Computing the costs for all values of d' is not (much) more expensive than computing only one of these values, and once all cost values have been determined, we can cache them and re-use them if they are needed again later during other parts of the computation of the heuristic value for the current state.

In fact, the similarity to shortest path problems is not superficial but runs quite deeply. If we ignore the recursive calls for computing cost values of dependent variables, *compute-costs* is basically an implementation of Dijkstra's algorithm for the single-source shortest path problem on domain transition graphs. The only difference to the "regular" algorithm lies in the fact that we do not know the cost for using an arc in advance. Transitions of derived variables have a base cost of 0 and transitions of fluents have a base cost of 1, but in addition to the base cost, we must pay the cost for achieving the conditions associated with a transition. However, the cost for achieving a given condition $v' = e'$ depends on the current value e of that state variable at the time the transition is taken. Thus, we can only compute the real cost for a transition once we know the values of the dependent state variables in the relevant situation.

Of course, there are many different ways of taking transitions through domain transition graphs, all potentially leading to different values for the dependent state variables. When we first introduced the causal graph heuristic, we showed that deciding plan existence for the local subtasks is NP-complete [58], so we are content with an approach that does not lead to a complete planning algorithm, as long as it works well for the subtasks we face in practice.

The approach we have chosen is to achieve each value of state variable v in the local subtask for v and d as quickly as possible, following a greedy policy. In the context of the Dijkstra algorithm, this means that we start by finding the cheapest possible plan to make a transition from d to some other value d'. Once we have found the cheapest possible plan $\pi_{d'}$, we commit to it, annotating the vertex d' of the domain transition graph with the local state obtained by applying plan $\pi_{d'}$ to the current state. In the next step, we look for the cheapest possible plan to achieve another value d'', by either considering transitions that start from the initial value d, or by considering transitions that continue the plan $\pi_{d'}$ by moving to a neighbour of d'. This process is iterated until all vertices of the domain transition graph have been reached or no further progress is possible.

Our implementation follows Dijkstra's algorithm (Fig. 11.2). We have implemented the priority queue as a vector of buckets for maximal speed and use a cache to avoid generating the same $cost_v(d, d')$ value twice for the same state. In addition to this, we use a global cache that is shared throughout the whole planning process so that we need to compute the values $cost_v(d, d')$ for variables v with few ancestors in the pruned causal graph only once. (Note that $cost_v(d, d')$ only depends on the current values of the ancestors of v.)

---

algorithm compute-costs(Π, s, v, d):
    Let $\mathcal{V}'$ be the set of immediate predecessors of v
      in the pruned causal graph of Π.
    Let DTG be the pruned domain transition graph of v.
    $cost_v(d, d) := 0$
    $cost_v(d, d') := \infty$ for all $d' \in \mathcal{D}_v \setminus \{d\}$
    $local\text{-}state_d := s$ restricted to $\mathcal{V}'$
    $unreached := \mathcal{D}_v$
    while $unreached$ contains a value $d' \in \mathcal{D}_v$ with $cost_v(d, d') < \infty$:
        Choose such a value $d' \in unreached$ minimizing $cost_v(d, d')$.
        $unreached := unreached \setminus \{d'\}$
        for each transition t in DTG from d' to some $d'' \in unreached$:
            $transition\text{-}cost := 0$ if v is derived; 1 if v is a fluent
            for each pair $v' = e'$ in the condition of t:
                $e := local\text{-}state_{d'}(v')$
                call compute-costs(Π, s, v', e).
                $transition\text{-}cost := transition\text{-}cost + cost_{v'}(e, e')$
            if $cost_v(d, d') + transition\text{-}cost < cost_v(d, d'')$:
                $cost_v(d, d'') := cost_v(d, d') + transition\text{-}cost$
                $local\text{-}state_{d''} := local\text{-}state_{d'}$
                for each pair $v' = e'$ in the condition of t:
                    $local\text{-}state_{d''}(v') := e'$

Fig. 11.2. Fast Downward's implementation of the causal graph heuristic: the *compute-costs* algorithm for computing the estimates $cost_v(d, d')$ for all values $d' \in \mathcal{D}_v$ in a state s of an MPT Π

Apart from these and some other technical considerations, Fig. 11.2 gives an accurate account of Fast Downward's implementation of the causal graph heuristic. For more details, including complexity considerations and a worked-out example, we refer to the original description of the algorithm [58].

11.2.3 States with Infinite Heuristic Value

We noted that Fast Downward uses an incomplete planning algorithm for determining solutions to local planning tasks. Therefore, there can be states s with $cost_v(s(v), s_\star(v)) = \infty$ even though the goal condition $v = s_\star(v)$ can still be reached. This means that we cannot trust infinite values returned by the causal graph heuristic. In our experience, states with infinite heuristic evaluation from which it is still possible to reach the goal are rare, so we indeed treat such states as *dead ends*.

If it turns out that *all* states at the search frontier are dead ends, we cannot make further progress with the causal graph heuristic. In this case, we use a sound dead-end detection routine to verify the heuristic assessment. If it turns out that all frontier states are indeed dead ends, then we report the task as unsolvable. Otherwise, search is restarted with the FF heuristic (cf. Sect. 11.3), which is sound for purposes of dead-end detection. (In practice, we have never observed the causal graph heuristic to fail on a solvable task. Therefore, the fallback mechanism is only used for some unsolvable tasks in the MICONIC-10-FULLADL domain which are not recognized by our dead-end detection technique.)

The dead-end detection routine has been originally developed for STRIPS-like tasks. However, extending it to full MPTs is easy; in fact, no changes to the core algorithm are required, as it works at the level of domain transition graphs and is still sound when applied to tasks with conditional effects and axioms. Since it is not a central aspect of Fast Downward, we do not discuss it here, referring to our earlier work instead [58].

11.2.4 Helpful Transitions

Inspired by Hoffmann's very successful use of *helpful actions* within the FF planner [68], we have extended our algorithm for computing the causal graph heuristic so that in addition to the heuristic estimate, it also generates a set of applicable operators considered useful for steering search towards the goal.

To compute helpful actions in FF, Hoffmann's algorithm generates a plan for the relaxed planning task defined by the current search state and considers those operators *helpful* which belong to the relaxed plan and are applicable in the current state.

Our approach follows a similar idea. After computing the heuristic estimate $cost_v(s(v), s_\star(v))$ for a variable v for which a goal condition is defined, we look into the domain transition graph of v to trace the path of transitions leading from $s(v)$ to $s_\star(v)$ that gave rise to the cost estimate. In particular, we consider the first transition on this path, starting at $s(v)$. If this transition corresponds to an applicable operator, we consider that operator a *helpful transition* and continue to check the next goal. If the transition does not correspond to an applicable operator because it has associated conditions of the form $v' = e'$ which are not currently satisfied, then we recursively look for helpful transitions in the domain transition graph of each such variable v', checking the path that was generated during the computation of $cost_{v'}(s(v'), e')$.

The recursive process continues until we have found all helpful transitions. Unlike the case for FF, where helpful actions can be found for all non-goal states, we might not find any helpful transition at all. It may be the case that a transition does not correspond to an applicable operator even though it has no associated conditions; this can happen when some operator preconditions are not represented in the pruned domain transition graph due to cycles in the causal graph. Even so, we have found helpful transitions to be a useful tool in guiding our best-first search algorithms.

11.3 The FF Heuristic

The *FF heuristic* is named after Hoffmann's planning algorithm of the same name, in the context of which it was originally introduced [68]. It is based on the notion of *relaxed planning tasks* that ignore *negative interactions*. In the context of MPTs, ignoring negative interactions means that we assume that each state variable can hold several values simultaneously. An operator effect or axiom that sets a variable v to a value d in the original task corresponds to an effect or axiom that *adds* the value d to the range of values assumed by v in the relaxed task. A condition $v = d$ in the original task corresponds to a condition requiring d to be an element of the set of values currently assumed by v in the relaxed task.

It is easy to see that applying some operator in a solvable relaxed planning task can never render it unsolvable. It can only lead to more operators being applicable and more goals being true, if it has any significant effect at all. For this reason, relaxed planning tasks can be solved efficiently, even though optimal solutions are still NP-hard to compute [19]. A plan for the relaxation of a planning task is called a *relaxed plan* for that task.

The FF heuristic estimates the goal distance of a world state by generating a relaxed plan for the task of reaching the goal from this world state. The number of operators in the generated plan is then used as the heuristic estimate. Our implementation of the FF heuristic does not necessarily generate the same, or even an equally long, relaxed plan as FF. In our experiments, this did not turn out to be problematic, as both implementations appear to be equally informative.

While the FF heuristic was originally introduced for ADL domains, extending it to tasks involving derived predicates is straight-forward. One possible extension is to simply assume that each derived predicate is initially set to its default value $\perp$ and treat axioms as relaxed operators of cost 0. In a slightly more complicated, but also more accurate approach, derived variables are initialized to their actual value in a given world state, allowing the relaxed planner to achieve the value $\perp$ (or other values) by applying the transitions of the extended domain transition graph of the derived variable. We have followed the second approach.

In addition to heuristic estimates, the FF heuristic can also be exploited for restricting or biasing the choice of operators to apply in a given world state s. The set of *helpful actions* of s consists of all those operators of the relaxed plan computed for s that are applicable in that state. As mentioned in Sect. 11.1, Fast Downward can be configured to treat helpful actions as preferred operators.

There is a wealth of work on the FF heuristic in the literature, so we do not discuss it further. For a more thorough treatment, we point to the references [63–65, 68].

11.4 Greedy Best-First Search in Fast Downward

Fast Downward uses *greedy best-first search with a closed list* as its default search algorithm. We assume that the reader is familiar with the algorithm and refer to the literature for details [105].

Greedy best-first search in Fast Downward differs from the textbook algorithm in two ways. First, it can treat helpful transitions computed by the causal graph heuristic or helpful actions computed by the FF heuristic as *preferred operators*. Second, it performs *deferred heuristic evaluation* to reduce the influence of large branching factors. We now turn to describing these two search enhancements.

11.4.1 Preferred Operators

To make use of helpful transitions computed by the causal graph heuristic or helpful actions computed by the FF heuristic, our variant of greedy best-first search supports the use of so-called *preferred operators*. The set of preferred operators of a given state is a subset of the set of applicable operators for this state. Which operators are considered preferred depends on the settings for the search component, as discussed earlier. The intuition behind preferred operators is that a randomly picked successor state is more likely to be closer to the goal if it is generated by a preferred operator, in which case we call it a *preferred successor*. Preferred successors should be considered before non-preferred ones on average.

Our search algorithm implements this preference by maintaining two separate open lists, one containing *all* successors of expanded states and one containing *preferred* successors exclusively. The search algorithm alternates between expanding a regular successor and a preferred successor. On even iterations it will consider the one open list, on odd iterations the other. No matter which open list a state is taken from, all its successors are placed in the first open list, and the preferred successors are additionally placed in the second open list. (Of course we could limit the first open list to only contain non-preferred successors; however, typically the total number of successors is vast and the number of preferred successors is tiny. Therefore, it is cheaper to add all successors to the first open list and detect duplicates upon expansion than scan through the list of successors determining for each element whether or not it is preferred.)

Since the number of preferred successors is smaller than the total number of successors, this means that preferred successors are typically expanded much earlier than others. This is especially important in domains where heuristic guidance is weak and a lot of time is spent exploring plateaus. When faced with plateaus, Fast Downward's open lists operate in a first-in-first-out fashion. (In other words: For a constant heuristic function, our search algorithm behaves like breadth-first search.) Preferred operators typically offer much better chances of escaping from plateaus since they lead to significantly lower effective branching factors.

11.4.2 Deferred Heuristic Evaluation

Upon expanding a state s, the textbook version of greedy best-first search computes the heuristic evaluation of all successor states of s and sorts them into the open list accordingly. This can be wasteful if s has many successors and heuristic evaluations are costly, two conditions that are often true for heuristic search approaches to planning. This is where our second modification comes into play.

If a successor with a better heuristic estimate than s is generated early and leads to a promising path towards the goal, we would like to avoid generating the other successors. Let us assume that s has 1000 successors, and that s', the 10th successor of s being generated, has a better heuristic estimate than s. Furthermore, let us assume that the goal can be reached from s' on a path with non-increasing heuristic estimates. Then we would like to avoid computing heuristic values for the 990 later successors of s altogether, focusing on s' instead.

Deferred heuristic evaluation achieves this by *not* computing heuristic estimates for the successors of an expanded state s immediately. Instead, the successors of s are placed in the open list together with the heuristic estimate *of state s*, and their own heuristic estimates are only computed when and if they are expanded, at which time it is used for sorting *their* successors into the open list, and so on. In general, each state is sorted into the open list according to the heuristic evaluation of its parent, with the initial state being an exception. In fact, we do not need to put the successor state itself into the open list, since we do not require its representation before we want to evaluate its heuristic estimate. Instead, we save memory by storing only a reference to the parent state and the operator transforming the parent state into the successor state in the open list.

It might not be clear how this approach can lead to significant savings in time, since deferred evaluation also means that information is only available later. The potential savings become most apparent when considering deferred heuristic evaluation together with the use of preferred operators: If an improving successor s' of a state s is reached by a preferred operator, it is likely that it will be expanded (via the second open list) long before most other successors – or even most siblings – of s. In the situation described above, where there exists a non-increasing path from s' to the goal, heuristic evaluations will never be computed for most successors of s. In fact, deferred heuristic evaluation can significantly improve search performance even when preferred operators are not used, especially in tasks where branching factors are large and the heuristic estimates are informative.

At first glance, deferred heuristic evaluation might appear related to another technique for reducing the effort of expanding a node within a best-first search algorithm, namely A* with Partial Expansion [114]. However, this algorithm is designed for reducing the *space* requirements of best-first search at the expense of additional heuristic evaluations: When expanding a node, A* with Partial Expansion computes the heuristic value of *all* successors, but only stores those in the open queue whose heuristic values fall below a certain *relevance threshold*. In later iterations, it might turn out that the threshold was chosen too low, in which case the node needs to be re-expanded and the heuristic values of its successors re-evaluated. In general, A* with Partial Expansion will never compute fewer heuristic estimates than standard A*, but it will usually require less memory.

However, for heuristic search approaches to planning (and certainly for Fast Downward), it is usually the case that heuristic evaluations are so costly in time that memory for storing open and closed lists is not a limiting factor. We are thus willing to trade off memory with time in the opposite way: Deferred heuristic evaluation normally leads to more node expansions and higher space requirements than standard best-first search because the heuristic values used for guiding the search are less informative (they evaluate the predecessor of a search node rather than the node itself). However, heuristic computations are only required for nodes that are actually removed from the open queue rather than for all nodes on the fringe, and the latter are usually significantly more numerous.

11.5 Multi-heuristic Best-First Search

As an alternative to greedy best-first search, Fast Downward supports an extended algorithm called *multi-heuristic best-first search*. This algorithm differs from greedy best-first search in its use of multiple heuristic estimators, based on our observation that different heuristic estimators have different weaknesses. It may be the case that a given heuristic is sufficient for directing the search towards the goal except for one part of the plan, where it gets stuck on a plateau. Another heuristic might have similar characteristics, but get stuck in another part of the search space.

Various ways of combining heuristics have been proposed in the literature, typically adding together or taking the maximum of the individual heuristic estimates. We believe that it is often beneficial *not* to combine the different heuristic estimates into a single numerical value. Instead, we maintain a *separate* open list for each heuristic estimator, which is sorted according to the respective heuristic. The search algorithm alternates between expanding a state from each open list. Whenever a state is expanded, estimates are calculated according to *each* heuristic, and the successors are put into each open list.

When Fast Downward is configured to use multi-heuristic best-first search, it computes estimates both for the causal graph heuristic and FF heuristic, maintaining two open lists. Of course, the approach can be combined with the use of preferred operators; in this case, the search algorithm maintains four open lists, as each heuristic distinguishes between normal and preferred successors.

11.6 Focused Iterative-Broadening Search

The *focused iterative-broadening search* algorithm is the most experimental piece of Fast Downward's search arsenal. In its present form, the algorithm is unsuitable for many planning domains, especially those containing comparatively few different goals. Yet we think that it might contain the nucleus for a successful approach to domain-independent planning which is very different to most current methods, so we include it for completeness and as a source of inspiration.

The algorithm is intended as a first step towards developing search techniques that emphasize the idea of using heuristic criteria locally, for limiting the set of operators to apply, rather than globally, for choosing which states to expand from a global set of open states. We made first experiments in this direction after observing the large boost in performance that can be obtained by using preferred operators in heuristic search. The algorithm performed surprisingly well in some of the standard benchmark domains, while performing badly in most others.

As the name suggests, the algorithm *focuses* the search by concentrating on one goal at a time, and by restricting its attention to operators which are supposedly important for reaching that goal:

Definition 11.6.1. *Modification Distances*
Let Π be an MPT, let o be an operator of Π, and let v be a variable of Π.

The **modification distance** *of o with respect to v is defined as the minimum, over all variables v' that occur as affected variables in the effect list of o, of the distance from v' to v in $CG(\Pi)$.*

For example, operators that modify v directly have a modification distance of 0 with respect to v, operators that modify variables which occur in preconditions of operators modifying v have a modification distance of 1, and so on. We assume that in order to change the value of a variable, operators with a low modification distance with respect to this variable are most useful.

Figure 11.3 shows the *reach-one-goal* procedure for achieving a single goal of an MPT. For the time being, assume that the *cond* parameter is always $\emptyset$. The procedure makes use of the assumption that high modification distance implies low usefulness in two ways. First, operators with high modification distance with respect to the goal variable are considered to have a higher associated cost, and are hence applied less frequently. Second, operators whose modification distance is beyond a certain threshold are forbidden completely. Instead of choosing a threshold a priori, the algorithm first tries to find a

---

algorithm reach-one-goal(Π, v, d, *cond*):
    for each $\vartheta \in \{0, 1, \ldots, max\text{-}threshold\}$:
        Let $\mathcal{O}_\vartheta$ be the set of operators of Π whose modification distance
        with respect to v is at most ϑ.
        Assign the cost c to each operator $o \in \mathcal{O}_\vartheta$ with modification
        distance c with respect to v.
        Call the *uniform-cost-search* algorithm with a closed list, using
        operator set O_ϑ, to find a state satisfying $\{v = d\} \cup$ *cond*.
        return the plan if uniform-cost-search succeeded.

Fig. 11.3. The *reach-one-goal* procedure for reaching a state with $v = d$. The value *max-threshold* is equal to the maximal modification distance of any operator with respect to v

solution with the lowest possible threshold of 0, increasing the threshold by 1 whenever the previous search has failed. The *uniform-cost-search* algorithm mentioned in Fig. 11.3 is the standard textbook method [105].

Although we were ignorant of this fact at the time our algorithm was conceived, the core idea of *reach-one-goal* is not new: Ginsberg and Harvey [49] present a search technique called *iterative broadening*, which is also based on the idea of repeatedly doing a sequence of uninformed searches with an ever-growing set of operators. Their work demonstrates the superiority of iterative broadening over standard depth-bounded search both empirically and analytically under the reasonable assumption that the choices made at each branching point are equally important. (See the original analysis for a precise definition of "equally important" [49]. While Ginsberg and Harvey's assumption is certainly not valid in practice, we find it much more convincing than the competing model where goal states are uniformly distributed across the search fringe.) The original iterative broadening algorithm applies to scenarios without any knowledge of the problem domain, so it chooses the set of operators which may be applied at every search node randomly, rather than using heuristic information from the causal graph as in our case. However, Ginsberg and Harvey already discuss the potential incorporation of heuristics into the operator selection. The introduction of operator costs (in the form of modification distances) is new, but it is a fairly straightforward extension where heuristic information is available.

The focused iterative-broadening search algorithm is based on the *reach-one-goal* method; the idea is to achieve the goals of the planning task one after the other, by using the *reach-one-goal* algorithm as the core subroutine for satisfying individual goals. Since it is not obvious what a good order of achieving the goals would be, one invocation of *reach-one-goal* is started for each goal in parallel. With each one-goal solver focusing on the (supposedly) relevant operators for reaching its particular goal, there is hope that the number of states considered before a goal is reached is small. Once one of the

---

algorithm reach-one-goal(Π, v, d, *cond*):

    for each $\vartheta \in \{0, 1, \ldots, max\text{-}threshold\}$:

        Let $\mathcal{O}_\vartheta$ be the set of operators of Π whose modification distance
        with respect to v is at most ϑ and which do not affect
        any state variable occurring in *cond*.

        Assign the cost c to each operator $o \in \mathcal{O}_\vartheta$ with modification
        distance c with respect to v.

        Call the *uniform-cost-search* algorithm with a closed list, using
        operator set $\mathcal{O}_\vartheta$, to find a state satisfying $\{v = d\} \cup$ *cond*.

        return the plan if uniform-cost-search succeeded.

    for each $\vartheta \in \{0, 1, \ldots, max\text{-}threshold\}$:

        Let $\mathcal{O}_\vartheta$ be the set of operators of Π whose modification distance
        with respect to v is at most ϑ.

        Assign the cost c to each operator $o \in \mathcal{O}_\vartheta$ with modification
        distance c with respect to v.

        Call the *uniform-cost-search* algorithm with a closed list, using
        operator set $\mathcal{O}_\vartheta$, to find a state satisfying $\{v = d\} \cup$ *cond*.

        return the plan if uniform-cost-search succeeded.

Fig. 11.4. The *reach-one-goal* procedure for reaching a state with $v = d$ (corrected)

one-goal solvers reaches its goal, the resulting plan is reported and all sub-searches are stopped. The overall search algorithm commits to this part of the plan; the situation in which the first goal has been reached is considered a new initial state.

From this situation, we try to satisfy the second goal, by once more starting parallel invocations of *reach-one-goal* for each possible second goal. Of course, this can lead to a situation where the search algorithm oscillates between goals, first achieving goal a, then abandoning it in favour of goal b, without any sign of making real progress. Therefore, we demand that *reach-one-goal* achieves the second goal *in addition* to the one we reached first, by setting the *cond* argument accordingly. Once two goals have been reached, the sub-searches are again stopped, sub-searches for the third goal are started, and so on, until all goals have been reached.

In some sense, our focusing technique is similar to the beam search algorithm [88], which also performs a fixed number of concurrent searches to avoid committing to a particular path in the search space too early. Beam search uses a heuristic function to evaluate which branches of search should be abandoned and where new branches should be spawned. While focused iterative-broadening search does not appear to use heuristic evaluations at first glance, the number of satisfied goals of a state is used as an evaluation criterion in essentially the same way. One important difference to beam search is our use of modification distances relative to a particular goal, which means that the different "beams" explore the state space in qualitatively different ways.

There is one final twist: To motivate *reach-one-goal* not to needlessly wander away from satisfied goals, we forbid applying operators that undo any of the previously achieved goals in *cond*. This is an old idea called *goal protection* [74]. It is well-known that protecting goals renders a search algorithm incomplete, even in state spaces where all operators are reversible and local search approaches like focused iterative-broadening search would be otherwise complete. In particular, search must fail in planning tasks which are not *serializable* [83]. Therefore, if the first solution attempt fails, the algorithm restarts without goal protection. The complete procedure is shown in Fig. 11.4, which concludes our discussion of Fast Downward's search component.

12

Experiments

This chapter evaluates the performance of Fast Downward on the IPC benchmark suite. The following Sect. 12.1 describes and motivates the design of the experiment. The results of the experiments are reported in Sect. 12.2 STRIPS domains from IPC1–3), Sect. 12.3 (ADL domains from IPC1–3) and Sect. 12.4 (domains from IPC4). Conclusions are drawn in Sect. 12.5.

12.1 Experiment Design

To evaluate the performance of Fast Downward, and specifically the differences between the various configurations of the search component, we have performed a number of experiments on the IPC benchmarks. The purpose of these experiments is to compare Fast Downward to the state of the art in PDDL planning, and to contrast the performance of the different search algorithms of Fast Downward (greedy best-first search with and without preferred operators, multi-heuristic best-first search with and without preferred operators, and focused iterative-broadening search).

To clearly state the purpose of our experiments, let us also point out two areas worthy of study that we do *not* choose to investigate here:

- We do not compare the causal graph heuristic to other heuristics, such as the FF or HSP heuristics. Such a comparison would require evaluating the different heuristics within otherwise identical planning systems. We have performed such an experiment in a separate publication [58] and do not restate its outcome here. The experiments presented in this chapter are concerned with the *complete* Fast Downward planning system, rather than just its heuristic function.
- We do not give a final answer to the question *why* Fast Downward performs well or badly in the domains we analyse. Where we do observe bad performance, we try to give a plausible explanation for this, but we do not conduct a full-blown study of heuristic quality in the spirit of Hoffmann's

work on the FF and h^+ heuristics [65]. We do believe that much could be learned from such an investigation, and we consider this an interesting avenue of research for future work.

Given that our aim in this chapter is to evaluate the Fast Downward planner as a whole, there are a number of algorithmic questions which we do not address. For example, one might wonder what (if any) speedup can be obtained by using successor generators over simpler methods which test each operator for applicability whenever a node is expanded. Another question concerns the extent to which deferred heuristic evaluation affects search performance. To keep this chapter at a reasonable length, we do not discuss either of these questions here. However, we have conducted experiments addressing them. The short summary is that successor generators speed up search by up to two orders of magnitude in extreme cases like the largest SATELLITE tasks, but have little impact on performance most of the time. Deferred heuristic evaluation is very beneficial in some domains, with speedups of more than one order of magnitude being common, is somewhat beneficial in the majority of domains, with speedups between 2 and 4, and is very rarely detrimental to performance. Details on these experiments are reported in an electronic appendix to the article in which we first reported these experiments [59].

12.1.1 Benchmark Set

The benchmark set we use consists of all propositional planning tasks from the fully automated tracks of the first four international planning competitions hosted at AIPS 1998, AIPS 2000, AIPS 2002 and ICAPS 2004 – i. e., the set of planning domains discussed in Part I. The set of benchmark domains is shown in Fig. 12.1. Altogether, the benchmark suite comprises 1442 tasks. (The numbers in Fig. 12.1 add up to 1462, but the 20 SATELLITE instances that were introduced for IPC3 were also part of the benchmark set of IPC4, so we only count them once.)

We distinguish between three classes of domains:

– *STRIPS domains:* These domains do not feature derived predicates or conditional effects, and all conditions appearing in goal and operators are conjunctions of positive literals.
– *ADL domains:* These domains make use of conditional effects in their operators and/or contain more general conditions than simple conjunctions in their goals and operators. However, they do not require axioms.
– *PDDL2.2 domains:* These domains use the full range of propositional PDDL2.2, including those features present in ADL domains and axioms.

At IPC4, some domains were presented in different *formulations*, meaning that the same real-world task was encoded in several different ways. Participants were asked to only work on one formulation per domain, being able to choose their preferred formulation for a given domain freely. For example,

Competition	Domain	Class	# Tasks
IPC1 (AIPS 1998)	ASSEMBLY	ADL	30
	GRID	STRIPS	5
	GRIPPER	STRIPS	20
	LOGISTICS	STRIPS	35
	MOVIE	STRIPS	30
	MYSTERY	STRIPS	30
	MYSTERYPRIME	STRIPS	35
IPC2 (AIPS 2000)	BLOCKSWORLD	STRIPS	35
	FREECELL	STRIPS	60
	LOGISTICS	STRIPS	28
	MICONIC-10-STRIPS	STRIPS	150
	MICONIC-10-SIMPLEADL	ADL	150
	MICONIC-10-FULLADL	ADL	150
	SCHEDULE	ADL	150
IPC3 (AIPS 2002)	DEPOTS	STRIPS	22
	DRIVERLOG	STRIPS	20
	FREECELL	STRIPS	20
	ROVERS	STRIPS	20
	SATELLITE	STRIPS	20
	ZENOTRAVEL	STRIPS	20
IPC4 (ICAPS 2004)	AIRPORT	STRIPS	50
	PROMELA-OPTICALTELEGRAPH	PDDL2.2	48
	PROMELA-PHILOSOPHERS	PDDL2.2	48
	PIPESWORLD-NOTANKAGE	STRIPS	50
	PIPESWORLD-TANKAGE	STRIPS	50
	PSR-SMALL	STRIPS	50
	PSR-MIDDLE	PDDL2.2	50
	PSR-LARGE	PDDL2.2	50
	SATELLITE	STRIPS	36

Fig. 12.1. The benchmark set

the AIRPORT domain was available in a STRIPS formulation and an ADL formulation.

However, the organizers did not strictly follow the rule of considering different encodings of the same real-world task different *formulations*, rather than different domains proper. Namely, for the PSR-MIDDLE and PROMELA domains, encodings with and without axioms were available, and these were considered as different domains on the grounds that the encodings without axioms were much larger and hence likely more difficult to solve. We apply the formulation vs. encoding view more strictly and thus only consider one PSR-MIDDLE domain and one domain for each of the two PROMELA variants, PROMELA-PHILOSOPHERS and PROMELA-OPTICALTELEGRAPH.

Of the IPC1 benchmark set, all tasks are solvable except for 11 MYS-
TERY instances. Of the IPC2 benchmark set, all tasks are solvable except
for 11 MICONIC-10-FULLADL instances. All IPC3 benchmarks are solvable.
For IPC4, we have not checked all instances of the PIPESWORLD-TANKAGE
domain, but we assume that all are tasks are solvable.

If run in any of the heuristic search modes, Fast Downward proves the
unsolvability of the unsolvable MYSTERY and MICONIC-10-FULLADL tasks
by using the dead-end detection routine described in our earlier article on the
causal graph heuristic [58], or in some cases in the MICONIC-10-FULLADL
domain by exhaustively searching all states with a finite FF heuristic. Of
course, if an unsolvable task is proved unsolvable by the planner, we report
this as a "successfully solved" instance in the experimental results.

12.1.2 Experiment Setup

As discussed in Chap. 11, there are eleven possible configurations of Fast
Downward's search component. However, not all of them are equally reasonable. For example, if we use FF's helpful actions, it would seem wasteful not to
use the FF heuristic estimate, since these two are calculated together. Therefore, for the greedy best-first search setup, we exclude configurations where FF
helpful actions are always computed. For the multi-heuristic best-first search
setup, we exclude configurations where only one type of preferred operators
is considered, but not the other, since this would seem to be a very arbitrary
choice. This leaves us with six different configurations of the planner:

1. **G**: Use greedy best-first search without preferred operators.
2. **G^P**: Use greedy best-first search with helpful transitions as preferred operators.
3. **G^{P+}**: Use greedy best-first search with helpful transitions as preferred
   operators. Use helpful actions as preferred operators in states with no
   helpful transitions.
4. **M**: Use multi-heuristic best-first search without preferred operators.
5. **M^P**: Use multi-heuristic best-first search with helpful transitions and helpful actions as preferred operators.
6. **F**: Use focused iterative-broadening search.

We apply each of these planner configurations to each of the 1442 benchmark tasks, using a computer with a 3.066 GHz Intel Xeon CPU – the same
machine that was used at IPC4 – and set a memory limit of 1 GB and a
timeout of 300 seconds.

To compare Fast Downward to the state of the art, we try to solve each
benchmark with the best-performing planners from the literature. Unfortunately, this involves some intricacies: some planners are not publicly available,
and others only cover a restricted subset of PDDL2.2. For this reasons, we
partition the benchmark domains into three sets depending on which planners
are available for comparison.

Domain	Task	Config.	Preparation	Search
FREECELL (IPC2)	probfreecell-10-1	M + P	9.30 s	298.64 s
GRID	prob05	M	10.04 s	291.01 s
MYSTERYPRIME	prob14	M	22.38 s	291.67 s
PSR-LARGE	p30-s179-n30-13-f30	G + P	43.43 s	265.29 s
SATELLITE (IPC4)	p33-HC-pfile13	M + P	180.74 s	169.09 s

Fig. 12.2. Tasks which could be solved by some configuration of Fast Downward with a search timeout of 300 seconds, but not with a total processing timeout of 300 seconds. The column "preparation" shows the combined time for translation and knowledge compilation

12.1.3 Translation and Knowledge Compilation vs. Search

Of course, the results we report for Fast Downward include the time spent in all three components of the planner: translation, knowledge compilation, and search. Therefore, in the following presentation of results, we only consider a task solved if the *total* processing time is below 300 seconds. However, we have also investigated which tasks can be solved with a timeout of 300 seconds for the *search* component alone, allowing the other components to use an arbitrary amount of resources. It turns out that this only makes a difference in five cases, most of which could have been solved in a total time below 310 seconds (Fig. 12.2). Only in one of these five cases, a SATELLITE instance of exorbitant size, did search take less time than the other two phases combined. These results show that the search component is the only time-critical part of Fast Downward in practice. Therefore, we do not report separate performance results for the individual components.

12.2 STRIPS Domains from IPC1–3

We now start presenting the experimental results. We abstain from listing runtimes for individual planning tasks due to the prohibitively large amount of data. (Again, the detailed data is provided in the electronic appendix to the article in which these experiments were first presented [59].) Instead, we report the following summarizing information:

- Tables showing the number of tasks *not solved* by each planner within the 300 second timeout. Here, we present individual results for each domain.
- Graphs showing the number of tasks solved in a given time by each planner. Here, we do not present separate results for each domain, as this would require too many graphs.

We do not discuss plan lengths here; our observations in this regard are similar to those made for the original implementation of the causal graph heuristic [58].

Domain	G	G^P	G^{P+}	M	M^P	F	Any	CG	FF	LPG
BLOCKSWORLD (35)	0	0	0	0	0	17	0	0	4	0
DEPOTS (22)	12	13	13	12	8	11	7	14	3	0
DRIVERLOG (20)	2	0	0	1	0	1	0	3	5	0
FREECELL IPC2 (60)	4	4	12	11	12	40	3	2	3	55
FREECELL IPC3 (20)	0	0	5	1	2	14	0	0	2	19
GRID (5)	1	2	1	1	0	4	0	1	0	1
GRIPPER (20)	0	0	0	0	0	0	0	0	0	0
LOGISTICS IPC1 (35)	1	0	0	4	0	26	0	0	0	4
LOGISTICS IPC2 (28)	0	0	0	0	0	0	0	0	0	0
MICONIC-10-STRIPS (150)	0	0	0	0	0	0	0	0	0	0
MOVIE (30)	0	0	0	0	0	0	0	0	0	0
MYSTERY (30)	1	2	1	0	0	13	0	1	12	15
MYSTERYPRIME (35)	0	0	0	2	0	14	0	1	3	7
ROVERS (20)	2	0	0	0	0	2	0	3	0	0
SATELLITE IPC3 (20)	1	0	0	0	0	6	0	0	0	0
ZENOTRAVEL (20)	0	0	0	0	0	0	0	0	0	0
Total (550)	24	21	32	32	22	148	10	25	32	101

Fig. 12.3. Number of unsolved tasks for the STRIPS domains from IPC1–3

Figure 12.3 shows the number of unsolved tasks for each of the STRIPS domains from IPC1–3. Figures 12.4 and 12.5 show the number of tasks solved by each planner within a given time bound between 0 and 300 seconds. In addition to the six configurations of Fast Downward under consideration, the table includes four other columns.

Under the heading "Any", we include results for a hypothetical meta-planner that guesses the best of the six configuration of Fast Downward for each input task and then executes Fast Downward with this setting. Under the heading "CG", we report the results for our first implementation of the causal graph heuristic [58]. (Apart from missing support for ADL and axioms, CG is very similar to Fast Downward using greedy best-first search and no preferred operators (configuration **G**). The translation and knowledge compilation components are essentially identical. The older search component mainly differs from Fast Downward in that it does not use deferred heuristic evaluation.) Finally, "FF" and "LPG" refer to the well-known planners [44,68] which won the fully-automated tracks of IPC2 and IPC3. They were chosen for comparison on this benchmark set because they showed the best performance by far of all publicly available planners we experimented with. For LPG, which uses a randomized search strategy, we attempted each task five times and report the median result.

The results show excellent performance of Fast Downward on this set of benchmarks. Compared to CG, which was already shown to solve more tasks than FF and LPG on this benchmark set [58], we get another slight improve-

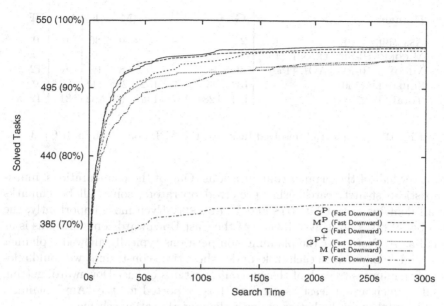

Fig. 12.4. Number of tasks solved vs. runtime for the STRIPS domains from IPC1–3. This graph shows the results for the various configurations of Fast Downward

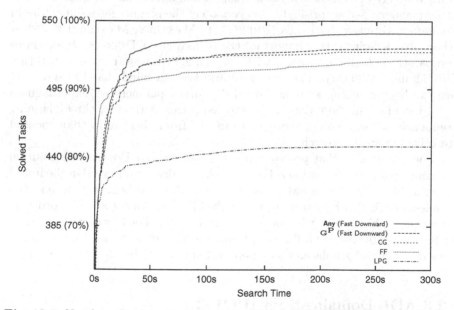

Fig. 12.5. Number of tasks solved vs. runtime for the STRIPS domains from IPC1–3. This graph shows the results for CG, FF and LPG and the hypothetical "Any" planner which always chooses the best configuration of Fast Downward. The result for greedy best-first search with helpful transitions is repeated for ease of comparison with Fig. 12.4

Domain	G	G^{P}	G$^{\mathrm{P+}}$	M	M^{P}	F	Any	FF
ASSEMBLY (30)	28	27	25	3	0	30	0	0
MICONIC-10-SIMPLEADL (150)	0	0	0	0	0	0	0	0
MICONIC-10-FULLADL (150)	9	8	9	9	8	90	6	12
SCHEDULE (150)	134	93	93	132	28	113	25	0
Total (480)	171	128	127	144	36	233	31	12

Fig. 12.6. Number of unsolved tasks for the ADL domains from IPC1–3

ment for half of the planner configurations. One of the configurations, multi-heuristic best-first search using preferred operators, solves all benchmarks in all domains except DEPOTS and FREECELL. Even more importantly, the number of tasks not solved by any of the Fast Downward configurations is as small as 10. Note that the planning competitions typically allowed a planner to spend 30 minutes on each task; under these time constraints, we could allocate five minutes to each of the six configurations of Fast Downward, getting results which are at least as good as those reported for the "Any" planner. Results might even be better under a cleverer allocation scheme.

Even the configuration using focused iterative-broadening search performs comparatively well on these benchmarks, although it cannot compete with the other planners. Not surprisingly, this version of the planner faces difficulties in domains with many dead ends (FREECELL, MYSTERY, MYSTERYPRIME) or where goal ordering is very important (BLOCKSWORLD, DEPOTS). It also fares comparatively badly in domains with very large instances, namely LOGISTICS (IPC1) and SATELLITE. The reader should keep in mind that FF and LPG are excellent planning systems; of all the other planners we experimented with, including all those that were awarded prizes at the first three planning competitions, none solved more benchmarks from this group than focused iterative-broadening search.

The one domain that proves quite resistant to Fast Downward's solution attempts in any configuration is DEPOTS. As we already observed in the initial experiments with the causal graph heuristic [58], we believe that one key problem here is that Fast Downward, unlike FF, does not use any goal ordering techniques, which are very important in this domain. The fact that the domain includes a BLOCKSWORLD-like subproblem is also an issue, as it gives rise to very dense causal graphs as we demonstrated in Sect. 10.3.3.

12.3 ADL Domains from IPC1–3

Second, we present results for the ADL domains of the first three planning competitions. This is a much smaller group than the previous, including only four domains. This time, we cannot consider CG or LPG, since neither CG nor the publicly available version of LPG supports ADL domains. Therefore,

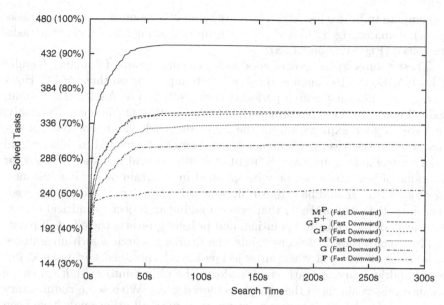

Fig. 12.7. Number of tasks solved vs. runtime for the ADL domains from IPC1–3. This graph shows the results for the various configurations of Fast Downward

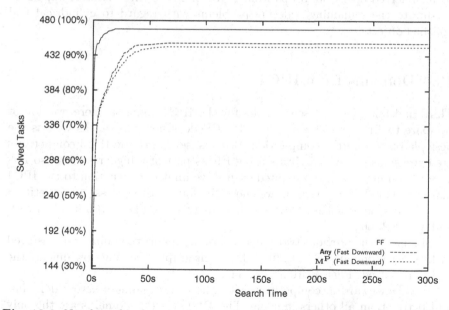

Fig. 12.8. Number of tasks solved vs. runtime for the ADL domains from IPC1–3. This graph shows the results for FF and the hypothetical "Any" planner which always chooses the best configuration of Fast Downward. The result for multi-heuristic best-first search with preferred operators is repeated for ease of comparison with Fig. 12.7

we compare to FF exclusively. Again, we report the number of unsolved tasks in each domain (Fig. 12.6) and present graphs showing how quickly the tasks are solved (Figs. 12.7 and 12.8).

These results do not look as good as for the first group of domains. Results in both MICONIC-10 domains are good, even improving on those of FF. However, greedy best-first search performs very badly in the ASSEMBLY domain, and all configurations perform badly in the SCHEDULE domain. Currently, we have no good explanation for the ASSEMBLY behaviour. For the SCHEDULE domain, the weak performance again seems to be related to missing goal ordering techniques: In many SCHEDULE tasks, several goals are defined for the same object which can only be satisfied in a certain order. For instance, for objects that should be cylindrical, polished and painted, these three goals must be satisfied in precisely that order: making an object cylindrical reverts the effects of polishing and painting, and polishing reverts the effect of painting. Not recognising these constraints, the heuristic search algorithm assumes to be close to the goal when an object is already polished and painted but not cylindrical, and is loathe to transform the object into cylindrical shape because this would undo the already achieved goals. With some rudimentary manual goal ordering, ignoring painting goals until all other goals have been satisfied, the number of tasks not solved by multi-heuristic best-first search with preferred operators drops from 28 to 3. These three failures appear to be due to the remaining ordering problems with regard to cylindrical and polished objects.

12.4 Domains from IPC4

Third and finally, we present results for the IPC4 domains. Here, we do not compare to FF: for these benchmarks, FF does not perform as well as the best planners from the competition. Besides, several of the IPC4 competitors are extensions of FF or hybrids using FF as part of a bigger system, so FF-based planning is well-represented even if we limit our attention to the IPC4 planners. For this comparison, we chose the four most successful competition participants besides Fast Downward, namely LPG-TD, SGPlan, Macro-FF and YAHSP [66].

Similar to the previous two experiments, we report the number of unsolved tasks in each domain (Fig. 12.9) and present graphs showing how quickly the tasks are solved (Figs. 12.10 and 12.11).

Fast Downward is competitive with the other planners across domains, and better than all others in some. The PIPESWORLD domains are the only ones in which any of the other planners is noticeably better than the two competition versions of Fast Downward. This is the case for YAHSP in both PIPESWORLD domains and for SGPlan in PIPESWORLD-NOTANKAGE. The PIPESWORLD domain is not very hierarchical in nature; this might be a domain where the decomposition approach of the causal graph heuristic is not

Domain	G	G^P	G^{P+}	M	M^P	F	Any
AIRPORT (50)	28	30	17	18	14	0	0
PIPESWORLD-NOTANKAGE (50)	24	25	23	14	7	10	7
PIPESWORLD-TANKAGE (50)	36	36	36	34	17	34	14
PROMELA-OPTICALTELEGRAPH (48)	48	47	48	47	46	13	13
PROMELA-PHILOSOPHERS (48)	0	0	0	16	0	21	0
PSR-SMALL (50)	0	0	0	0	0	1	0
PSR-MIDDLE (50)	0	0	0	0	0	22	0
PSR-LARGE (50)	22	20	22	23	22	39	20
SATELLITE IPC4 (36)	8	0	0	8	3	22	0
Total (432)	166	158	146	160	109	162	54

Domain	FD	FDD	LPG-TD	Macro-FF	SG-Plan	YA-HSP
AIRPORT (50)	0	0	7	30	6	17
PIPESWORLD-NOTANKAGE (50)	11	7	10	12	0	0
PIPESWORLD-TANKAGE (50)	34	19	29	29	20	13
PROMELA-OPTICALTELEGRAPH (48)	22	22	37	31	29	36
PROMELA-PHILOSOPHERS (48)	0	0	1	36	0	19
PSR-SMALL (50)	0	0	2	50	6	3
PSR-MIDDLE (50)	0	0	0	19	4	50
PSR-LARGE (50)	22	22	50	50	39	50
SATELLITE IPC4 (36)	0	3	1	0	6	0
Total (432)	89	73	137	257	110	188

Fig. 12.9. Number of unsolved tasks for the IPC4 domains. Results for the various configurations of Fast Downward are listed in the upper part, results for the competition participants in the lower part. "FD" and "FDD" denote the versions of Fast Downward that participated in IPC4 under the names "Fast Downward" and "Fast Diagonally Downward" (cf. Sect. 11.1)

very appropriate. The results of the heuristic search configurations in the PROMELA-OPTICALTELEGRAPH domain are extremely bad and require further investigation.

Interestingly, focused iterative-broadening search performs very well on some of the benchmarks from this suite. One of the reasons for this is that in many of the tasks of the IPC4 suite, there are many individual goals which are easy to serialize and can be solved mostly independently. (Indeed, we have devised an additional experiment which shows that if this property is artificially violated by a simple goal reformulation, the performance of the algorithm degrades quickly; again, consult see the electronic appendix of the original publication for details [59].) Comparing the configuration **G** to **G^{P+}** and especially **M** to **M^P**, we also observe that using preferred operators is very useful for these benchmarks, even more so than in the two previous experiments.

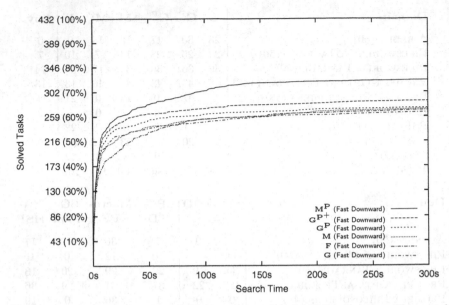

Fig. 12.10. Number of tasks solved vs. runtime for the IPC4 domains. This graph shows the results for the various configurations of Fast Downward

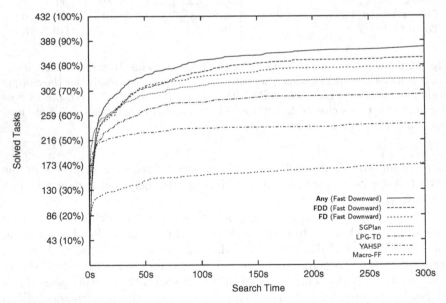

Fig. 12.11. Number of tasks solved vs. runtime for the IPC4 domains. This graph shows the results for the hypothetical "Any" planner which always chooses the best configuration of Fast Downward, the competition configurations of Fast Downward and the best four other participants

As a final remark, we observe that if we implemented the "Any" meta-planner by calling the six Fast Downward configurations in a round-robin fashion, we would obtain a planning system that could solve all but 54 of the IPC4 benchmarks within a $6 \cdot 5 = 30$ minute timeout. This is almost on par with the top performer of IPC4, Fast Diagonally Downward, which solved all but 52 of the IPC4 benchmarks under the same timeout. Thus, this is a benchmark set for which exploring different planner configurations definitely pays off.

12.5 Conclusions from the Experiment

What do we learn from the experimental results? Our first conclusion is that Fast Downward is clearly competitive with the state of the art. This is especially true for the configuration using multi-heuristic best-first search with preferred operators ($\mathbf{M^P}$), which outperforms all competing planning systems both on the set of STRIPS domains from IPC1–3 and on the domains from IPC4. If it were not for the issues with the SCHEDULE domain, the same would be true for the remaining group of benchmarks, the ADL domains from IPC1–3.

With regard to the second objective of the investigation, evaluating the relative strengths of the different planner configurations, the $\mathbf{M^P}$ configuration emerges as a clear-cut winner. In 23 out of 29 domains, no other configuration solves more tasks, and unlike the other configurations, there is only one domain (PROMELA-OPTICALTELEGRAPH) in which it performs very badly. We conclude that both multi-heuristic best-first search and the use of preferred operators are promising extensions to heuristic planners.

This is particularly true for preferred operators. Indeed, after the $\mathbf{M^P}$ configuration, the two variants of greedy best-first search with preferred operators show the next best overall performance, both in terms of the number of domains where they are among the top performers and in terms of the total number of tasks solved. Comparing $\mathbf{G}$ to $\mathbf{G^P}$, there are ten domains in which the variant using preferred operators solves more tasks than the one not using them; the opposite is true in five domains. Comparing $\mathbf{M}$ to $\mathbf{M^P}$, the difference is even more striking, with the preferred operator variant outperforming the other in fifteen domains, while being worse in two (in both of which it only solves one task less). These are convincing arguments for the use of preferred operators.

13

Discussion

In this final chapter, we provide a summary of Part II (Sect. 13.1), identify and discuss the major and minor contributions (Sects. 13.2 and 13.3) highlighting opportunities for further research, and conclude (Sect. 13.4).

13.1 Summary

As a motivating starting point for this part, we explained that planning tasks often exhibit a simpler structure if expressed with *multi-valued state variables*, rather than the traditional propositional representations. We then introduced *Fast Downward*, a planning system based on the idea of converting tasks into a multi-valued formalism and exploiting the causal information underlying such encodings.

Fast Downward processes PDDL planning tasks in three stages. The first of these stages, *translation*, automatically transforms a PDDL task into an equivalent multi-valued planning task with a nicer causal structure, using invariant synthesis techniques for grouping related propositions and efficient grounding algorithms for instantiating operators and axioms. We then discussed the second stage, *knowledge compilation*, demonstrating in depth what kind of knowledge the planner extracts from the task representation, discussing *causal graphs*, *domain transition graphs*, *successor generators* and *axiom evaluators*. During our discussion of Fast Downward's *search* component, we introduced its heuristic search algorithms, which use the technique of *deferred heuristic evaluation* to reduce the number of states for which a heuristic goal distance estimate must be computed. In addition to greedy best-first search, Fast Downward employs the *multi-heuristic best-first search* algorithm to usefully integrate the information of two heuristic estimators, namely the *causal graph heuristic* and *FF heuristic*. Both heuristic search algorithms can utilize preference information about operators. We also introduced Fast Downward's experimental *focused iterative-broadening search* algorithm, which is based on the idea of pruning the set of operators to only consider those successor states which are likely to lead towards a specific goal.

We thus tried to give a complete account of the Fast Downward planning system's approach to solving multi-valued planning tasks, including its motivation, architecture, and algorithmic foundations. In the previous chapter, we studied its empirical behaviour, showing good performance across the whole range of propositional benchmarks from the previous planning competitions.

13.2 Major Contributions

Among the novel techniques we introduced, there are two contributions which we consider to be of central importance and which we would like to emphasize.

13.2.1 Multi-valued Representations

The first of these major contributions is the use of multi-valued state variables for PDDL-style planning. We believe that multi-valued representations are much more structured and hence much more amenable to automated reasoning – be it for the purposes of heuristic evaluation, task decomposition, or other aspects of planning, such as goal ordering or extraction of landmarks.

One issue that would be worth studying further is the suitability of multi-valued representations for other planning algorithms. There exists a host of planning algorithms for which such representations appear to be more natural (or more efficient) than binary ones. One example is symbolic state-space exploration with binary decision diagrams (BDDs), the original MIPS planning algorithm [35]. It might seem somewhat unintuitive that multi-valued state variables are a natural representation for *binary* decision diagrams, but it is typically much more efficient to use an encoding that encodes a set of 15 mutually exclusive propositions with four binary variables encoding an integer from the set $\{0, \ldots, 15\}$ than using 15 separate BDD variables. Indeed, such an encoding is critical to the performance of BDD exploration in the case of MIPS.

An approach closely related to symbolic exploration with BDDs is planning based on Boolean satisfiability (SAT). Indeed, similar encodings can be used for either problem. For SAT-based planning, however, there are reasons to believe that it is more promising to introduce individual variables for each proposition of the planning task and encode mutual exclusion constraints explicitly [103]. Still, *arriving* at such mutual exclusion constraints, which *are* critical to performance, requires invariant synthesis techniques similar to the ones we have presented, an algorithmic chore that would be largely unnecessary if multi-valued representations were given as a base input. Given a multi-valued representation, the mutual exclusivity of different value assignments to the same variable would be immediately apparent, and we hypothesize that these mutual exclusions alone already account for a large percentage of *all* mutual exclusion constraints in a typical planning task.

Closely related to SAT planning, multi-valued representations are very beneficial for planning approaches that compile planning tasks into integer programs (IPs). Compared to an IP representation based on a binary task encoding, one can dramatically reduce the number of integer variables by representing a group of mutually exclusive propositions as a single number. The planning algorithm by van den Briel et al. uses Fast Downward's translator to obtain such a succinct encoding, and the authors report that this results in a significant speedup of the integer program solver [111].

Multi-valued representations are also a natural fit for planning approaches based on constraint satisfaction, although we are not aware of relevant work in this area.

So clearly, multi-valued representations are of potentially great benefit to many approaches to planning, not limited to the particular heuristic search technique presented here. However, much experience about the usefulness of such representations remains anecdotal, and a more formal study of this topic is needed. One interesting issue in this context is the question when one should prefer a given representation over another one. We illustrate this issue with an example from the BLOCKSWORLD domain. In addition to the standard binary encoding, there are at least three "obvious" multi-valued representations:

- For each block, encode what is *below* it (another block, the table, or thin air if it is being held). Moreover, encode for each block whether or not it is clear.
- For each block, encode what is *above* it (another block, the robot arm if it is being held, or thin air if it is a clear block not being held). Moreover, encode for each block whether or not it is on the table.
- For each block, encode what is below it *and* what is above it.

Our translation algorithm always generates either the first or second encoding. The actual choice is made arbitrarily, because both encodings are considered equally succinct by our greedy mutex group selection algorithm (Fig. 9.9). What is important here is that the translation algorithm never generates the third, redundant representation due to the intuition that redundant representations are often detrimental to performance. However, this is not universally true; for example in constraint satisfaction, redundant representations are often found to be beneficial because they allow additional constraint propagation. Using a modified version of the Fast Downward translator within the above-mentioned IP-based planning system, van den Briel reports that redundant representations are sometimes, but not always beneficial (personal communications).

We believe that a detailed comparison of such representations, using different planning approaches, would be of considerable worth. In addition to comparing redundant and non-redundant multi-valued representations, such a study should also analyse trivial translations from PDDL using binary domains for each state variable as well as hand-tailored representations, to provide a clear picture whether or not multi-valued representations are useful at all for a given combination of planning approach and planning domain, and to see where the limits of automatic translations lie. For the planning approach we presented here, it is fairly obvious that a binary representation will not fare very well due to the much less hierarchical causal graphs, as we illustrated in Chap. 8. Anecdotal empirical evidence strongly suggests that non-binary representations greatly contribute to the informedness of the causal graph heuristic, and that they are of great importance to BDD- and IP-based approaches. However, as of now, many issues here remain poorly understood.

13.2.2 Task Decomposition Heuristics

The second key contribution in Fast Downward is the use of hierarchical task decompositions within a heuristic planning framework. Applying such task decompositions to domain-independent planning has a considerable potential, but since the research of Knoblock [79] and Bacchus and Yang [9], little work has been published on this topic. With Fast Downward, we hope to renew interest in this approach, which we believe to be a very promising ground for further advances in automated planning.

The purely hierarchical decompositions we have pursued in this work are not the only kinds of task decompositions one could envisage. While the hierarchical approach does have the benefit of admitting very efficient heuristic computations, it does show its weaknesses for tasks with very highly connected causal graphs, as in the DEPOTS domain. In principle, it would thus be very appealing to consider local subtasks, defined by small subgraphs of the causal graph, *without* previously eliminating causal cycles. Initial experiments in this direction have shown that it is difficult to achieve this goal without losing the performance of Fast Downward's heuristic estimator, but perhaps better heuristic accuracy can outweigh worse per-state performance in many cases. This is a very interesting area for further investigation.

Another question of interest is whether it is possible to improve performance significantly by considering *other* subtasks than those induced by star-shaped subgraphs of the causal graph. For example, one might consider larger subtasks, maybe spanning three levels of the causal graph rather than just two, or subtasks that group state variables that co-occur as effects of many operators. Another option would be merging pairs of tightly interrelated state variables into single state variables prior to search by building the cross product of their domain transition graphs. Such an approach could be very beneficial in the SCHEDULE domain, where merging all state variables that pertain to the same physical object could easily circumvent the goal ordering problems we discussed in the previous chapter. We should point out that merging state variables in such a fashion is very closely related to Edelkamp's work on using *pattern databases* for heuristic planning [32].

Returning to the causal graph heuristic itself, we believe that an investigation of search space topology under that heuristic, following Hoffmann's work on the FF heuristic [63–65], would be of considerable interest. Intuition suggests that the quality of the heuristic should be closely related to the causal structure of the tasks in a given domain. The more connected the causal graph of a task is, the more conditions will be ignored by the causal graph heuristic, and one would expect that this would have a detrimental effect on heuristic accuracy. In future work, we hope to design an experiment that can clearly show such ties between connectedness measures for causal graphs and topological properties of the corresponding heuristic search space.

13.3 Minor Contributions

Before we conclude, let us also briefly discuss some of the minor contributions of this part, and discuss some potential follow-up research pertaining to them.

One such contribution is the development of the invariant synthesis algorithm presented in Sect. 9.4. We observed that there are a number of related algorithms suggested by other researchers and explained why we nevertheless devised a new technique rather than reusing one of the established ones. The chief reasons for a new algorithm were the need to support PDDL2.2 and the need to efficiently deal with very large numbers of schematic operators. Given the considerable amount of parallel developments in this area, we believe that it is time for a comparison study of invariant synthesis algorithms, providing some details about their relative efficiency and power (for example measured by the fraction of unreachable search states pruned by the invariants generated by a given algorithm).

A second minor contribution is the efficient grounding algorithm presented in Sect. 9.5. Our Horn exploration algorithm was motivated by observed performance shortcomings of more simple approaches, like systematically instantiating all parameters of schematic operators and axioms in all possible ways. In this case, it is doubtful whether a detailed performance study for grounding algorithms would be all that useful, given that there are only few domains where the grounding phase is the bottleneck of the planner. On the other hand, some evidence that there is need for efficient grounding algorithms is obscured by the fact that planning domains are commonly reformulated *by the human domain designer* to cater for the shortcomings of existing grounding algorithms. For example, in the PDDL formulation of the PIPESWORLD domain, the conceptually atomic operation of pushing a batch into a pipeline segment is split into two separate schematic actions push-start and push-end by the domain designer to facilitate grounding. In a similar vein, an application study by Boddy et al. suggests that current planners have limitations in grounding operators with many parameters [15], which prompted the authors of the study to perform similar splits of conceptually atomic operations into different operators, to be applied sequentially.

Third, we believe that the search enhancements presented in Chap. 11 are applicable beyond the scope of Fast Downward. The potential usefulness of preferred operators within best-first search should not require further discussion; the results presented in the previous chapter clearly show their benefit across a wide number of planning domains.

Our results also clearly show that multi-heuristic best-first search is a viable approach, although more experiments are needed here to analyse its suitability in other application domains, for other heuristic evaluation functions, and in particular to compare it to other approaches for combining heuristic estimates, such as taking the sum or maximum of different estimates, or simply running searches with different estimators in parallel. One theoretical consideration in favour of multi-heuristic best-first search is that its choices are invariant under strictly monotonic transformations of individual heuristic estimators. For example, if we are given two heuristic estimators, one of which has a systematic error causing it to overestimate all heuristic distances by a factor of 5, then this estimator would have a dominating influence under search schemes that base their decision on the maximum or sum of the two heuristic estimates. Multi-heuristic best-first search, on the other hand, is robust against such systematic errors because estimates by different heuristic estimators are never compared to each other.

Finally, deferred heuristic evaluation is another search enhancement that should be studied in some more detail, for example in artificially constructed search spaces with varying branching factors and varying heuristic accuracy. We expect that the technique should lead to significant speedups in tasks with wide branching factors and heuristic estimators of good quality, and that this effect should be reinforced when using preferred operators (provided that the preferred operators generated by the heuristic are actually, on average, superior choices to non-preferred ones).

13.4 Going Further

This concludes the discussion in this chapter, and thus also in this volume. As a final remark, we believe that similar techniques to the ones we presented for the case of multi-valued planning tasks – identifying local subtasks based on causal graph structure, and solving them by traversing domain transition graphs – may also be of some use for solving *numerical* planning tasks, expressed in "level 2" of the PDDL language [42].

Clearly, going beyond finite domains to state variables which can assume any value from the infinite set of natural numbers (or integers, or rational numbers) poses significant challenges – not least the fact that the planning problem for such tasks is undecidable [56]. On the other hand, we already observed in Chap. 1 that even propositional PDDL planning algorithms are *practically incomplete*, and still the last decade has seen truly remarkable progress in this area.

References

1. Allen, J., Hendler, J., Tate, A. (eds.): Readings in Planning. Morgan Kaufmann, San Francisco (1990)
2. Amir, E., Engelhardt, B.: Factored planning. In: Gottlob, G., Walsh, T. (eds.) Proceedings of the 18th International Joint Conference on Artificial Intelligence (IJCAI'03), pp. 929–935. Morgan Kaufmann, San Francisco (2003)
3. Andrews, M.: Hardness of buy-at-bulk network design. In: Proceedings of the 45th Annual IEEE Symposium on Foundations of Computer Science (FOCS 2004), pp. 115–124. IEEE Computer Society Press, Los Alamitos (2004)
4. Andrews, M., Zhang, L.: Bounds on fiber minimization in optical networks with fixed fiber capacity. In: Makki, K., Knightly, E.W. (eds.) Proceedings of the 24th Annual Joint Conference of the IEEE Computer and Communications Societies (INFOCOM 2005), pp. 409–419. IEEE Computer Society Press, Los Alamitos (2005)
5. Andrews, M., Zhang, L.: Minimizing maximum fiber requirement in optical networks. Journal of Computer and Systems Sciences 72, 118–131 (2006)
6. Arora, S.: Polynomial time approximation schemes for Euclidean traveling salesman and other geometric problems. Journal of the ACM 45(5), 753–782 (1998)
7. Ausiello, G., Crescenzi, P., Gambosi, G., Kann, V., Marchetti-Spaccamela, A., Protasi, M.: Complexity and Approximation. Springer, Heidelberg (1999)
8. Bacchus, F.: The AIPS'00 planning competition. AI Magazine 22(3), 47–56 (2001)
9. Bacchus, F., Yang, Q.: Downward refinement and the efficiency of hierarchical problem solving. Artificial Intelligence 71(1), 43–100 (1994)
10. Bäckström, C., Nebel, B.: Complexity results for SAS$^+$ planning. Computational Intelligence 11(4), 625–655 (1995)
11. Ball, M., Magnanti, T., Monma, C., Nemhauser, G. (eds.): Network Models. Handbooks in Operations Research and Management Science, vol. 8. Elsevier, Amsterdam (1995)
12. Biundo, S., Fox, M. (eds.): ECP 1999. LNCS, vol. 1809. Springer, Heidelberg (2000)
13. Biundo, S., Myers, K., Rajan, K. (eds.): Proceedings of the Fifteenth International Conference on Automated Planning and Scheduling (ICAPS 2005). AAAI Press, Menlo Park (2005)

14. Blum, A., Furst, M.L.: Fast planning through planning graph analysis. Artificial Intelligence 90(1–2), 281–300 (1997)
15. Boddy, M., Gohde, J., Haigh, T., Harp, S.: Course of action generation for cyber security using classical planning. In: Biundo, S., Myers, K., Rajan, K. (eds.) Proceedings of the Fifteenth International Conference on Automated Planning and Scheduling (ICAPS 2005), pp. 12–21. AAAI Press, Menlo Park (2005)
16. Bonet, B., Geffner, H.: Planning as heuristic search. Artificial Intelligence 129(1), 5–33 (2001)
17. Brafman, R.I., Domshlak, C.: Structure and complexity in planning with unary operators. Journal of Artificial Intelligence Research 18, 315–349 (2003)
18. Brafman, R.I., Domshlak, C.: Factored planning: How, when and when not. In: Proceedings of the Twenty-First National Conference on Artificial Intelligence (AAAI-2006), pp. 809–814. AAAI Press, Menlo Park (2006)
19. Bylander, T.: The computational complexity of propositional STRIPS planning. Artificial Intelligence 69(1–2), 165–204 (1994)
20. Bylander, T.: A probabilistic analysis of propositional STRIPS planning. Artificial Intelligence 81(1–2), 241–271 (1996)
21. Cesta, A., Borrajo, D. (eds.): Pre-proceedings of the Sixth European Conference on Planning (ECP'01), Toledo, Spain (2001)
22. Chalasani, P., Motwani, R.: Approximating capacitated routing and delivery problems. SIAM Journal on Computing 28(6), 2133–2149 (1999)
23. Cheeseman, P., Kanefsky, B., Taylor, W.M.: Where the really hard problems are. In: Mylopoulos, J., Reiter, R. (eds.) Proceedings of the 12th International Joint Conference on Artificial Intelligence (IJCAI'91), pp. 331–337. Morgan Kaufmann, San Francisco (1991)
24. Chien, S., Kambhampati, S., Knoblock, C.A. (eds.): Proceedings of the Fifth International Conference on Artificial Intelligence Planning and Scheduling (AIPS 2000). AAAI Press, Menlo Park (2000)
25. Cormen, T.H., Leiserson, C.E., Rivest, R.L.: Introduction to Algorithms. MIT Press, Cambridge (1990)
26. Crescenzi, P.: A short guide to approximation preserving reductions. In: Proceedings of the 12th Annual IEEE Conference on Computational Complexity (CCC'97), pp. 262–273. IEEE Computer Society Press, Los Alamitos (1997)
27. Cresswell, S., Fox, M., Long, D.: Extending TIM domain analysis to handle ADL constructs. In: AIPS '02 Workshop on Knowledge Engineering Tools and Techniques for A.I. Planning (2002)
28. Domshlak, C., Brafman, R.I.: Structure and complexity in planning with unary operators. In: Ghallab, M., Hertzberg, J., Traverso, P. (eds.) Proceedings of the Sixth International Conference on Artificial Intelligence Planning and Scheduling (AIPS 2002), pp. 34–43. AAAI Press, Menlo Park (2002)
29. Domshlak, C., Dinitz, Y.: Multi-agent off-line coordination: Structure and complexity. In: Cesta, A., Borrajo, D. (eds.) Pre-proceedings of the Sixth European Conference on Planning (ECP'01), Toledo, Spain, pp. 277–288 (2001)
30. Dowling, W.F., Gallier, J.H.: Linear-time algorithms for testing the satisfiability of propositional Horn formulae. Journal of Logic Programming 1(3), 367–383 (1984)
31. Drescher, M.: Approximationseigenschaften von Transportproblemen in der Handlungsplanung. Diplomarbeit, Albert-Ludwigs-Universität Freiburg (2005)

32. Edelkamp, S.: Planning with pattern databases. In: Cesta, A., Borrajo, D. (eds.) Pre-proceedings of the Sixth European Conference on Planning (ECP'01), Toledo, Spain, pp. 13–24 (2001)

33. Edelkamp, S.: Limits and possibilities of PDDL for model checking software. In: Edelkamp, S., Hoffmann, J. (eds.) Proceedings of the ICAPS-03 Workshop on the Competition: Impact, Organisation, Evaluation, Benchmarks (2003)

34. Edelkamp, S., Helmert, M.: Exhibiting knowledge in planning problems to minimize state encoding length. In: Biundo, S., Fox, M. (eds.) ECP 1999. LNCS, vol. 1809, pp. 135–147. Springer, Heidelberg (2000)

35. Edelkamp, S., Helmert, M.: On the implementation of MIPS. In: Paper presented at the Fifth International Conference on Artificial Intelligence Planning and Scheduling, Workshop on Model-Theoretic Approaches to Planning, Breckenridge, Colorado, 14 April (2000)

36. Edelkamp, S., Helmert, M.: The model checking integrated planning system (MIPS). AI Magazine 22(3), 67–71 (2001)

37. Edelkamp, S., Hoffmann, J.: Quo vadis, IPC-4? — Proposals for the classical part of the 4th International Planning Competition. In: Edelkamp, S., Hoffmann, J. (eds.) Proceedings of the ICAPS-03 Workshop on the Competition: Impact, Organisation, Evaluation, Benchmarks (2003)

38. Edelkamp, S., Hoffmann, J.: PDDL2.2: The language for the classical part of the 4th International Planning Competition. Technical Report 195, Albert-Ludwigs-Universität Freiburg, Institut für Informatik (2004)

39. Erol, K., Nau, D.S., Subrahmanian, V.S.: Complexity, decidability and undecidability results for domain-independent planning. Artificial Intelligence 76(1–2), 65–88 (1995)

40. Fikes, R.E., Nilsson, N.J.: STRIPS: A new approach to the application of theorem proving to problem solving. Artificial Intelligence 2, 189–208 (1971)

41. Fox, M., Long, D.: The automatic inference of state invariants in TIM. Journal of Artificial Intelligence Research 9, 367–421 (1998)

42. Fox, M., Long, D.: PDDL2.1: An extension to PDDL for expressing temporal planning domains. Journal of Artificial Intelligence Research 20, 61–124 (2003)

43. Garey, M.R., Johnson, D.S.: Computers and Intractability — A Guide to the Theory of NP-Completeness. W.H. Freeman, New York (1979)

44. Gerevini, A., Saetti, A., Serina, I.: Planning through stochastic local search and temporal action graphs in LPG. Journal of Artificial Intelligence Research 20, 239–290 (2003)

45. Gerevini, A., Schubert, L.: Inferring state constraints for domain-independent planning. In: Rich, C., Mostow, J. (eds.) Proceedings of the Fifteenth National Conference on Artificial Intelligence (AAAI-98), pp. 905–912. AAAI Press, Menlo Park (1998)

46. Gerevini, A., Schubert, L.: Discovering state constraints for planning: DIS-COPLAN. Technical Report 2005-09-48, Department of Electronics for Automation, University of Brescia, 2005.

47. Gerevini, A., Serina, I.: LPG: A planner based on local search for planning graphs with action costs. In: Ghallab, M., Hertzberg, J., Traverso, P. (eds.) Proceedings of the Sixth International Conference on Artificial Intelligence Planning and Scheduling (AIPS 2002), pp. 13–22. AAAI Press, Menlo Park (2002)

48. Ghallab, M., Hertzberg, J., Traverso, P. (eds.): Proceedings of the Sixth International Conference on Artificial Intelligence Planning and Scheduling (AIPS 2002). AAAI Press, Menlo Park (2002)

49. Ginsberg, M.L., Harvey, W.D.: Iterative broadening. Artificial Intelligence 55, 367–383 (1992)
50. Giunchiglia, E., Muscettola, N., Nau, D. (eds.): Proceedings of the Thirteenth International Conference on Automated Planning and Scheduling (ICAPS 2003). AAAI Press, Menlo Park (2003)
51. Gørtz, I.L.: Hardness of preemptive finite capacity dial-a-ride. IMADA Preprints 4, Syddansk Universitet, 2006.
52. Gupta, N., Nau, D.S.: On the complexity of blocks-world planning. Artificial Intelligence 56(2–3), 223–254 (1992)
53. Hearn, R.A., Demaine, E.D.: PSPACE-completeness of sliding-block puzzles and other problems through the nondeterministic constraint logic model of computation. Theoretical Computer Science 343(1–2), 72–96 (2005)
54. Helmert, M.: On the complexity of planning in transportation and manipulation domains. Diplomarbeit, Albert-Ludwigs-Universität Freiburg (2001)
55. Helmert, M.: On the complexity of planning in transportation domains. In: Cesta, A., Borrajo, D. (eds.) Pre-proceedings of the Sixth European Conference on Planning (ECP'01), Toledo, Spain, pp. 349–360 (2001)
56. Helmert, M.: Decidability and undecidability results for planning with numerical state variables. In: Ghallab, M., Hertzberg, J., Traverso, P. (eds.) Proceedings of the Sixth International Conference on Artificial Intelligence Planning and Scheduling (AIPS 2002), pp. 303–312. AAAI Press, Menlo Park (2002)
57. Helmert, M.: Complexity results for standard benchmark domains in planning. Artificial Intelligence 143(2), 219–262 (2003)
58. Helmert, M.: A planning heuristic based on causal graph analysis. In: Zilberstein, S., Koehler, J., Koenig, S. (eds.) Proceedings of the Fourteenth International Conference on Automated Planning and Scheduling (ICAPS 2004), pp. 161–170. AAAI Press, Menlo Park (2004)
59. Helmert, M.: The Fast Downward planning system. Journal of Artificial Intelligence Research 26, 191–246 (2006)
60. Helmert, M.: New complexity results for classical planning benchmarks. In: Long, D., Smith, S.F., Borrajo, D., McCluskey, L. (eds.) Proceedings of the Sixteenth International Conference on Automated Planning and Scheduling (ICAPS 2006), pp. 52–61. AAAI Press, Menlo Park (2006)
61. Helmert, M.: Solving Planning Tasks in Theory and Practice. PhD thesis, Albert-Ludwigs-Universität Freiburg (2006)
62. Helmert, M., Mattmüller, R., Röger, G.: Approximation properties of planning benchmarks. In: Proceedings of the 17th European Conference on Artificial Intelligence (ECAI 2006), pp. 585–589 (2006)
63. Hoffmann, J.: Local search topology in planning benchmarks: An empirical analysis. In: Nebel, B. (ed.) Proceedings of the 17th International Joint Conference on Artificial Intelligence (IJCAI'01), pp. 453–458. Morgan Kaufmann, San Francisco (2001)
64. Hoffmann, J.: Local search topology in planning benchmarks: A theoretical analysis. In: Ghallab, M., Hertzberg, J., Traverso, P. (eds.) Proceedings of the Sixth International Conference on Artificial Intelligence Planning and Scheduling (AIPS 2002), pp. 92–100. AAAI Press, Menlo Park (2002)
65. Hoffmann, J.: Where 'ignoring delete lists' works: Local search topology in planning benchmarks. Journal of Artificial Intelligence Research 24, 685–758 (2005)

66. Hoffmann, J., Edelkamp, S.: The deterministic part of IPC-4: An overview. Journal of Artificial Intelligence Research 24, 519–579 (2005)
67. Hoffmann, J., Edelkamp, S., Thiébaux, S., Englert, R., dos Santos Liporace, F., Trüg, S.: Engineering benchmarks for planning: the domains used in the deterministic part of IPC-4. Journal of Artificial Intelligence Research 26, 453–541 (2006)
68. Hoffmann, J., Nebel, B.: The FF planning system: Fast plan generation through heuristic search. Journal of Artificial Intelligence Research 14, 253–302 (2001)
69. Hoffmann, J., Nebel, B.: RIFO revisited: Detecting relaxed irrelevance. In: Cesta, A., Borrajo, D. (eds.) Pre-proceedings of the Sixth European Conference on Planning (ECP'01), Toledo, Spain, pp. 325–336 (2001)
70. Holzmann, G.J.: The model checker SPIN. IEEE Transactions on Software Engineering 23(5), 279–295 (1997)
71. Jonsson, P., Bäckström, C.: Incremental planning. In: Ghallab, M., Milani, A. (eds.) New Directions in AI Planning: EWSP '95 — 3rd European Workshop on Planning. Frontiers in Artificial Intelligence and Applications, vol. 31, pp. 79–90. IOS Press, Amsterdam (1995)
72. Jonsson, P., Bäckström, C.: State-variable planning under structural restrictions: Algorithms and complexity. Artificial Intelligence 100(1–2), 125–176 (1998)
73. Jonsson, P., Bäckström, C.: Tractable plan existence does not imply tractable plan generation. Annals of Mathematics and Artificial Intelligence 22(3), 281–296 (1998)
74. Joslin, D., Roach, J.: A theoretical analysis of conjunctive-goal problems. Artificial Intelligence 41(1), 97–106 (1989) Research Note
75. Kautz, H., McAllester, D., Selman, B.: Encoding plans in propositional logic. In: Aiello, L.C., Doyle, J., Shapiro, S.C. (eds.) Proceedings of the Fifth International Conference on Principles of Knowledge Representation and Reasoning (KR'96), pp. 374–384. Morgan Kaufmann, San Francisco (1996)
76. Kautz, H., Selman, B.: Planning as satisfiability. In: Neumann, B. (ed.) Proceedings of the 10th European Conference on Artificial Intelligence (ECAI 92), pp. 359–363. John Wiley, Chichester (1992)
77. Kautz, H., Selman, B.: Pushing the envelope: Planning, propositional logic, and stochastic search. In: Proceedings of the Thirteenth National Conference on Artificial Intelligence (AAAI-96), pp. 1194–1201. AAAI Press, Menlo Park (1996)
78. Kautz, H., Selman, B.: Unifying SAT-based and graph-based planning. In: Dean, T. (ed.) Proceedings of the Sixteenth International Joint Conference on Artificial Intelligence (IJCAI'99), pp. 318–325. Morgan Kaufmann, San Francisco (1999)
79. Knoblock, C.A.: Automatically generating abstractions for planning. Artificial Intelligence 68(2), 243–302 (1994)
80. Koehler, J., Hoffmann, J.: On reasonable and forced goal orderings and their use in an agenda-driven planning algorithm. Journal of Artificial Intelligence Research 12, 338–386 (2000)
81. Koehler, J., Nebel, B., Hoffmann, J., Dimopoulos, Y.: Extending planning graphs to an ADL subset. In: Steel, S. (ed.) ECP 1997. LNCS, vol. 1348, pp. 273–285. Springer, Heidelberg (1997)

82. Koehler, J., Schuster, K.: Elevator control as a planning problem. In: Chien, S., Kambhampati, S., Knoblock, C.A. (eds.) Proceedings of the Fifth International Conference on Artificial Intelligence Planning and Scheduling (AIPS 2000), pp. 331–338. AAAI Press, Menlo Park (2000)

83. Korf, R.E.: Planning as search: A quantitative approach. Artificial Intelligence 33(1), 65–88 (1987)

84. Lifschitz, V.: On the semantics of STRIPS. In: Georgeff, M., Lansky, A. (eds.) Reasoning about Actions and Plans, pp. 1–9. Morgan Kaufmann, San Francisco (1987)

85. Long, D., Fox, M.: Automatic synthesis and use of generic types in planning. In: Chien, S., Kambhampati, S., Knoblock, C.A. (eds.) Proceedings of the Fifth International Conference on Artificial Intelligence Planning and Scheduling (AIPS 2000), pp. 196–205. AAAI Press, Menlo Park (2000)

86. Long, D., Fox, M.: The 3rd International Planning Competition: Results and analysis. Journal of Artificial Intelligence Research 20, 1–59 (2003)

87. Long, D., et al.: The AIPS-98 planning competition. AI Magazine 21(2), 13–33 (2000)

88. Lowerre, B.T.: The HARPY Speech Recognition System. PhD thesis, Computer Science Department, Carnegie Mellon University, Pittsburgh, Pennsylvania (1976)

89. Mattmüller, R.: Approximatives Planen in der Grid-Domäne. Studienarbeit, Albert-Ludwigs-Universität Freiburg (2005)

90. McDermott, D.: Using regression-match graphs to control search in planning. Artificial Intelligence 109(1–2), 111–159 (1999)

91. McDermott, D.: The 1998 AI Planning Systems competition. AI Magazine 21(2), 35–55 (2000)

92. Milidiú, R.L., dos Santos Liporace, F., de Lucena, C.J.P.: Pipesworld: Planning pipeline transportation of petroleum derivatives. In: Edelkamp, S., Hoffmann, J. (eds.) Proceedings of the ICAPS-03 Workshop on the Competition: Impact, Organisation, Evaluation, Benchmarks (2003)

93. Nebel, B., Dimopoulos, Y., Koehler, J.: Ignoring irrelevant facts and operators in plan generation. In: Steel, S. (ed.) ECP 1997. LNCS, vol. 1348, pp. 338–350. Springer, Heidelberg (1997)

94. Newell, A., Simon, H.A.: GPS: A program that simulates human thought. In: Feigenbaum, E.A., Feldman, J. (eds.) Computers and Thought, pp. 279–293. Oldenbourg (1963)

95. Papadimitriou, C.H.: Computational Complexity. Addison-Wesley, Reading (1994)

96. Papadimitriou, C.H., Yannakakis, M.: Optimization, approximation, and complexity classes. Journal of Computer and System Sciences 43, 425–440 (1991)

97. Penberthy, J.S., Weld, D.S.: UCPOP: A sound, complete, partial order planner for ADL. In: Nebel, B., Rich, C., Swartout, W. (eds.) Proceedings of the Third International Conference on Principles of Knowledge Representation and Reasoning (KR'92), pp. 103–114. Morgan Kaufmann, San Francisco (1992)

98. Pessoa, A.A.: Planning the transportation of multiple commodities in bidirectional pipeline networks. In: Fleischer, R., Trippen, G. (eds.) ISAAC 2004. LNCS, vol. 3341, pp. 766–777. Springer, Heidelberg (2004)

99. Pólya, G.: How to Solve It: A New Aspect of Mathematical Method. Princeton University Press, Princeton (1945)

100. Rintanen, J.: A planning algorithm not based on directional search. In: Cohn, A.G., Schubert, L., Shapiro, S.C. (eds.) Proceedings of the Sixth International Conference on Principles of Knowledge Representation and Reasoning (KR'98), pp. 617–624. Morgan Kaufmann, San Francisco (1998)

101. Rintanen, J.: An iterative algorithm for synthesizing invariants. In: Kautz, H., Porter, B. (eds.) Proceedings of the Seventeenth National Conference on Artificial Intelligence (AAAI-2000), pp. 806–811. AAAI Press, Menlo Park (2000)

102. Rintanen, J.: Phase transitions in classical planning: an experimental study. In: Zilberstein, S., Koehler, J., Koenig, S. (eds.) Proceedings of the Fourteenth International Conference on Automated Planning and Scheduling (ICAPS 2004), pp. 101–110. AAAI Press, Menlo Park (2004)

103. Rintanen, J.: Compact representation of sets of binary constraints. In: Proceedings of the 17th European Conference on Artificial Intelligence (ECAI 2006), pp. 143–147 (2006)

104. Rintanen, J., Heljanko, K., Niemelä, I.: Parallel encodings of classical planning as satisfiability. Technical Report 198, Albert-Ludwigs-Universität Freiburg, Institut für Informatik (2004)

105. Russell, S., Norvig, P.: Artificial Intelligence — A Modern Approach. Prentice-Hall, Englewood Cliffs (2003)

106. Sacerdoti, E.D.: Planning in a hierarchy of abstraction spaces. Artificial Intelligence 5, 115–135 (1974)

107. Selman, B.: Near-optimal plans, tractability, and reactivity. In: Doyle, J., Sandewall, E., Torasso, P. (eds.) Proceedings of the Fourth International Conference on Principles of Knowledge Representation and Reasoning (KR'94), pp. 521–529. Morgan Kaufmann, San Francisco (1994)

108. Slaney, J., Thiébaux, S.: Blocks World revisited. Artificial Intelligence 125, 119–153 (2001)

109. Tenenberg, J.D.: Abstraction in planning. In: Allen, J.F., Kautz, H.A., Pelavin, R.N., Tenenberg, J.D. (eds.) Reasoning About Plans, chapter 4, pp. 213–283. Morgan Kaufmann, San Mateo (1991)

110. Thiébaux, S., Cordier, M.-O.: Supply restoration in power distribution systems — a benchmark for planning under uncertainty. In: Cesta, A., Borrajo, D. (eds.) Pre-proceedings of the Sixth European Conference on Planning (ECP'01), Toledo, Spain, pp. 85–95 (2001)

111. van den Briel, M., Vossen, T., Kambhampati, S.: Reviving integer programming approaches for AI planning: A branch-and-cut framework. In: Biundo, S., Myers, K., Rajan, K. (eds.) Proceedings of the Fifteenth International Conference on Automated Planning and Scheduling (ICAPS 2005), pp. 310–319. AAAI Press, Menlo Park (2005)

112. Weld, D.S.: Recent advances in AI planning. AI Magazine 20(2), 93–123 (1999)

113. Williams, B.C., Nayak, P.P.: A reactive planner for a model-based executive. In: Pollack, M.E. (ed.) Proceedings of the 15th International Joint Conference on Artificial Intelligence (IJCAI'97), pp. 1178–1195. Morgan Kaufmann, San Francisco (1997)

114. Yoshizumi, T., Miura, T., Ishida, T.: A* with partial expansion for large branching factor problems. In: Kautz, H., Porter, B. (eds.) Proceedings of the Seventeenth National Conference on Artificial Intelligence (AAAI-2000), pp. 923–929. AAAI Press, Menlo Park (2000)

115. Zilberstein, S., Koehler, J., Koenig, S. (eds.): Proceedings of the Fourteenth International Conference on Automated Planning and Scheduling (ICAPS 2004). AAAI Press, Menlo Park (2004)

Index

abstraction, 164
achievability definition, 214
action, *see* operator
  helpful action, 224, 229, **230**, 231,
    242
active vertex, 209
approximation, 16
  c-approximation, 16, 17
  constant factor approximation, 59–68
  f-approximation, 16
  p-approximation, 17
approximation algorithm, 13, **15–16**
  deterministic, 15
  exponential-time, 15
  for MINIMUM FEEDBACK VERTEX
    SET, 80
  non-deterministic, 15
  polynomial-time, 15
approximation class, **16–18**, 26
  APX, 17
  exp-APX, **17**, 28
  EXPO, **17**, 24
  EXPS, 17
  FPTAS, **17**, 27
  NPO, 17
  NPS, 17
  PO, 16
  poly-APX, 17
  PS, **17**, 28
  PTAS, 17
APX, *see* approximation class
axiom, 172, 174, 178, 197, 211
  applicability rule, 192
  body, 172, 174

condition, 174
effect rule, 192
evaluator, 208, 221–222
head, 172, 174
layer, 174, 202
schematic axiom, 172

benchmark domain, 31–37
  AIRPORT, 35, 40, **108–111**
  ASSEMBLY, 32, **113–117**
  BLOCKSWORLD, 31, 34, 117
  DEPOTS, 34, 40, **85–88**
  DRIVERLOG, 34, 40, **103–107**
  FREECELL, 34, **117–126**
  GRID, 33, 40, **98–102**
  GRIPPER, 32, 33, 40, **75–76**
  LOGISTICS, 31, 33, 40, **78–83**
  MICONIC-10, 34, 40, **88–93**
  MICONIC-10-FULLADL, 90
  MICONIC-10-SIMPLEADL, 90
  MICONIC-10-STRIPS, 88
  MOVIE, 33, **126**
  MYSTERY, 33, 40, **76–77**
  MYSTERYPRIME, 33, 40, **76–77**
  PROMELA-OPTICALTELEGRAPH, 136
  PROMELA-PHILOSOPHERS, 135
  PIPESWORLD, 35, **127–132**
  PIPESWORLD-NOTANKAGE, 129
  PROMELA, 35, **132–138**
  PSR, 36, **138–142**
  ROVERS, 35, 40, **93–98**
  SATELLITE, 35, **142–145**
  SCHEDULE, 34, **145–149**
  ZENOTRAVEL, 35, 40, **83–85**

body
  of a Horn clause, 191
  of an axiom, see axiom
bounded plan existence, see plan

canonical model, see logic program
causal graph, 160, 164–168, 208,
    213–220
  acyclic causal graph, 214
  heuristic, see heuristic
  pruned causal graph, 217
condition, 177, 201
  in domain transition graph, 209
  transition condition, 214
  trivially false condition, 214
cost
  estimate, 226
  operator cost, see operator
  plan cost, 22
covered facts, 181

deferred heuristic evaluation, 223,
    232–233
derived predicate, see predicate
derived value, 174
derived variable, see variable
domain, see planning domain
domain transition graph, 160, 207,
    208–213
  extended domain transition graph,
    212
  pruned domain transition graph, 217
  strongly connected domain transition
    graph, 214
  weighted domain transition graph,
    see weight
downward refinement property, 215

effect, 172, 179, 200–201
  co-occurring effect, 214
  conditional effect, 172
  conjunctive effect, 172
  of MPT operator, 174
  rule, 192
  simple effect, 172
  trigger rule, 192
  universal effect, 172
exp-APX, see approximation class
EXPO, see approximation class
EXPS, see approximation class

Fast Downward, 168–170
fluent
  predicate, see predicate
  variable, see variable
FPTAS, see approximation class

goal, 174
  formula, 172
  protection, 237
  rule, 192
  state, 21
graph
  causal graph, see causal graph
  dependency graph, 214
  reducible graph, 50
  state transition graph, see state
  stretchable graph, 50
  weight-expandable graph, 50
grounding, 190–197

HAMILTONIAN PATH, 54
head
  of a Horn clause, 191
  of an axiom, see axiom
heuristic
  causal graph heuristic, 224–229
  FF heuristic, 230
Horn
  exploration, 191–197
  positive Horn clause, 191

invariant, 180–190
  monotonicity invariant, 181
  monotonicity invariant candidate, 181
  threatened candidate, 183

join rule, 193

knowledge compilation, 207–222

layering property, 174
local subtask, 225
logic program
  canonical model, 191, 195–197
  normal form, 193
  positive logic program, 191
  stratified logic program, 202

measure, 14, 22
  function, 14
  optimal measure, 16

minimization problem, 14–15
MINIMUM FEEDBACK VERTEX SET, 78
    approximation algorithm, 80
MINIMUM SET COVER, 15, 19, 62
MINIMUM VERTEX COVER, 15, 19, 56
modification distance, 234
MPT, 173–175, 197–202
multi-valued planning task, see MPT

node
    generator node, 220
    selector node, 220
normal form
    for positive logic programs, 193
normalization, 176–180
NPO, see approximation class
NPS, see approximation class

operator, 21
    applicability, 21
    applicability rule, 192
    cheap operator, 27
    cost, 21, 22
    generated operators, 221
    MPT operator, 174
    preferred operator, 223, 224, 231–232
    schematic operator, 172
    too heavy, 184
    unbalanced operator, 184
optimization problem, 13–21
ordered monotonicity property, 164, 215

partial variable assignment, 173
PDDL, 4, 168, **171**
    task, 171
        grounded task, 190
        normalized task, 179
performance ratio, 16
phase transition, 6
plan, 22
    abstract plan, 164, 215
    existence problem, 23, 25, 26, 28, 29,
        52, 53, 69
    bounded plan existence, 23–25, 28,
        29, 69
    for MPTs, 175
    reasonable plan, 46–47
    relaxed plan, 230
    short optimal plan, 26
    short plan, 26

planning
    domain, 22
    incomplete planning technique, 5
    multi-valued planning task, see MPT
    problem, **23**, 24, 28, 29
    task, 22
        relaxed planning task, 190, 191, 230
PO, see approximation class
poly-APX, see approximation class
predicate
    constant predicate, 198
    derived predicate, 172, 198
    fluent predicate, 172, 198
projection rule, 193
PS, see approximation class
PTAS, see approximation class

reducibility, see reduction
reduction, 18–21, 51
    AP-reduction, 19, 65
    approximation-preserving, see
        AP-reduction
    Karp reduction, 18, 25, 29
    OP-reduction, 20, 29, 47, 51, 67
    optimization-preserving reduction,
        see OP-reduction
    polynomial many-one reduction, see
        Karp reduction
relaxed planning task, see planning task
roadmap, 43
roadmap graph, 41
route planning, 43–45, 51, 53, 54, 62,
        65, 67
    IPC domains, 75–111
    ROUTE domain, 44
    ROUTE task, 43

search, 223–236
    focused iterative-broadening search,
        224, 234–236
    greedy best-first search, 223, 231–233
    multi-heuristic best-first search, 223,
        233–234
semantics
    add-after-delete semantics, 186
    consistent effect semantics, 186
separability definition, 214
solution, 14, 15
    function, 14
    optimal solution, 16

solvability, 14, 16, 46, 191, 230
state, 21
    extended state, 174
    goal state, *see* goal
    initial state, 21, 172, 174, 199
    local state, 133
    partial state, *see* partial variable
        assignment
    reduced state, 174
    space, 21, 23, 209
    transition graph, 21, **175**
    variable, *see* variable
stratifiability, 172, 202
successor
    generator, 208, **220–221**
    preferred successor, 231–232
    state, 220

transition, 133, 209, 227
    condition, *see* condition
    domain transition graph, *see* domain
        transition graph
    dominated transition, 217
    helpful transition, 224, 229, 245
    read transition, 133
    write transition, 133
translation, 169, 171–205
transportation planning, 39–45, 47, 51,
    52, 54, 56, 59
    IPC domains, 75–111
    TRANSPORT domain, 41–43
    TRANSPORT task, 41

TRAVELLING SALESPERSON PROBLEM,
    54, 55

unit cost model, 22, 61, 62, 67
upward solution property, 164

variable
    affected variable, 174, 213
    counted variable, 181
    derived variable, 173, 212, 222
    fluent variable, 173
    free variable, 172
    multi-valued state variable, 161, 168,
        182, 198, 254
    partial assignment, *see* partial
        variable assignment
    selection, 198
    selection variable, 221
    splitting variable, 166
    state variable, 173, 208, 209
    static variable, 166
    symmetrically reversible variable, 166
variable-unique
    atom, 193
    rule, 193

weight, 41, 50, 181, 183
    expandability, 50
    expansion, 55, 66
    in causal graphs, 217
    in domain transition graphs, 209

Lecture Notes in Artificial Intelligence (LNAI)

Vol. 4929: M. Helmert, Understanding Planning Tasks. XIV, 270 pages. 2008.

Vol. 4897: M. Baldoni, T.C. Son, M.B. van Riemsdijk, M. Winikoff (Eds.), Declarative Agent Languages and Technologies V. X, 245 pages. 2008.

Vol. 4885: M. Chetouani, A. Hussain, B. Gas, M. Milgram, J.-L. Zarader (Eds.), Advances in Nonlinear Speech Processing. XI, 284 pages. 2007.

Vol. 4874: J. Neves, M.F. Santos, J.M. Machado (Eds.), Progress in Artificial Intelligence. XVIII, 704 pages. 2007.

Vol. 4869: F. Botana, T. Recio (Eds.), Automated Deduction in Geometry. X, 213 pages. 2007.

Vol. 4850: M. Lungarella, F. Iida, J. Bongard, R. Pfeifer (Eds.), 50 Years of Artificial Intelligence. X, 399 pages. 2007.

Vol. 4845: N. Zhong, J. Liu, Y. Yao, J. Wu, S. Lu, K. Li (Eds.), Web Intelligence Meets Brain Informatics. XI, 516 pages. 2007.

Vol. 4840: L. Paletta, E. Rome (Eds.), Attention in Cognitive Systems. XI, 497 pages. 2007.

Vol. 4830: M.A. Orgun, J. Thornton (Eds.), AI 2007: Advances in Artificial Intelligence. XIX, 841 pages. 2007.

Vol. 4828: M. Randall, H.A. Abbass, J. Wiles (Eds.), Progress in Artificial Life. XII, 402 pages. 2007.

Vol. 4827: A. Gelbukh, Á.F. Kuri Morales (Eds.), MICAI 2007: Advances in Artificial Intelligence. XXIV, 1234 pages. 2007.

Vol. 4826: P. Perner, O. Salvetti (Eds.), Advances in Mass Data Analysis of Signals and Images in Medicine, Biotechnology and Chemistry. X, 183 pages. 2007.

Vol. 4819: T. Washio, Z.-H. Zhou, J.Z. Huang, X. Hu, J. Li, C. Xie, J. He, D. Zou, K.-C. Li, M.M. Freire (Eds.), Emerging Technologies in Knowledge Discovery and Data Mining. XIV, 675 pages. 2007.

Vol. 4811: O. Nasraoui, M. Spiliopoulou, J. Srivastava, B. Mobasher, B. Masand (Eds.), Advances in Web Mining and Web Usage Analysis. XII, 247 pages. 2007.

Vol. 4798: Z. Zhang, J.H. Siekmann (Eds.), Knowledge Science and Engineering and Management. XVI, 669 pages. 2007.

Vol. 4795: F. Schilder, G. Katz, J. Pustejovsky (Eds.), Annotating, Extracting and Reasoning about Time and Events. VII, 141 pages. 2007.

Vol. 4790: N. Dershowitz, A. Voronkov (Eds.), Logic for Programming, Artificial Intelligence, and Reasoning. XIII, 562 pages. 2007.

Vol. 4788: D. Borrajo, L. Castillo, J.M. Corchado (Eds.), Current Topics in Artificial Intelligence. XI, 280 pages. 2007.

Vol. 4775: A. Esposito, M. Faundez-Zanuy, E. Keller, M. Marinaro (Eds.), Verbal and Nonverbal Communication Behaviours. XII, 325 pages. 2007.

Vol. 4772: H. Prade, V.S. Subrahmanian (Eds.), Scalable Uncertainty Management. X, 277 pages. 2007.

Vol. 4766: N. Maudet, S. Parsons, I. Rahwan (Eds.), Argumentation in Multi-Agent Systems. XII, 211 pages. 2007.

Vol. 4755: V. Corruble, M. Takeda, E. Suzuki (Eds.), Discovery Science. XI, 298 pages. 2007.

Vol. 4754: M. Hutter, R.A. Servedio, E. Takimoto (Eds.), Algorithmic Learning Theory. XI, 403 pages. 2007.

Vol. 4737: B. Berendt, A. Hotho, D. Mladenic, G. Semeraro (Eds.), From Web to Social Web: Discovering and Deploying User and Content Profiles. XI, 161 pages. 2007.

Vol. 4733: R. Basili, M.T. Pazienza (Eds.), AI*IA 2007: Artificial Intelligence and Human-Oriented Computing. XVII, 858 pages. 2007.

Vol. 4724: K. Mellouli (Ed.), Symbolic and Quantitative Approaches to Reasoning with Uncertainty. XV, 914 pages. 2007.

Vol. 4722: C. Pelachaud, J.-C. Martin, E. André, G. Chollet, K. Karpouzis, D. Pelé (Eds.), Intelligent Virtual Agents. XV, 425 pages. 2007.

Vol. 4720: B. Konev, F. Wolter (Eds.), Frontiers of Combining Systems. X, 283 pages. 2007.

Vol. 4702: J.N. Kok, J. Koronacki, R. Lopez de Mantaras, S. Matwin, D. Mladenič, A. Skowron (Eds.), Knowledge Discovery in Databases: PKDD 2007. XXIV, 640 pages. 2007.

Vol. 4701: J.N. Kok, J. Koronacki, R. Lopez de Mantaras, S. Matwin, D. Mladenič, A. Skowron (Eds.), Machine Learning: ECML 2007. XXII, 809 pages. 2007.

Vol. 4696: H.-D. Burkhard, G. Lindemann, R. Verbrugge, L.Z. Varga (Eds.), Multi-Agent Systems and Applications V. XIII, 350 pages. 2007.

Vol. 4694: B. Apolloni, R.J. Howlett, L. Jain (Eds.), Knowledge-Based Intelligent Information and Engineering Systems, Part III. XXIX, 1126 pages. 2007.

Vol. 4693: B. Apolloni, R.J. Howlett, L. Jain (Eds.), Knowledge-Based Intelligent Information and Engineering Systems, Part II. XXXII, 1380 pages. 2007.

Vol. 4692: B. Apolloni, R.J. Howlett, L. Jain (Eds.), Knowledge-Based Intelligent Information and Engineering Systems, Part I. LV, 882 pages. 2007.

Vol. 4687: P. Petta, J.P. Müller, M. Klusch, M. Georgeff (Eds.), Multiagent System Technologies. X, 207 pages. 2007.

Vol. 4682: D.-S. Huang, L. Heutte, M. Loog (Eds.), Advanced Intelligent Computing Theories and Applications. XXVII, 1373 pages. 2007.

Vol. 4676: M. Klusch, K.V. Hindriks, M.P. Papazoglou, L. Sterling (Eds.), Cooperative Information Agents XI. XI, 361 pages. 2007.

Vol. 4667: J. Hertzberg, M. Beetz, R. Englert (Eds.), KI 2007: Advances in Artificial Intelligence. IX, 516 pages. 2007.

Vol. 4660: S. Džeroski, L. Todorovski (Eds.), Computational Discovery of Scientific Knowledge. X, 327 pages. 2007.

Vol. 4659: V. Mařík, V. Vyatkin, A.W. Colombo (Eds.), Holonic and Multi-Agent Systems for Manufacturing. VIII, 456 pages. 2007.

Vol. 4651: F. Azevedo, P. Barahona, F. Fages, F. Rossi (Eds.), Recent Advances in Constraints. VIII, 185 pages. 2007.

Vol. 4648: F. Almeida e Costa, L.M. Rocha, E. Costa, I. Harvey, A. Coutinho (Eds.), Advances in Artificial Life. XVIII, 1215 pages. 2007.

Vol. 4635: B. Kokinov, D.C. Richardson, T.R. Roth-Berghofer, L. Vieu (Eds.), Modeling and Using Context. XIV, 574 pages. 2007.

Vol. 4632: R. Alhajj, H. Gao, X. Li, J. Li, O.R. Zaïane (Eds.), Advanced Data Mining and Applications. XV, 634 pages. 2007.

Vol. 4629: V. Matoušek, P. Mautner (Eds.), Text, Speech and Dialogue. XVII, 663 pages. 2007.

Vol. 4626: R.O. Weber, M.M. Richter (Eds.), Case-Based Reasoning Research and Development. XIII, 534 pages. 2007.

Vol. 4617: V. Torra, Y. Narukawa, Y. Yoshida (Eds.), Modeling Decisions for Artificial Intelligence. XII, 502 pages. 2007.

Vol. 4612: I. Miguel, W. Ruml (Eds.), Abstraction, Reformulation, and Approximation. XI, 418 pages. 2007.

Vol. 4604: U. Priss, S. Polovina, R. Hill (Eds.), Conceptual Structures: Knowledge Architectures for Smart Applications. XII, 514 pages. 2007.

Vol. 4603: F. Pfenning (Ed.), Automated Deduction – CADE-21. XII, 522 pages. 2007.

Vol. 4597: P. Perner (Ed.), Advances in Data Mining. XI, 353 pages. 2007.

Vol. 4594: R. Bellazzi, A. Abu-Hanna, J. Hunter (Eds.), Artificial Intelligence in Medicine. XVI, 509 pages. 2007.

Vol. 4585: M. Kryszkiewicz, J.F. Peters, H. Rybinski, A. Skowron (Eds.), Rough Sets and Intelligent Systems Paradigms. XIX, 836 pages. 2007.

Vol. 4578: F. Masulli, S. Mitra, G. Pasi (Eds.), Applications of Fuzzy Sets Theory. XVIII, 693 pages. 2007.

Vol. 4573: M. Kauers, M. Kerber, R. Miner, W. Windsteiger (Eds.), Towards Mechanized Mathematical Assistants. XIII, 407 pages. 2007.

Vol. 4571: P. Perner (Ed.), Machine Learning and Data Mining in Pattern Recognition. XIV, 913 pages. 2007.

Vol. 4570: H.G. Okuno, M. Ali (Eds.), New Trends in Applied Artificial Intelligence. XXI, 1194 pages. 2007.

Vol. 4565: D.D. Schmorrow, L.M. Reeves (Eds.), Foundations of Augmented Cognition. XIX, 450 pages. 2007.

Vol. 4562: D. Harris (Ed.), Engineering Psychology and Cognitive Ergonomics. XXIII, 879 pages. 2007.

Vol. 4548: N. Olivetti (Ed.), Automated Reasoning with Analytic Tableaux and Related Methods. X, 245 pages. 2007.

Vol. 4539: N.H. Bshouty, C. Gentile (Eds.), Learning Theory. XII, 634 pages. 2007.

Vol. 4529: P. Melin, O. Castillo, L.T. Aguilar, J. Kacprzyk, W. Pedrycz (Eds.), Foundations of Fuzzy Logic and Soft Computing. XIX, 830 pages. 2007.

Vol. 4520: M.V. Butz, O. Sigaud, G. Pezzulo, G. Baldassarre (Eds.), Anticipatory Behavior in Adaptive Learning Systems. X, 379 pages. 2007.

Vol. 4511: C. Conati, K. McCoy, G. Paliouras (Eds.), User Modeling 2007. XVI, 487 pages. 2007.

Vol. 4509: Z. Kobti, D. Wu (Eds.), Advances in Artificial Intelligence. XII, 552 pages. 2007.

Vol. 4496: N.T. Nguyen, A. Grzech, R.J. Howlett, L.C. Jain (Eds.), Agent and Multi-Agent Systems: Technologies and Applications. XXI, 1046 pages. 2007.

Vol. 4483: C. Baral, G. Brewka, J. Schlipf (Eds.), Logic Programming and Nonmonotonic Reasoning. IX, 327 pages. 2007.

Vol. 4482: A. An, J. Stefanowski, S. Ramanna, C.J. Butz, W. Pedrycz, G. Wang (Eds.), Rough Sets, Fuzzy Sets, Data Mining and Granular Computing. XIV, 585 pages. 2007.

Vol. 4481: J. Yao, P. Lingras, W.-Z. Wu, M.S. Szczuka, N.J. Cercone, D. Ślęzak (Eds.), Rough Sets and Knowledge Technology. XIV, 576 pages. 2007.

Vol. 4476: V. Gorodetsky, C. Zhang, V.A. Skormin, L. Cao (Eds.), Autonomous Intelligent Systems: Multi-Agents and Data Mining. XIII, 323 pages. 2007.

Vol. 4460: S. Aguzzoli, A. Ciabattoni, B. Gerla, C. Manara, V. Marra (Eds.), Algebraic and Proof-theoretic Aspects of Non-classical Logics. VIII, 309 pages. 2007.

Vol. 4457: G.M.P. O'Hare, A. Ricci, M.J. O'Grady, O. Dikenelli (Eds.), Engineering Societies in the Agents World VII. XI, 401 pages. 2007.

Vol. 4456: Y. Wang, Y.-m. Cheung, H. Liu (Eds.), Computational Intelligence and Security. XXIII, 1118 pages. 2007.

Vol. 4455: S. Muggleton, R. Otero, A. Tamaddoni-Nezhad (Eds.), Inductive Logic Programming. XII, 456 pages. 2007.

Vol. 4452: M. Fasli, O. Shehory (Eds.), Agent-Mediated Electronic Commerce. VIII, 249 pages. 2007.

Vol. 4451: T.S. Huang, A. Nijholt, M. Pantic, A. Pentland (Eds.), Artifical Intelligence for Human Computing. XVI, 359 pages. 2007.

Vol. 4442: L. Antunes, K. Takadama (Eds.), Multi-Agent-Based Simulation VII. X, 189 pages. 2007.